ADVANCED METHODS OF

PHYSIOLOGICAL SYSTEM MODELING

Volume 2

ADVANCED METHODS OF
PHYSIOLOGICAL SYSTEM MODELING

Volume 2

Edited by
Vasilis Z. Marmarelis
University of Southern California
Los Angeles, California

Published in cooperation with
BIOMEDICAL SIMULATIONS RESOURCE
University of Southern California, Los Angeles
by
PLENUM PRESS
New York and London

Library of Congress Cataloging in Publication Data

Workshop on Advanced Methods of Physiological System Modeling (3rd: 1988: Marina del Rey, Calif.)
    Advanced methods of physiological system modeling, volume 2 / edited by Vasilis Z. Marmarelis.
        p.   cm.
    Based, in part, on papers presented at the Third Workshop on Advanced Methods of Physiological System Modeling, held May 27–28, 1988, in Marina del Rey, California, and supported by the Division of Research Resources, National Institutes of Health
    Includes bibliographical references.
    ISBN 978-1-4613-9791-5   ISBN 978-1-4613-9789-2 (eBook)
    DOI 10.1007/ 978-1-4613-9789-2
    1. Physiology—Mathematical models—Congresses. 2. Nonlinear functional analysis—Congresses. I. Marmarelis, Vasilis Z. II. University of Southern California. Biomedical Simulations Resource. III. Title.
    [DNLM: 1. Models, Biological—congresses. 2. Physiology—congresses. 3. Systems Analysis—congresses. QT 4 W926a 1988]
    QP33.6.M36W67   1988
    612'.001'5118—dc20
    DNLM/DLC   89-16239
    for Library of Congress   CIP

Based, in part, on papers presented at the Third Workshop on Advanced Methods of Physiological System Modeling, held May 27–28, 1988, in Marina del Rey, California, and supported by the Division of Research Resources, National Institutes of Health

© 1989 Plenum Press, New York
Softcover reprint of the hardcover 1st edition 1989
A Division of Plenum Publishing Corporation
233 Spring Street, New York, N.Y. 10013

PREFACE

This volume is the second in a series of publications sponsored by the Biomedical Simulations Resource (BMSR) at the University of Southern California that report on recent research developments in the area of physiological systems modeling and analysis of physiological signals. As in the first volume of this series, the work reported herein is concerned with the development of advanced methodologies and their novel application to problems of biomedical interest, with emphasis on nonlinear aspects of physiological function. The term "advanced methodologies" is used to indicate that the scope of this work extends beyond the ordinary type of analysis used by most investigators in this area, which is confined primarily in the linear domain. As the importance of nonlinearities in understanding the complex mechanisms of physiological function is increasingly recognized, the need for effective and practical methodologies that address the issue of nonlinear dynamics in life sciences becomes more and more pressing. The publication of these volumes and the workshops, organized by the BMSR on the same subject, are two key activities in our efforts to promote and intensify research in this area, foster interaction and collaboration among interested investigators, and disseminate recent results throughout the biomedical community. In the three years since the establishment of the BMSR, we have seen our efforts produce gratifying results in terms of increased interest and research output by the approximately one hundred participants of our workshops and contributors to our volumes. Twice as many investigators worldwide have received the benefit of our first volume. The enthusiastic support of this pioneering group of investigators has vindicated our efforts and has fulfilled the expectations of our sponsor, the Biomedical Research Technology Program of the Division of Research Resources at the National Institutes of Health, whose support and foresight has made these activities possible.

The present volume contains 16 articles that report recent results in the study of nonlinear dynamics in the context of physiological systems. These articles cover several diverse topics ranging in scope from methodological issues to quantitative interpretations of intriguing aspects of physiological function. In expanding the scope of the previous volume, which was primarily focused on issues related to the Volterra–Wiener approach, this volume includes three articles on chaotic dynamics and fractal dimensions. This is a subject of emerging research interest that shares the same object of study with the Volterra–Wiener approach. One of the most significant accomplishments of our workshops was in bringing together experts in those two approaches and facilitate interaction aimed at bridging the two previously segregated communities. This interaction has led to interesting comparisons of the two approaches that promise to enrich them both conceptually and methodologically. We intend to continue fostering this interaction with the expectation of mutual benefit, that may lead to a more global viewpoint about nonlinear dynamics in living systems and the effective analysis of actual data. In addition to these two major directions of research, a notable contribution by Prof. D.R. Brillinger charts a promising route for the analysis of non–Gaussian physiological signals by use of high–order spectra and maximum likelihood estimation. In terms of physiological domains covered by these applications, the majority of studies is focused on the function of the nervous system, although the developed methodologies can be readily adapted to a variety of physiological domains. It is my hope that this volume will contribute to further advances and broader dissemination of this area of research in order to entice a greater number of competent investigators in the use of these "cutting–edge" methodologies.

I wish to thank all the authors for their excellent contributions to this volume and their enthusiastic participation in this exciting and ambitious journey. I know that they all share my conviction that our collective efforts will eventually transform the landscape of scientific research in this area and help unravel the tangled questions of nonlinear physiological function. My sincere thanks are also due to Ms. Gabriele Larmon, my Administrative Assistant, and Ms. Diane Lord for their valuable help in putting this volume together and their assiduous efforts in running effectively the various activities of the BMSR, including our annual workshops. Finally, I wish to express my gratitude to our sponsor, the Biomedical Research Technology Program of the DRR/NIH, whose financial support under Grant No. RR–01861 and intellectual foresight has made the establishment of the BMSR and the publication of this volume possible.

Vasilis Z. Marmarelis, Ph.D.

Professor of Biomedical and Electrical Engineering
Director of the Biomedical Simulations Resource
University of Southern California
Los Angeles, California

CONTENTS

VOLTERRA–WIENER ANALYSIS OF A CLASS OF NONLINEAR FEEDBACK SYSTEMS AND APPLICATION TO SENSORY BIOSYSTEMS

Vasilis Z. Marmarelis

Departments of Biomedical and Electrical Engineering
University of Southern California
Los Angeles, CA 90089-1451

INTRODUCTION

This article addresses an issue that sits at the confluence of two important, yet marginally explored, problems: the Volterra–Wiener expansions of nonlinear differential equations and the Wiener (white noise) analysis of nonlinear feedback systems. These two problems can be addressed in the same methodological context, as explained below. The importance of this issue derives from the desire to relate parametric (i.e., differential equations) models with nonparametric (i.e., Volterra–Wiener functional expansions) models for nonlinear dynamic systems whose internal structure and functional organization are inadequately known to the investigator to allow the development of an explicit mathematical model on the basis of physical/chemical principles ("black box" formulation). This problem has been addressed extensively in the linear case (linear realization theory) and practical methods have been developed for this purpose. However, this problem has received little attention in the nonlinear case, not for lack of importance but because of its considerable complexity. Notable efforts have been made for development of a "nonlinear realization theory" (e.g., Rugh, 1981) but no practical methods are currently available that can be employed by investigators who study nonlinear physiological "black box" systems in order to arrive at parametric (nonlinear differential equation) models from input–output experimental data. On the other hand, many investigators have followed the Wiener approach to obtain nonparametric models (in the form of Volterra–Wiener expansions) of nonlinear physiological systems in recent years.

The appeal of the Wiener approach is that it does not require prior knowledge of the internal structure/organization of the system, and yields models that are "true to the data" (i.e., they do not reflect subjective biases of the investigator) using well–established methodologies. The Wiener approach has its own practical limitations but those are now well–understood and thoroughly investigated (Marmarelis & Marmarelis, 1978). The main advantage of obtaining an equivalent parametric model is its compactness which, in turn, affords greater interpretability of the model in terms of the relevant scientific issues. For instance, it has been observed that several sensory systems undergo a gradual transition from an overdamped to an underdamped dynamic mode (e.g., retinal horizontal cells) or the resonance frequency of a band–pass characteristic shifts downward (e.g., auditory nerve fibers) when the power of the input signal (stimulus) increases. How are these changes mediated physiologically? Are they the result of nonlinear feedback, as has been previously postulated, and if so, how can we determine the form of this nonlinear feedback? The answers to these questions may be provided more readily by a nonlinear parametric model that accurately describes the input–output relation under various stimulus conditions. We must note that an equally useful model for this purpose may be in the form of a "block structured" model, i.e., a model structured from simple block components (e.g., linear filters and static nonlinearities) with simple interconnections (e.g., cascade, parallel, feedback etc.).

With this motivation in mind, we consider a nonlinear ordinary differential equation of the form:

$$L(D)y + f(y, Dy, \ldots, D^r y) = M(D)x \qquad (1)$$

where D is the differential operator $\frac{d(\cdot)}{dt}$, $x(t)$ and $y(t)$ are the system input and output respectively, $f(\cdot)$ is a continuous function of its arguments, and $L(D)$, $M(D)$ are polynomials in D (L is of degree higher than r, and M is of lower degree than L). This equation is equivalent to the block structured system of Fig. 1. If the function $f(\cdot)$ is nonlinear then the system of Fig. 1 is a nonlinear (negative) feedback system with a linear forward subsystem and a linear subsystem in pre–cascade.

In the following section, we will study the Volterra–Wiener expansion of a subset of this class of nonlinear feedback systems and derive some analytical expressions that may help interpret nonparametric measurements (Wiener kernels) in the aforementioned parametric and/or block–structured modeling context. In the subsequent section, computer simulations are used to illustrate our theoretical findings and demonstrate analogies in the observed behavior of this class of nonlinear feedback systems with experimental observations from the visual and auditory system that have been previously reported. It is hoped that these results will begin to re–orient our thinking towards the possible development of physiologically interpretable models of this type from nonparametric experimental measurements (Wiener kernels).

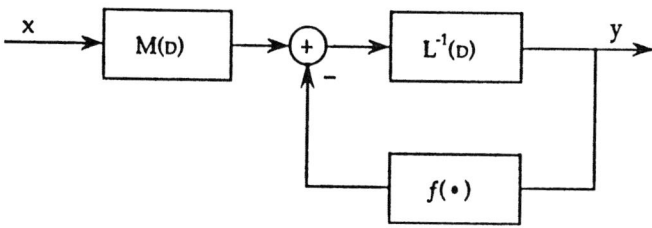

Fig. 1. Block–structured model of nonlinear feedback system described by the non-linear differential Eq. (1).

VOLTERRA–WIENER EXPANSIONS OF A CLASS OF NONLINEAR DIFFERENTIAL EQUATIONS AND NONLINEAR FEEDBACK SYSTEMS

As indicated above, the study of nonlinear feedback systems can be cast in the context of Volterra–Wiener expansions of nonlinear differential equations. To demonstrate this, we consider the class of nonlinear feedback dynamic systems described by Eq. (1) (or Fig. 1) when the function $f(\cdot)$ has only two arguments (y and Dy) and is represented by a power series expansion:

$$f(y,\ Dy) = \sum_{\substack{i=0}}^{\infty} \sum_{\substack{j=0 \\ i+j \geq 2}}^{\infty} c_{i,j} y^i (Dy)^j \tag{2}$$

where the coefficients of the nonlinear terms are of small magnitude, i.e., $|c_{i,j}| \leq \varepsilon \ll 1$ for all i and j. Note that the polynomial $L(D)$ is assumed of degree higher than first (i.e., it involves at least the second derivative of y) and the polynomial $M(D)$ is of lower degree than $L(D)$ by at least two degrees. The nonlinear terms are of degree two and higher since linear terms may be absorbed into $L(D)$, and must be thought as terms of a Taylor (or Weierstrass) expansion of an analytic (or continuous) function of y and its derivative.

Volterra series expansion

For the region of stable solutions of this equation, there exists a Volterra functional expansion (Volterra, 1930; Barrett, 1965; Rugh, 1981):

$$y(t) = \sum_{n=1}^{\infty} \int_0^{\infty} \cdots \int_0^{\infty} k_n(\tau_1, \ldots, \tau_n)\, x(t - \tau_1) \cdots x(t - \tau_n)\, d\tau_1 \cdots d\tau_n \tag{3}$$

which represents the system output in terms of a series of multiple convolution integrals of the input. The kernel functions $\{k_n\}$ characterize the dynamics of the nonlinear system and are called the Volterra kernels of the system. They are symmetric functions of their arguments, i.e., attain the same value for all permutations of given $(\tau_1, \cdots, \tau_n)$ values.

A method of "generalized harmonic balance" can be used to derive analytically the Volterra kernels that correspond to the system described by Eqs. (1) and (2). This method is based on ideas found in (Barrett, 1963; Bedrosian & Rice, 1971; Korenberg, 1973; Rugh, 1981) and has been briefly outlined in (Marmarelis, 1982, 1989). According to this method, the m-th order Volterra kernel can be evaluated in the m-dimensional Laplace domain by considering generalized harmonic inputs:

$$x_m(t) = e^{s_1 t} + \cdots + e^{s_m t} \tag{4}$$

where s_i are distinct complex (Laplace) variables. Substitution of the input $x_m(t)$ into Eq. (3) results in the output expression:

$$y_m(t) = \sum_{n=1}^{\infty} \sum_{j_1=1}^{m} \cdots \sum_{j_n=1}^{m} e^{(s_{j_1} + \cdots + s_{j_n})t} \int \cdots \int k_n(\tau_1, \ldots, \tau_n) e^{-s_{j_1} \tau_1 - \cdots - s_{j_n} \tau_n} d\tau_1 \cdots d\tau_n$$

$$= \sum_{n=1}^{\infty} \sum_{j_1=1}^{m} \cdots \sum_{j_n=1}^{m} e^{(s_{j_1}+\cdots s_{j_n})t} K_n(s_{j_1},\ldots,s_{j_n}) \tag{5}$$

and its r-th derivative:

$$D^r y_m(t) = \sum_{n=1}^{\infty} \sum_{j_1=1}^{m} \cdots \sum_{j_n=1}^{m} (s_{j_1}+\cdots+s_{j_n})^r e^{(s_{j_1}+\cdots+s_{j_n})t} K_n(s_{j_1},\ldots,s_{j_n}) \tag{6}$$

where K_n is the n-dimensional Laplace transform of the kernel k_n.

Note that the output terms that contain the complex exponential with all distinct complex frequencies of the input (i.e., terms of the form $e^{(s_1+\cdots+s_m)t}$) are terms associated with the m-th order kernel.

Substitution of the output expression and its derivatives given by Eqs. (5) and (6) into Eq. (1) yields an expression that can be used for the evaluation of the m-th order kernel $K_m(s_1,\ldots,s_m)$ on the basis of harmonic balance, i.e., by selecting only those terms in the equation that contain the complex exponential $e^{(s_1+\cdots+s_m)t}$. The contributions of these terms in the equation must balance out, according to the harmonic balance approach.

Let us follow this approach for increasing values of m. For $m = 1$ we have:

$$y_1(t) = \sum_{n=1}^{\infty} e^{n s_1 t} K_n(s_1,\ldots,s_1) \tag{7}$$

and,

$$D^r y_1(t) = \sum_{n=1}^{\infty} (n s_1)^r e^{n s_1 t} K_n(s_1,\ldots,s_1) \tag{8}$$

The nonlinear terms $y_1^i (Dy_1)^j$ yield:

$$
\begin{aligned}
y_1^i(t)[Dy_1(t)]^j = & \\
& \sum_{n_1=1}^{\infty} \cdots \sum_{n_i=1}^{\infty} e^{(n_1+\cdots+n_i)s_1 t} K_{n_1}(s_1,\ldots,s_1)\cdots K_{n_i}(s_1,\ldots,s_1) \\
& \sum_{n_1'=1}^{\infty} \cdots \sum_{n_j'=1}^{\infty} n_1' \cdots n_j' \, s_1^j e^{(n_1'+\cdots+n_j')s_1 t} K_{n_1'}(s_1,\ldots,s_1)\cdots K_{n_j'}(s_1,\ldots,s_1)
\end{aligned} \tag{9}
$$

Therefore, balancing terms containing $e^{s_1 t}$ we obtain:

$$L(s_1) K_1(s_1) e^{s_1 t} = M(s_1) e^{s_1 t} \tag{10}$$

where the polynomials $L(s_1)$ and $M(s_1)$ are as defined in Eq. (1).

Therefore:

$$K_1(s_1) = \frac{M(s_1)}{L(s_1)} \tag{11}$$

Note that the nonlinear terms in Eq. (1) do not contribute to K_1, i.e, the first–order Volterra kernel of the system represents strictly the linear portion of the nonlinear differential equation, as expected.

The second order kernel can be evaluated for m=2. Then :

$$y_2(t) = \sum_{n=1}^{\infty} \sum_{j_1=1}^{2} \cdots \sum_{j_n=1}^{2} e^{(s_{j_1}+\cdots+s_{j_n})t} K_n(s_{j_1}, \ldots, s_{j_n}) \tag{12}$$

$$D^r y_2(t) = \sum_{n=1}^{\infty} \sum_{j_1=1}^{2} \cdots \sum_{j_n=1}^{2} (s_{j_1}+\cdots+s_{j_n})^r e^{(s_{j_1}+\cdots+s_{j_n})t} K_n(s_{j_1}, \ldots, s_{j_n}) \tag{13}$$

Since the resulting expressions for $y_2^i(t)[Dy_2(t)]^j$ are rather unwieldy, we focus only on those terms that will contain the exponential $e^{(s_1+s_2)t}$. No such terms will be present in these expressions for $i+j > 2$. The only such terms will come from the expressions for $y_2^2(t)$, $y_2(t)Dy_2(t)$, $[Dy_2(t)]^2$ and, of course, from $y_2(t)$ and $D^r y_2(t)$. Observing that the kernels are symmetric functions of their arguments, we have:

$$2L(s_1 + s_2)K_2(s_1, s_2) + 2c_{2,0}K_1(s_1)K_1(s_2)$$
$$+c_{1,1}(s_1 + s_2)K_1(s_1)K_1(s_2) + 2c_{0,2}s_1 s_2 K_1(s_1)K_1(s_2) = 0 \tag{14}$$

Solving Eq. (14) for K_2, we obtain:

$$K_2(s_1, s_2) = -[c_{2,0} + c_{1,1}\frac{(s_1 + s_2)}{2} + c_{0,2}s_1 s_2]K_1(s_1)K_1(s_2)/L(s_1 + s_2) \tag{15}$$

Continuing on with $m = 3$, we see that terms containing the exponential $e^{(s_1+s_2+s_3)t}$ are contributed by $y_3(t)$, $D^r y_3(t)$ and $y_3^i(t)[Dy_3(t)]^j$ for $i+j = 2, 3$. Note, however, that the contributions from expressions for $i+j = 2$ are of order ε^2, since we have assumed that $|c_{i,j}| \leq \varepsilon \ll 1$. Thus, if we neglect these terms, the resulting third–order Volterra kernel is:

$$K_3(s_1, s_2, s_3) =$$
$$-[c_{3,0} + c_{2,1}\frac{(s_1 + s_2 + s_3)}{3} + c_{1,2}\frac{(s_1 s_2 + s_2 s_3 + s_3 s_1)}{3} + c_{0,3}s_1 s_2 s_3]$$
$$K_1(s_1)K_1(s_2)K_1(s_3)/L(s_1 + s_2 + s_3) \tag{16}$$

Generalizing this analysis and observing that for the m-th order harmonic balance of terms containing $e^{(s_1+\cdots+s_m)t}$ the only non-negligible terms (i.e., of order ε) will be contributed by the expressions for $y_m(t)$, $D^r y_m(t)$ and $y_m^i(t)[Dy_m(t)]^j$ for $i+j = m$, we have (for $m > 1$):

$$K_m(s_1, \ldots, s_m) =$$
$$-\left\{\sum_{n=0}^{m} \frac{(m-n)! n!}{m!} c_{m-n,n} R_{m,n}(s_1, \ldots, s_m)\right\} K_1(s_1) \cdots K_1(s_m) / L(s_1 + \cdots + s_m) \quad (17)$$

where $R_{m,n}(s_1, \cdots, s_m)$ denotes the sum of all distinct products $(s_{j_1} s_{j_2} \cdots s_{j_n})$ that can be formed with combinations of the indices $(j_1, j_2, \ldots, j_n)$ from the set $(1, 2, \ldots, m)$. Note that $R_{m,0} = 1$ by definition. Equation (17), in combination with Eq. (11), yields the approximate general expression for the Volterra kernels of this class of nonlinear systems, under the stated assumption of small magnitude coefficients for the nonlinear terms of Eq. (1).

Wiener series expansion

Let us now turn to the Wiener series expansion of this class of systems. System identification and modeling studies of a given "black-box" nonlinear system require estimation of the kernels from input-output data. This task is complicated by the fact that the Volterra functional terms are, in general, coupled for a given input and the problem requires solving a complicated set of simultaneous equations. To solve this estimation problem, Wiener proposed the orthogonalization of the Volterra functional expansion for a Gaussian white noise (GWN) input. This leads to decoupling of the functional terms (of the Wiener orthogonal expansion) and allows the estimation of the (Wiener) kernels one at a time.

This approach has been studied extensively in the last 30 years and the reader is referred to the original Wiener monograph (Wiener, 1958) and recent books on the subject (Marmarelis & Marmarelis, 1978; Schetzen, 1980; Rugh, 1981; Marmarelis, 1987) for details. A summary, necessary for the developments of this paper, is given below.

The Wiener orthogonal expansion is:

$$y(t) = \sum_{n=0}^{\infty} G_n[h_n; x(t'), t' \leq t] \quad (18)$$

where the n-th order Wiener functional

$$G_n[h_n; x(t'), t' \leq t] = \sum_{n=0}^{\infty} \sum_{m=0}^{[n/2]} \frac{(-1)^m \, n! \, P^m}{(n-2m)! \, m! \, 2^m} \int_0^\infty \cdots \int h_n(\tau_1, \ldots \tau_{n-2m}, \sigma_1, \sigma_1 \ldots \sigma_m, \, \sigma_m)$$
$$x(t - \tau_1) \cdots x(t - \tau_{n-2m}) \, d\tau_1 \cdots d\tau_{n-2m} d\sigma_1 \cdots d\sigma_m \quad (19)$$

contains the n-th order Wiener kernel h_n and the power level P of the GWN input.

The most widely used method for the estimation of Wiener kernels is the cross-correlation technique (Lee & Schetzen, 1965), according to which the n-th order Wiener kernel can be estimated from high-order input-output cross-correlations as:

$$h_n(\tau_1 \cdots \tau_n) = \frac{1}{n! \, P^n} E[y_n(t) \, x(t - \tau_1) \ldots x(t - \tau_n)] \quad (20)$$

where $y_n(t)$ is the n-th order output residual:

$$y_n(t) = y(t) - \sum_{m=0}^{n-1} G_m[h_m; x(t', t' \leq t)] \tag{21}$$

As indicated by Eqs. (20) and (21), the kernels are estimated successively, in ascending order. In practice the ensemble-average of Eq. (20) is evaluated as a time-average over the experimental input-output record under the assumption of system stationarity. Issues of actual implementation and estimation accuracy in this approach have been addressed extensively (Marmarelis & Marmarelis, 1978).

The Wiener kernels are, in general, different from the Volterra kernels of a given system (for which both expansions exist.) The n-th order Wiener kernel can be expressed in terms of the Volterra kernels of the same and higher order as (Marmarelis & Marmarelis, 1978; Marmarelis, 1987):

$$h_n(\tau_1, \ldots, \tau_n) = \sum_{m=0}^{\infty} \frac{(n+2m)! \, P^m}{n! \, m! \, 2^m} \int_0^{\infty} \cdots \int k_{n+2m}(\tau_1, .., \tau_n, \sigma_1, \sigma_1, .., \sigma_m, \sigma_m) d\sigma_1 .. d\sigma_m \tag{22}$$

Estimation of the Wiener kernels in the frequency domain is possible by evaluating high-order input-output cross-spectra (Brillinger, 1970; French & Butz, 1973):

$$H_n(j\omega_1, \ldots, j\omega_n) = \frac{1}{n! P^n} \Phi_{y_n x \cdots x}(j\omega_1, \ldots, j\omega_n) \tag{23}$$

where $\Phi_{y_n x \cdots x}$ is the n-th order cross-spectrum between the output residual y_n and the GWN input x. The relation between Volterra and Wiener kernels in the frequency domain is:

$$H_n(j\omega_1, \ldots, j\omega_n) = \sum_{m=0}^{\infty} \frac{(n+2m)! \, P^m}{n! \, m! \, 2^m \, (2\pi)^m}$$
$$\int_{-\infty}^{\infty} \cdots \int K_{n+2m}(j\omega_1, \ldots, j\omega_n, ju_1, -ju_1, \ldots, ju_m, -ju_m) du_1 \cdots du_m \tag{24}$$

For the specific class of systems described by Eq. (1), the Volterra kernels of order higher than first have the approximate form given by Eq. (17). Therefore, combining Eq. (17) with Eq. (24) we can obtain, in first approximation, the general expressions for the high-order Wiener kernels of this class of systems.

Equation (23) indicates that the result of (first-order) cross-spectral analysis is the first-order Wiener kernel in the frequency domain. This is commonly referred to as the (apparent) transfer function of the system, and yields a linearized model of the input-output relation:

$$\tilde{Y}(j\omega) = H_1(j\omega) X(j\omega) \tag{25}$$

for deterministic inputs, or:

$$\tilde{\Phi}_{yx}(j\omega) = H_1(j\omega)\Phi_{xx}(\omega) \qquad (26)$$

for stochastic inputs. This model is the best linear model (in output mean-square error sense) for GWN inputs.

Let us study how this model is affected by the presence of nonlinearities for various values of GWN input power level P. From Eq. (24) we have for $n = 1$:

$$H_1(j\omega) = K_1(j\omega) + \sum_{m=1}^{\infty} \frac{(2m+1)!P^m}{m!\,2^m\,(2\pi)^m}$$
$$\int_{-\infty}^{\infty} \cdots \int K_{2m+1}(j\omega, ju_1, -ju_1, \ldots, ju_m, -ju_m)du_1\cdots du_m \qquad (27)$$

which indicates that the (apparent) transfer function $H_1(j\omega)$ is a power series in P and depends on all odd-order Volterra kernels of the system. Note that $H_1(j\omega)$ coincides with $K_1(j\omega)$ (which represents the linear portion of the differential equation) for $P = 0$, as expected. Substituting $K_{2m+1}(\ )$ from the general expression of Eq. (17), we obtain the general (approximate) expression of $H_1(j\omega)$ for the considered class of systems:

$$H_1(j\omega) = K_1(j\omega) - \frac{K_1(j\omega)}{L(j\omega)} \sum_{m=1}^{\infty} \sum_{n=0}^{2m+1} \frac{(2m-n+1)!\,n!}{m!} \left(\frac{P}{4\pi}\right)^m c_{2m+1-n,n}$$
$$\int_{-\infty}^{\infty} \cdots \int R_{2m+1,n}(j\omega, ju_1, -ju_1, \ldots, ju_m, -ju_m)|K_1(u_1)\cdots K_1(u_m)|^2 du_1 \cdots du_m \quad (28)$$

Inspection of the function $R_{2m+1,n}(j\omega, ju_1, -ju_1, \ldots, ju_m, -ju_m)$, as defined following Eq. (17), indicates that its values for n even do not depend on ω whereas its values for n odd depend linearly on $(j\omega)$. This lead to the following interesting expression:

$$H_1(j\omega) = K_1(j\omega) - \frac{K_1(j\omega)}{L(j\omega)} \sum_{m=1}^{\infty} \frac{(P/2)^m}{m!}$$
$$\sum_{l=0}^{m}[(2m-2l+1)!\,(2l)!\,c_{2m-2l+1,2l} + (j\omega)(2m-2l)!\,(2l+1)!\,c_{2m-2l,2l+1}]\,Q_{m,l} \quad (29)$$

where,

$$Q_{m,l} = \frac{1}{(2\pi)^m} \int_{-\infty}^{\infty} \cdots \int R_{m,l}(u_1^2, \ldots, u_m^2)\,|K_1(u_1)\cdots K_1(u_m)|^2\,du_1 \cdots du_m \qquad (30)$$

Considering the definition of $R_{m,l}$, we see that the constants $Q_{m,l}$ depend on the Euclidian norms of $|K_1(u)|$ and $|uK_1(u)|$. For these quantities to be finite the degree of the polynomial $L(D)$ in Eq. (1) must be at least two degrees higher than the polynomial $M(D)$. Thus, Eq. (29) can be written as:

$$
\begin{aligned}
H_1(j\omega) &= K_1(j\omega) - \frac{K_1(j\omega)}{L(j\omega)}[A(P) + j\omega B(P)] \\
&= K_1(j\omega)[1 - \frac{A(P) + j\omega B(P)}{L(j\omega)}]
\end{aligned} \tag{31}
$$

where $A(P)$ and $B(P)$ are power series in P with coefficients dependent on $\{Q_{m,l}\}$ and $\{c_{i,j}\}$ for $(i+j)$ odd (i.e., i odd and j even for $A(P)$, and i even and j odd for $B(P)$ coefficients). Equation (31) indicates that $H_1(j\omega)$ is affected by the nonlinear terms of Eq. (1) for which $(i+j)$ is odd, and depends on the power level P of the GWN input. Therefore, the linearized model obtained through cross-spectral analysis may deviate considerably from the truly linear portion of the system (represented by K_1) depending on the values of A and B, which in turn depend on the input power level.

An analytical example

Consider as an example a system of this class described by the differential equation:

$$
L(D)y + c_{3,0}y^3 + c_{2,1}y^2(Dy) + c_{3,2}y^3(Dy)^2 + c_{0,5}(Dy)^5 = M(D)x \tag{32}
$$

where $|c_{i,j}| \ll 1$ and

$$
\begin{aligned}
L(D) &= a_2 D^2 + a_1 D + a_0 \tag{33} \\
M(D) &= b_0 \tag{34}
\end{aligned}
$$

Then, the only non-negligible quantities $\{Q_{m,l}\}$ in Eq. (29) are:

$$
Q_{1,0} = \frac{1}{2\pi} \int_{-\infty}^{\infty} |K_1(u_1)|^2 \, du_1 \tag{35}
$$

$$
Q_{2,1} = \frac{1}{4\pi^2} \int \int_{-\infty}^{\infty} (u_1^2 + u_2^2)|K_1(u_1)K_1(u_2)|^2 \, du_1 du_2 \tag{36}
$$

$$
Q_{2,2} = \frac{1}{4\pi^2} \int \int_{-\infty}^{\infty} u_1^2 u_2^2 |K_1(u_1)K_1(u_2)|^2 \, du_1 du_2 \tag{37}
$$

where,

$$
|K_1(u)|^2 = \frac{b_0^2}{[(a_0 - a_2 u^2)^2 + a_1^2 u^2]} \tag{38}
$$

Then, from Eqs. (29) and (31) we obtain:

$$
A(P) = 3c_{3,0}Q_{1,0}P + \frac{3}{2}c_{3,2}Q_{2,1}P^2 \tag{39}
$$

$$
B(P) = c_{2,1}Q_{1,0}P + 15c_{0,5}Q_{2,2}P^2 \tag{40}
$$

Therefore, the (apparent) transfer function of the linearized model of this system is:

$$H_1(s) = b_0 \frac{[a_2 s^2 + (a_1 - B)s + (a_0 - A)]}{(a_2 s^2 + a_1 s + a_0)^2} \tag{41}$$

For very small values of P, that make A and B negligible relative to a_0 and a_1 respectively, the measured transfer function $H_1(s)$ has the poles and zeroes of $K_1(s)$. However, for values of P for which A and B become significant relative to a_0 and a_1 respectively, two new zeroes emerge for $H_1(s)$ and its poles become double. These effects should become, in general, more pronounced as the value of P increases (i.e., the nonlinear effects are increasing in importance).

The case of static nonlinear feedback

A special case of considerable practical interest is when the nonlinearity involves only the output variable, y, and not its derivative, i.e.,

$$L(D)y + \varepsilon f(y) = M(D)x \tag{42}$$

where $|\varepsilon| \ll 1$. If the function $f()$ is analytic or can be approximated to an arbitrary degree of accuracy by a power series (note that the linear term is excluded since it can be absorbed into L):

$$f(y) = \sum_{n=2}^{\infty} \alpha_n y^n \tag{43}$$

then the resulting Volterra kernels are:

$$K_1(s) = \frac{M(s)}{L(s)} \tag{44}$$

$$K_n(s_1, \ldots, s_n) = -\varepsilon \alpha_n K_1(s_1) \ldots K_1(s_n)/L(s_1 + \ldots + s_n) \tag{45}$$

where terms of order ε^2 or higher have been considered negligible.

The first-order Wiener kernel in this case is (cf. Eq. (27)):

$$H_1(jw) = K_1(jw) \left\{ 1 - \frac{\varepsilon}{L(jw)} \sum_{m=1}^{\infty} \frac{(2m+1)!}{m!} \left(\frac{P\kappa}{2} \right)^m \alpha_{2m+1} \right\}$$

$$= K_1(jw) \left[1 - \frac{\varepsilon}{L(jw)} C_1(P) \right] \tag{46}$$

where,

$$\kappa = \frac{1}{2\pi} \int_{-\infty}^{\infty} |K_1(u)|^2 du \tag{47}$$

and the second-order Wiener kernel is:

$$
\begin{aligned}
H_2(jw_1, jw_2) &= -\varepsilon \frac{K_1(jw_1)K_1(jw_2)}{L(jw_1 + jw_2)} \sum_{m=o}^{\infty} \frac{(2m+2)!}{m!2} \left(\frac{P\kappa}{2}\right)^m \alpha_{2m+2} \\
&= -\varepsilon \frac{K_1(jw_1)K_1(jw_2)}{L(jw_1 + jw_2)} C_2(P) \tag{48}
\end{aligned}
$$

We observe that, as the input power level varies, the waveform of the first-order Wiener kernel changes but the second-order Wiener kernel remains unchanged in shape and changes only in scale. Note that the functions $C_1(P)$ and $C_2(P)$ are power series (or polynomials) in $(P\kappa)$ and characteristic of the system nonlinearities; furthermore the Wiener kernels approach their Volterra counterparts as the input power level diminishes (as expected).

These results indicate that, for a system with linear forward and weak nonlinear feedback (i.e., $|\varepsilon\alpha_i| \ll 1$), the first-order Wiener kernel in the time domain will be:

$$
h_1(\tau) = k_1(\tau) - \varepsilon C_1(P) \int_o^\tau k_1(\tau - \lambda)g(\lambda)d\lambda \tag{49}
$$

and the second-order Wiener kernel will be:

$$
h_2(\tau_1, \tau_2) = -\varepsilon C_2(P) \int_o^{min(\tau_1,\tau_2)} k_1(\tau_1 - \lambda)k_1(\tau_2 - \lambda)g(\lambda)d\lambda \tag{50}
$$

where $g(\lambda)$ is the inverse Fourier transform of $1/L(jw)$.

Effect of GWN input mean level

A companion issue to that of changing input power level is the effect of changing the mean level of the experimental GWN input. This is a rather common situation in experimental investigations of physiological systems, whereby different mean levels are used for the experimental input (with white-noise perturbations superimposed on them) in order to explore different ranges of the system function (Marmarelis & Naka, 1973; Sakai & Naka, 1987). The resulting kernels for each different mean level of the input will vary, in general, for a nonlinear system. In order to reconcile these different measurements, we can use a "reference mean level" μ_o and refer to its corresponding kernels $\{k_n^o\}$ the kernels $\{k_n^\mu\}$ obtained from different mean levels μ, according to the relation:

$$
k_n^\mu(\tau_1, \ldots, \tau_n) = \sum_{\ell=o}^{\infty} \frac{(n+\ell)!}{n!\,\ell!} (\mu - \mu_0)^\ell \int_o^\infty \cdots \int k_{n+\ell}^0(\tau_1, \ldots, \tau_n, \sigma_1, \ldots, \sigma_\ell)\, d\sigma_1 \cdots d\sigma_\ell \tag{51}
$$

The correspondence with the Wiener kernel measurements is given by the relation:

$$h_n^\mu(\tau_1,\ldots,\tau_n) \;=\; \sum_{m=0}^{\infty}\sum_{\ell=0}^{\infty}\frac{(n+2m+\ell)!}{n!\,m!\,\ell!}\left(\frac{P}{2}\right)^m(\mu-\mu_0)^\ell$$

$$\int_0^{\infty}\!\!\cdots\!\int k_{n+2m+\ell}^0(\tau_1,\ldots,\tau_n,\lambda_1,\lambda_1,\ldots,\lambda_m,\lambda_m,\sigma_1,\ldots,\sigma_\ell)d\lambda_1\cdots d\lambda_m\,d\sigma_1\cdots d\sigma_\ell \quad (52)$$

and in the frequency domain:

$$H_n^\mu(\omega_1,\ldots,\omega_n) \;=\; \sum_{m=0}^{\infty}\sum_{\ell=0}^{\infty}\frac{(n+2m+\ell)!}{n!\,m!\,\ell!}\left(\frac{P}{4\pi}\right)^m(\mu-\mu_0)^\ell$$

$$\int_{-\infty}^{\infty}\!\!\cdots\!\int K_{n+2m+\ell}^0(\omega_1,\ldots,\omega_n,u_1,-u_1,\ldots,u_m,-u_m,0,\ldots,0)\,du_1\cdots du_m \quad (53)$$

The first-order Wiener kernel for the class of nonlinear feedback systems discussed above (cf. Eq. (42)) is given in terms of the "reference" Volterra kernels (when $\mu_o = 0$) by the expression:

$$h_1^\mu(\tau) = k_1(\tau) - \varepsilon\int_o^\tau g(\lambda)k_1(\tau-\lambda)d\lambda\cdot\left\{\sum_{\substack{m=o\\ m+\ell\geq 1}}^{\infty}\sum_{l=o}^{\infty}\frac{(2m+\ell+1)!}{m!\ell!}\left(\frac{P\kappa}{2}\right)^m(\mu\gamma)^\ell\alpha_{2m+\ell+1}\right\}$$

$$(54)$$

where,

$$\gamma = \int_o^\infty k_1(\lambda)d\lambda \quad (55)$$

and all other parameters and functions are as defined before (k_1 is the zero-mean input first-order Volterra "reference" kernel). Note that the first-order Wiener kernel for $\mu \neq o$ is also affected by the even-order terms of the nonlinearity, unlike the case of $\mu = o$ where it is affected only by the odd-order terms of the nonlinearity.

In the following section we use computer simulations to demonstrate the effect of changing input power level and/or mean level on the waveform of the first-order Wiener kernel, and draw the analogy with changes observed in the first-order Wiener kernels of some sensory systems when the GWN input power level and/or mean level is varied experimentally.

COMPUTER SIMULATIONS OF SOME NONLINEAR FEEDBACK SYSTEMS

We demonstrate the theoretical results obtained in the previous section by simulating nonlinear feedback systems of the aforementioned class (cf. Eq. (42)) and observing the effects of various nonlinearities, input power levels and input mean levels on the estimated first–order Wiener kernels.

Cubic feedback systems

First, we consider a system with a low-pass forward linear subsystem (L) and a cubic negative feedback $(f = \varepsilon y^3)$ as shown in Fig. 1 (for $M \equiv 1$). For $|\varepsilon| \ll 1$, the first–order Wiener kernel is: (cf. Eq. (49) and note that $g \equiv k_1$ in this case)

$$h_1(\tau) = k_1(\tau) - 3\varepsilon P\kappa \int_o^\tau k_1(\lambda)k_1(\tau - \lambda)d\lambda \qquad (56)$$

where $k_1(\tau)$ is the impulse response function of the low–pass linear forward subsystem (as well as the first-order Volterra kernel of the overall system) shown in Fig. 2. For a GWN input with zero mean and unity power level, we compute the first-order Wiener kernel estimates (based on 4,096 input/output data points) for $\varepsilon = 0.004, 0.008, 0.016$ and 0.032. The resulting estimates are shown along with the estimate for $\varepsilon = 0$ (i.e., no cubic feedback) in Fig. 3. We observe a gradual decrease of damping (i.e., emergence of an "undershoot") of the kernel estimates and a gradual increase of their bandwidth as the degree of the cubic feedback (ε) increases. This is also demonstrated in Fig. 4, where the FFT magnitude functions of these kernel estimates are shown up to normalized frequency of 0.2 Hz (Nyquist frequency is 0.5 Hz). We observe the gradual transition from an overdamped to an underdamped mode as ε increases. To examine whether these results are in agreement with Eq. (56), we compute the differences of the Wiener kernel estimates $\{h_1\}$ for nonzero ε values from the kernel estimate for $\varepsilon = 0$ (i.e., the estimate of k_1). These differences are shown in Fig. 5, and they exhibit the scalar dependence on ε predicted by Eq. (56) as well as the similarity of waveform with the convolution of k_1 with itself, i.e.,

$$[k_1(\tau) - h_1(\tau)] = 3\varepsilon P\kappa \int_0^\tau k_1(\lambda)k_1(\tau - \lambda)d\lambda \qquad (57)$$

To demonstrate this similarity of waveform, we show in Fig.6 the convolution of the k_1 estimate with itself along with the properly normalized (i.e., by $3\varepsilon P\kappa$) kernel difference for the case of $\varepsilon = 0.004$. The normalized kernel differences for all the aforementioned values of ε are shown in Fig. 7. The agreement is better for the smaller values of ε and begins to break down as ε increases in value. This is expected because of the assumption of very small ε used in the theoretical derivations of the previous section. The observed disparity is two-fold: (1) deviation from linearity versus ε in the scaling of the kernel differences, and (2) slight deviations in the theoretically predicted waveform. This disparity is due to contributions of higher (than third) odd-order Volterra kernels which were neglected in our theoretical derivations as being negligible (i.e., of order ε^2 or higher in magnitude). We observe that the deviation in scaling amounts to a slight reduction in the linear scaling relation with ε (as expected since the fifth-order Volterra kernel, being the most important contributor to this deviation, is proportional to ε^2 - - i.e., it has opposite sign from the third-order

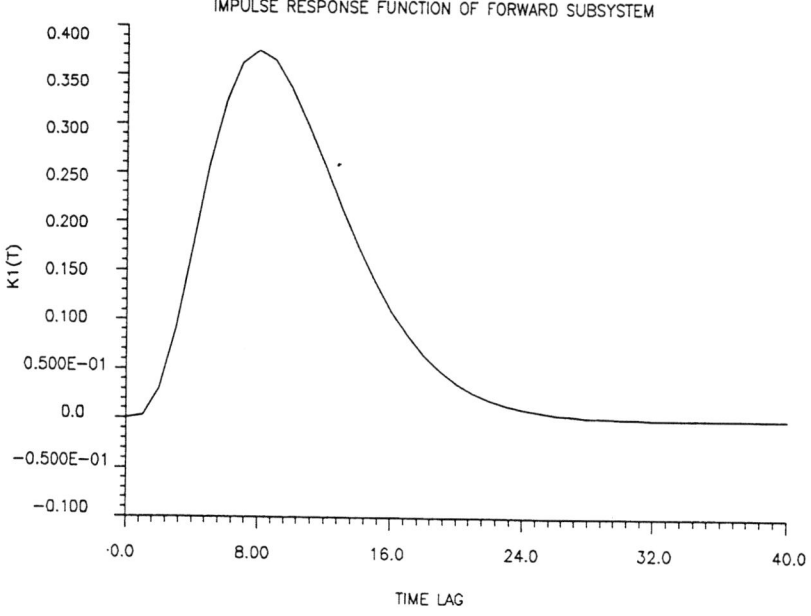

Fig. 2. Impulse response function of overdamped linear forward subsystem (L) used in the simulation example.

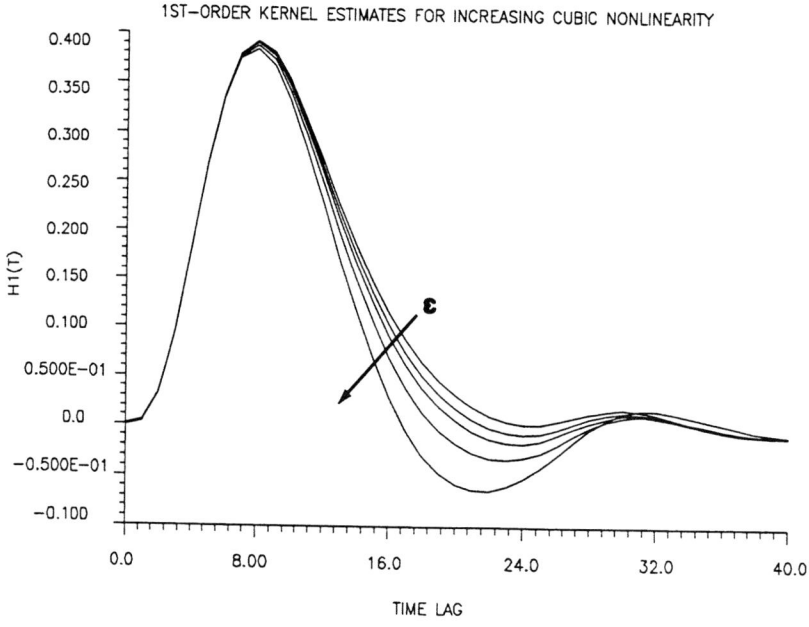

Fig. 3. First–order Wiener kernel estimates of system with cubic feedback nonlinearity and the forward subsystem shown in Fig. 2, for $\varepsilon = 0$, 0.004, 0.008, 0.016 and 0.032 ($P = 1$ in all cases).

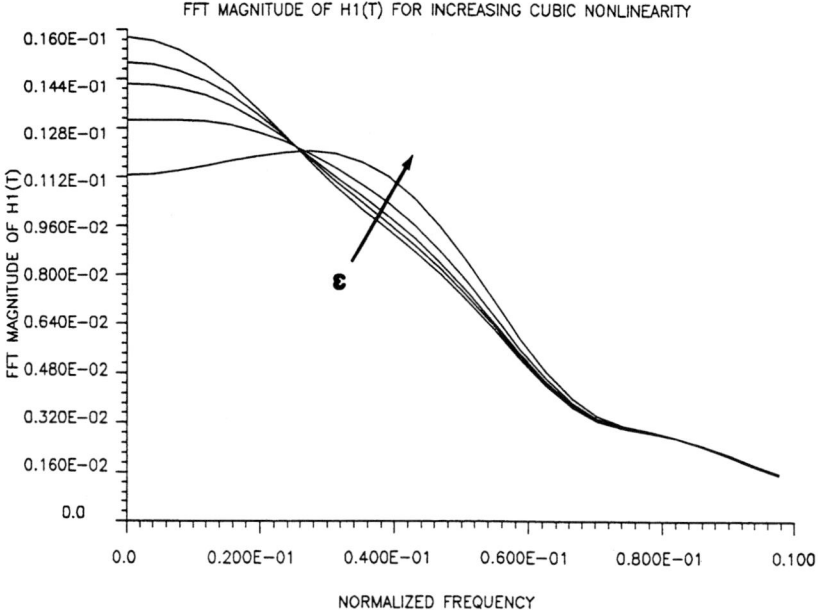

Fig. 4. FFT magnitudes of the Wiener kernel estimates shown in Fig. 3. Curves shown are for $\varepsilon = 0$, 0.004, 0.008, 0.016 and 0.032. Zero–frequency gain decreases and bandwidth increases for increasing ε.

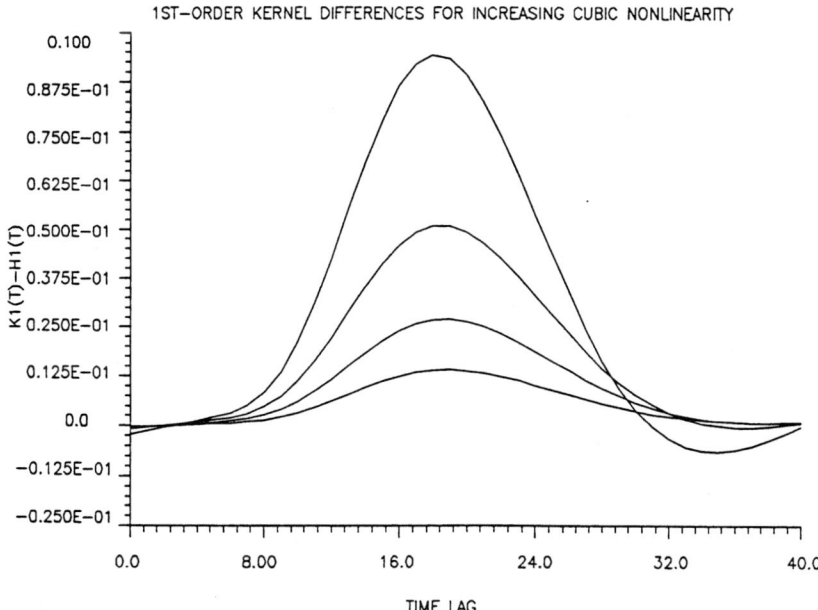

Fig. 5. Differences between first–order Wiener kernel estimates and Volterra kernel for $\varepsilon = 0.004$, 0.008, 0.016 and 0.032. Curves are increasing in magnitude for increasing $\varepsilon(P = 1$ in all cases).

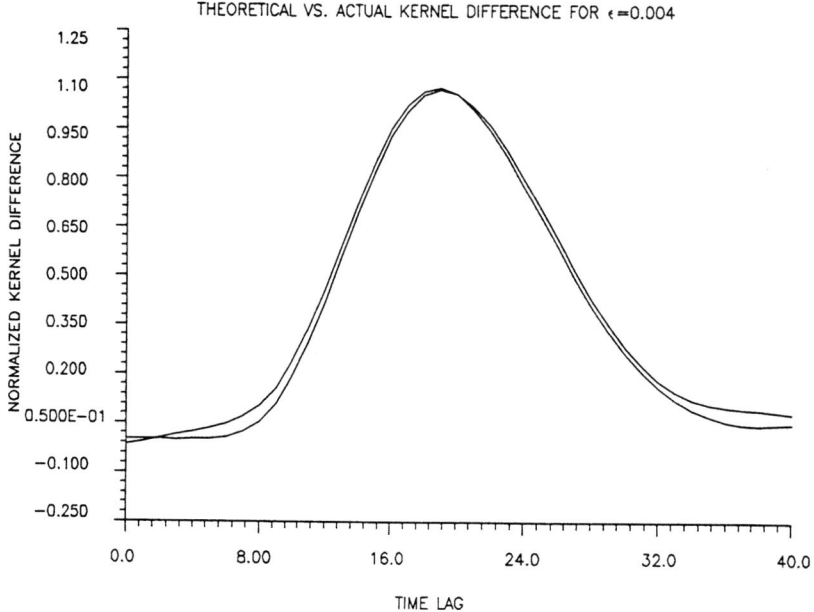

Fig. 6. Convolution of $k_1(\tau)$ with itself, superimposed on the first–order kernel difference shown in Fig. 5 for $\varepsilon = 0.004$ (normalized by $3\varepsilon P \kappa$) to demonstrate the validity of Eq. (57).

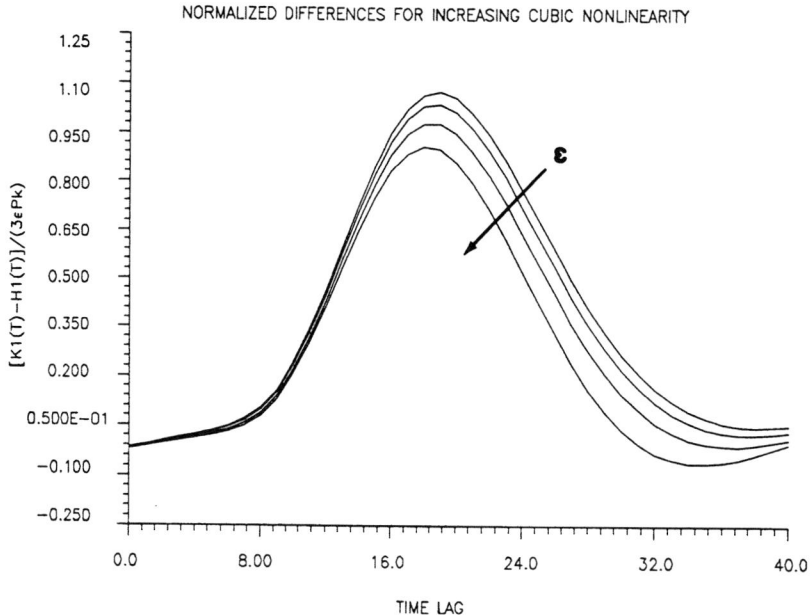

Fig. 7. First–order Wiener kernel differences for $\varepsilon = 0.004$, 0.008, 0.016 and 0.032, normalized by $3\varepsilon P\kappa$. We observe decreasing magnitude for increasing ε ($P = 1$ in all cases).

Volterra kernel on which the theoretical kernel difference of Eq. (57) is based). The deviation in waveform can be explained similarly by looking at the structure of the fifth-order Volterra kernel for this system. In the Laplace domain we have:

$$K_5(s_1, s_2, s_3, s_4, s_5) = \frac{3}{10}\varepsilon^2 K_1(s_1)K_1(s_2)K_1(s_3)K_1(s_4)K_1(s_5)$$
$$K_1(s_1 + s_2 + s_3 + s_4 + s_5) \sum_{j_1,j_2,j_3} K_1(s_{j_1} + s_{j_2} + s_{j_3}) \qquad (58)$$

where the summation extends over all ten combinations of (j_1, j_2, j_3) from the set of integers (1, 2, 3, 4, 5). The contribution of K_5 to H_1 in the Fourier domain is:

$$
\begin{aligned}
H_{1,5}(w) &= \frac{15P^2}{(2\pi)^2} \int^\infty \int_{-\infty} K_5(w, u_1, -u_1, u_2, -u_2) du_1 du_2 \\
&= \frac{9}{2}\left(\frac{P\varepsilon}{2\pi}\right)^2 K_1^2(w) \int^\infty \int_{-\infty} |K_1(u_1)|^2 |K_1(u_2)|^2 Q(w, u_1, u_2) du_1 du_2 \quad (59)
\end{aligned}
$$

where the function Q is the sum of ten terms, four of which have the form $K_1(w \pm u_1 \pm u_2)$, two are $K_1(w)$ and the remaining four are $K_1(\pm u_1)$, $K_1(\pm u_2)$. Further analytical manipulations of Eq. (59) lead to the following expression:

$$
\begin{aligned}
H_{1,5}(w) = 9(P\varepsilon)^2 K_1^2(w) \Big\{ &\kappa\nu + \frac{\kappa^2}{2}K_1(w) \\
&+ \frac{1}{(2\pi)^2}\int^\infty \int_{-\infty} |K_1(u_1)K_1(u_2)|^2 K_1(w - u_1 - u_2) du_1 du_2 \Big\} \quad (60)
\end{aligned}
$$

where κ was defined in Eq. (47) and,

$$\nu = \frac{1}{2\pi}\int_{-\infty}^\infty |K_1(u)|^2 Re\{K_1(u)\} du \qquad (61)$$

Conversion of Eq. (60) to the time domain yields:

$$
\begin{aligned}
h_{1,5}(\tau) &= 9(P\varepsilon)^2 \Big\{ \kappa\nu c_1(\tau) + \frac{\kappa^2}{2}\int_o^\infty k_1(\lambda)c_1(\tau - \lambda)d\lambda \\
&+ \int_o^\infty k_1(\lambda)a_1(\lambda)c_1(\tau - \lambda)d\lambda \Big\} \quad (62)
\end{aligned}
$$

where $a_1(\tau)$ and $c_1(\tau)$ are the autocorrelation and the convolution of $k_1(\tau)$ with itself, respectively. This rather complicated expression explains the deviations from the approximate relation of Eq. (56) when ε is increased. To illustrate the form of the three terms in the brackets of Eq. (62) for this example, we show them separately in Fig. 8 (traces 1, 2 and 3, respectively, in the order they appear in Eq. (62)). The

17

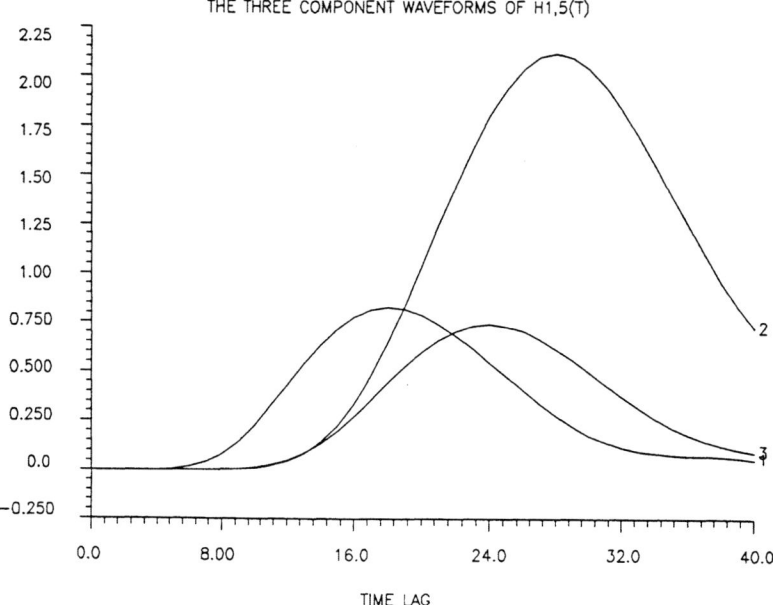

Fig. 8. Waveforms of the three components of $h_{1,5}(\tau)$ described by Eq. (62), normalized by $9(P\varepsilon)^2$. Traces 1, 2 and 3 correspond to the order the terms appear in Eq. (62).

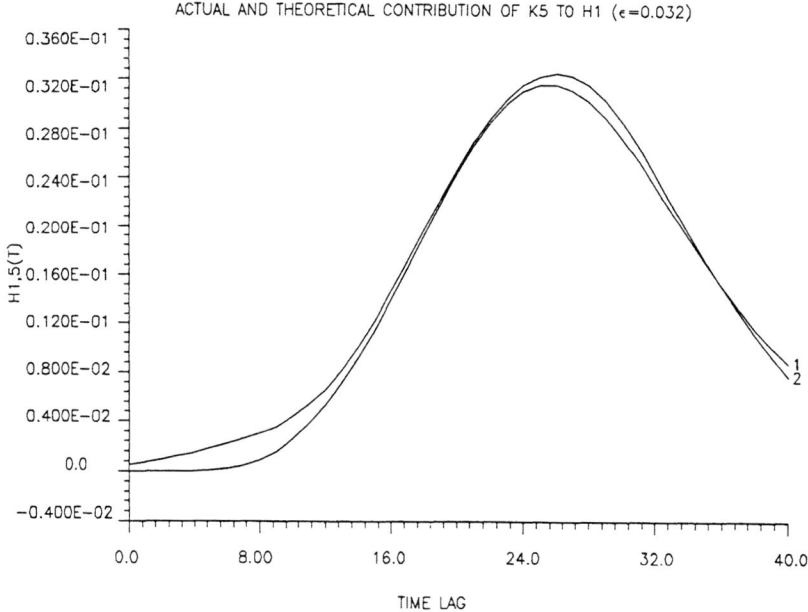

Fig. 9. The composite waveform $h_{1,5}(\tau)$ described by Eq. (62) (trace 1) superimposed on the actual deviation of kernel waveform due to k_5 (trace 2) for $\varepsilon = 0.032$.

18

composite waveform of $h_{1,5}(\tau)$ described by Eq. (62) is shown in Fig. 9 properly scaled along with the actual waveform deviation of $h_1(\tau)$ from the prediction of Eq. (56) for $\varepsilon = 0.032$. We observe good agreement between the two curves (kernel estimation errors notwithstanding) that corroborates our analytical derivations.

In the next round of simulations we kept the value of ε constant and varied the input power level P. The obtained results were in agreement with the theoretical expressions of Eq. (56) and Eq. (62) demonstrating that the role of P and ε is interchangeable insofar their effect on the first-order Wiener kernel. As indicated in our theoretical expressions this effect is determined by the product (εP).

This theoretical analysis and supporting computer simulations demonstrate the fact that the first-order Wiener kernel of a low-pass forward linear subsystem with cubic feedback makes a gradual transition from overdamped to underdamped mode (within the region of stable behavior) as the input power level increases (with a companion increase of bandwidth and decrease of zero-frequency gain).

Next we explore the effect of varying the (white noise) input mean level while keeping ε and P constant. We simulated the previous system for $\varepsilon = 0.001$ and $P=1$ using input mean levels $\mu = 0, 1, 2$ and 4, successively. The response amplitude histograms are shown in Fig. 10, and the cubic nonlinearity used in negative feedback is shown in Fig. 11. The obtained first-order Wiener kernel estimates are shown in Fig. 12. We observe that the changes in the waveform of the kernels as the input mean level increases are qualitatively similar to the ones induced by increasing input power level - - viz., increasing bandwidth and decreasing damping and sensitivity (zero-frequency gain). According to the general expression of Eq. (54), we have for this system:

$$h_1^\mu(\tau) = k_1(\tau) - 3\varepsilon[P\kappa + (\mu\gamma)^2] \int_0^\tau k_1(\lambda)k_1(\tau - \lambda)d\lambda \qquad (63)$$

Comparing Eq. (63) with Eq. (56), we see that, for small ε, the effect of increasing P is similar to the effect of increasing μ^2 (which also implies that the effect is the same for positive or negative μ) and the differential effect is proportional to κ and γ^2 respectively. The latter observation implies that the differential effect of changing P or μ^2 by the same amount is greater for mean level changes in the case of a low-pass forward subsystem, since:

$$\frac{\gamma^2}{\kappa} = \frac{\{\int_o^\infty k_1(\tau)d\tau\}^2}{\int_o^\infty k_1^2(\tau)d\tau} \geq 1 \qquad (64)$$

However, for underdamped or band-pass forward subsystems this differential effect may be reversed, since the ratio of Eq. (64) will be probably less than unity.

The qualitative changes in the first–order kernel waveform, that have been observed in these simulations as the GWN input power and mean levels change, have been also seen in experimental investigations of retinal horizontal cells (viz., the gradual transition from an overdamped to an underdamped mode). However, in the

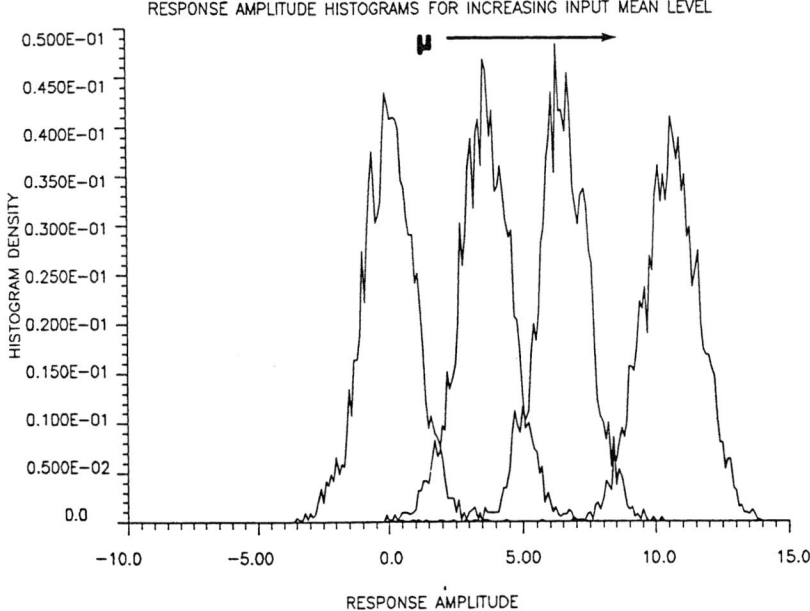

Fig. 10. Amplitude histograms of cubic feedback system responses for GWN inputs ($P = 1$) with mean levels $\mu = 0$, 1, 2 and 4.

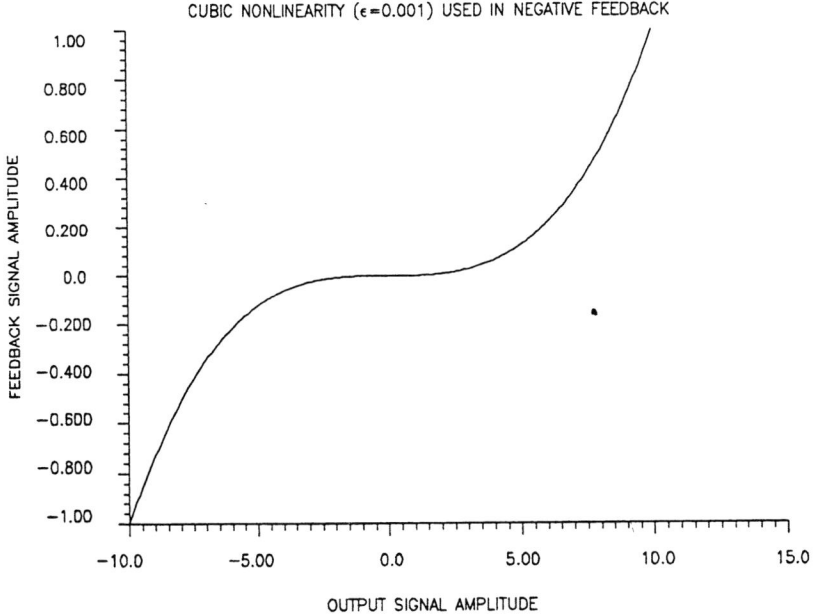

Fig. 11. Form of the cubic nonlinearity used in negative feedback ($\varepsilon = 0.001$).

20

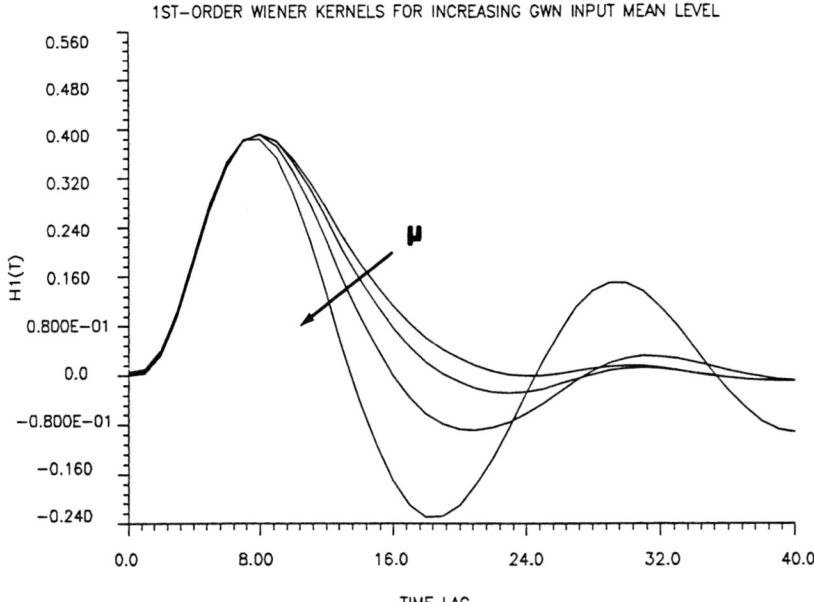

Fig. 12. First–order Wiener kernels for GWN input mean levels $\mu = 0$, 1, 2 and 4 ($P = 1$, $\varepsilon = 0.001$).

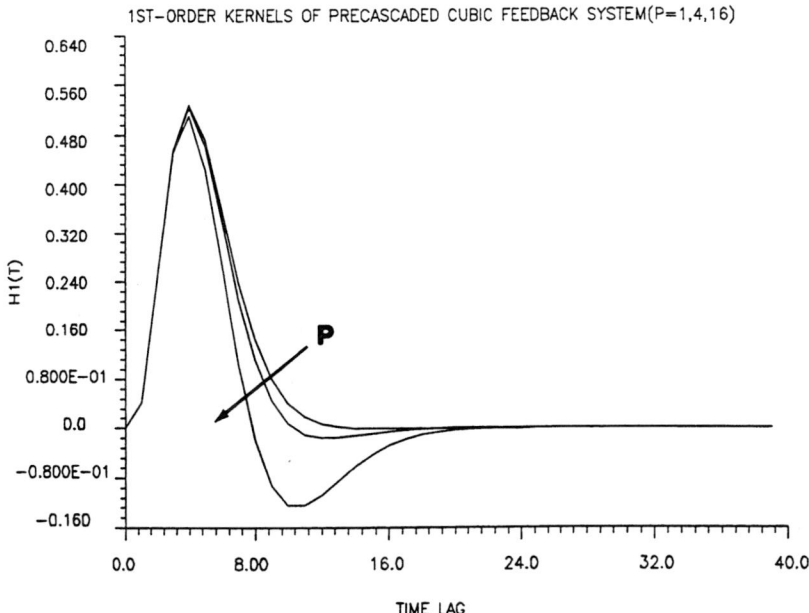

Fig. 13. First–order Wiener kernels of another cubic feedback system ($\varepsilon = 0.004$) with broader bandwidth for $P = 1$, 4 and 16.

case of retinal horizontal cells, the gradual reduction in the size of the kernel (zero-frequency gain) and peak–response (latency) time, as P and/or μ increase, is much more pronounced than in these simulations. To account for the greater reduction in kernel size, we may introduce a compressive (static) nonlinearity in cascade with the overall feedback system, that leads to an additional reduction of the gain of the overall cascade system as P and/or μ increase. On the other hand, a greater reduction in the peak–response (latency) time may require the introduction of another dynamic component in cascade with the feedback system.

To demonstrate the latter point, a similar feedback system is introduced in front of the previous one, which has a low–pass linear forward subsystem (of twice the bandwidth of the one shown in Fig. 2) and negative cubic feedback (with $\varepsilon = 0.004$). The first–order Wiener kernels of this feedback system are shown in Fig. 13 for $P = 1, 4$ and 16. The kernels of the cascade of the two feedback systems (the second one having $\varepsilon = 0.001$) are shown in Fig. 14 for $P = 1, 4$ and 16. We observe greater reduction of the kernel size and peak–response time as P increases, relative to the case of the second feedback system alone. As discussed before, the reduction in kernel size can be made greater by introducing a compressive (static) nonlinearity at the end of this cascade or in–between the two cascaded feedback systems.

The possibility of nonlinear feedback in the receptor-horizontal cell complex in the retina has been postulated in the past (Marmarelis & Naka, 1973) and the presented simulations seem to provide additional support for this hypothesis. Greater changes in first–order kernel waveform have been observed in experiments that used steady illumination for the surround (annulus) of the receptive field of the horizontal cell (Sakai & Naka, 1987) in addition to GWN stimulus at the center (spot). This is consistent with the hypothesis that surround stimulation makes a stronger contribution to the nonlinear feedback mechanism, possibly due to the integrative function of the horizontal cell over the surround of the receptive field, that triggers the nonlinear feedback to the centrally active receptors. The presented analysis offers the quantitative means for testing rigorously this hypothesis and for estimating the form of this nonlinear feedback on the basis of Eq. (54) using experimental kernel measurements obtained for various values of P and μ.

Another point of practical interest, that must be made in connection with this example, is the difference between the first–order kernel (Volterra or Wiener) and the system response to an impulse. This point is often the source of confusion due to biases engrained by linear system analysis. For a third–order system, such as in this example for small ε, the response to an impulse input $x(t) = A\delta(t)$ is:

$$
\begin{aligned}
r(t) &= Ak_1(t) + A^3 k_3(t, t, t) \\
&= Ak_1(t) - \varepsilon A^3 \int_0^t k_1(\lambda) k_1^3(t - \lambda) d\lambda
\end{aligned}
\tag{65}
$$

This is clearly different from the first–order Volterra kernel $k_1(t)$ and its Wiener counterpart given by Eq. (56). This fact is demonstrated in Fig. 15, for the aforementioned system, where the responses to impulses of strengths $A = 1, 2$ and 4 are shown (in all cases $\varepsilon = 0.032$). The waveforms of these responses become increasingly less damped as the impulse strength increases (cf. Eq. (65)), however, all of

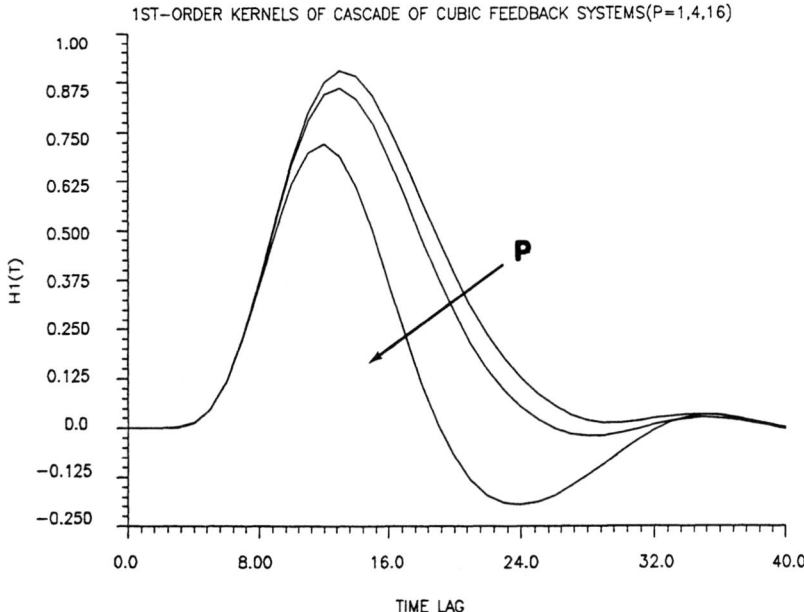

Fig. 14. First–order Wiener kernels of the cascade of two cubic feedback systems for $P = 1$, 4 and 16.

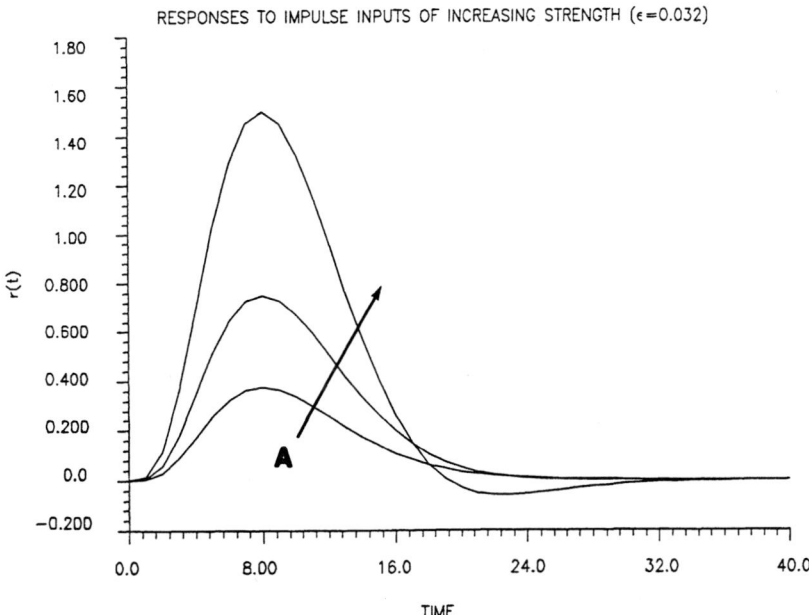

Fig. 15. Response of cubic feedback system ($\varepsilon = 0.032$) to impulsive inputs of strengths $A = 1$, 2 and 4.

23

them are considerably more damped than the first–order Wiener kernel in this case
($\varepsilon = 0.032$) for $P = 1$. This is demonstrated in Fig. 16, where the Wiener kernel is
shown along with the response to an impulse of strength $A = 4$ (normalized by A).
Note that in this simulation $P = 1$ corresponds to a GWN input variance of 1 (since
the GWN input bandwidth is 0.5 Hz) and, therefore, the GWN input amplitude
never exceeds 4 (the impulse strength) within the data–record used. Nonetheless,
the Wiener kernel is considerably less damped indicating the "energy sensitivity"
of the nonlinear feedback system. The latter point is further illustrated in Fig. 17,
where the system response to a unit pulse is shown ($\varepsilon = 0.032$). The on–set response
is highly underdamped and distinctly different from both the integrated first–order
kernel or impulse response (the "step response" of linear system theory). Observe
also the sharp difference between on–set and off–set response, characteristic of nonlinear system and so often seen in physiological systems. Note also that this system
becomes unstable when we double the amplitude of the pulse.

Finally, to demonstrate the changes in pulse response waveforms as the pulse
amplitude increases, we show in Fig. 18 the responses of the system (when $\varepsilon = 0.001$)
for pulse amplitudes of 1,2 and 4. The observed changes are qualitatively consistent
with the previous discussion, however, we cannot obtain the first–order Wiener kernel
or the response to an impulse by differentiating them in time as in the linear case.
For small values of ε, the response to a step function input $x(t) = Au(t)$ is:

$$s(t) = A \int_0^t k_1(\tau)d\tau - \varepsilon A^3 \int_0^t k_1(\tau) \left\{ \int_0^t k_1(\lambda - \tau)d\lambda \right\}^3 d\tau \qquad (66)$$

Note also that the steady–state value of the step response for various values of A is
given by:

$$L(0)y + \varepsilon y^3 = A \qquad (67)$$

in the region of stability of this system, where $L(0) = 1/K_1(0)$ for this system. The
steady–state values of the pulse response as a function of pulse amplitude are shown
in Fig. 19, for this system when $\varepsilon = 0.001$. Note that these values are different, in
general, from the mean response value when the GWN input has nonzero mean. For
small values of ε, this steady–state value is approximately (cf. Eq. (66)):

$$y_A \cong AK_1(0) - \varepsilon A^3 K_1^4(0) \qquad (68)$$

In the next series of simulations, we consider an underdamped linear forward
subsystem (its impulse response function shown in Fig. 20) with the same negative cubic feedback as before. The resulting first–order Wiener kernel estimates for
increasing GWN input power level (viz., $P = 1, 2$ and 4) are shown in Fig. 21 (for
$\varepsilon = 0.008$) along with the first–order Volterra kernel of the system (which is the same
as the impulse response function of the linear forward subsystem, shown in Fig. 20,
and the Wiener kernel for very small values of P). We observe a gradual deepening

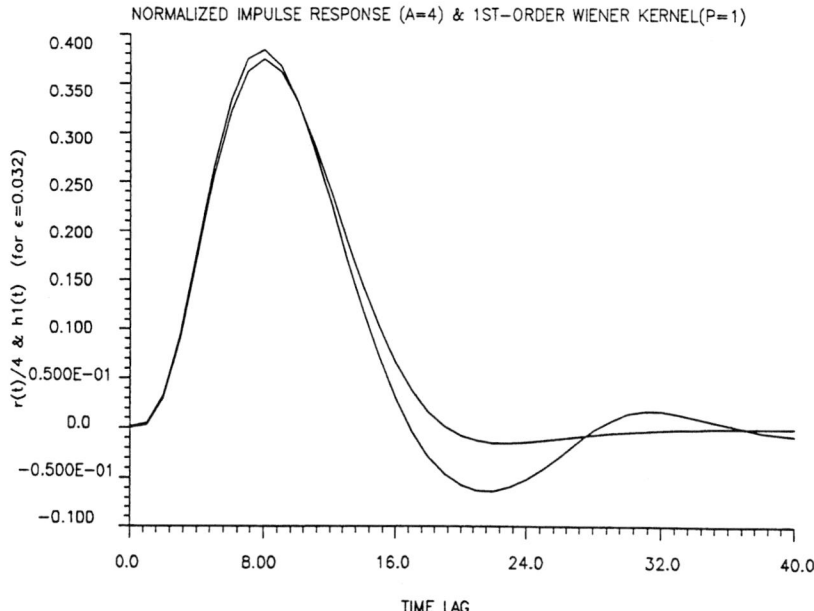

Fig. 16. Normalized impulse response ($A = 4$) and first–order Wiener kernel ($P = 1$) for cubic feedback system ($\varepsilon = 0.032$).

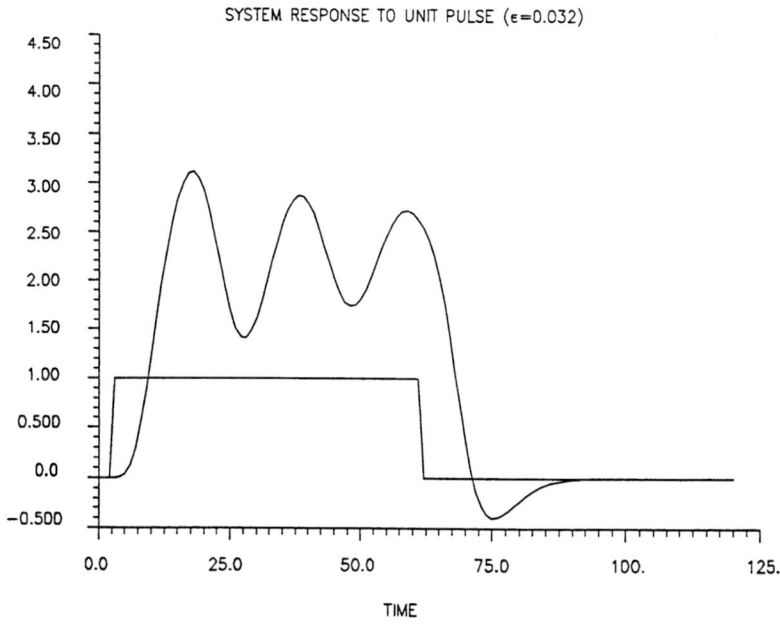

Fig. 17. Response of cubic feedback system ($\varepsilon = 0.032$) to unit pulse input. Observe the highly underdamped on–set response in contrast with the off–set response.

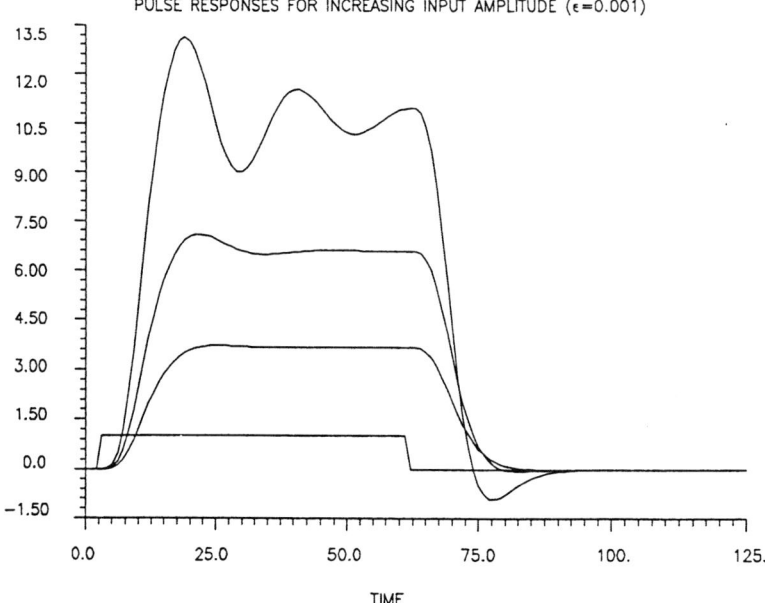

Fig. 18.   Responses of cubic feedback system ($\varepsilon = 0.001$) to pulse inputs of different amplitudes (1, 2 and 4). Observe the gradual decrease of damping and latency of the on–set response.

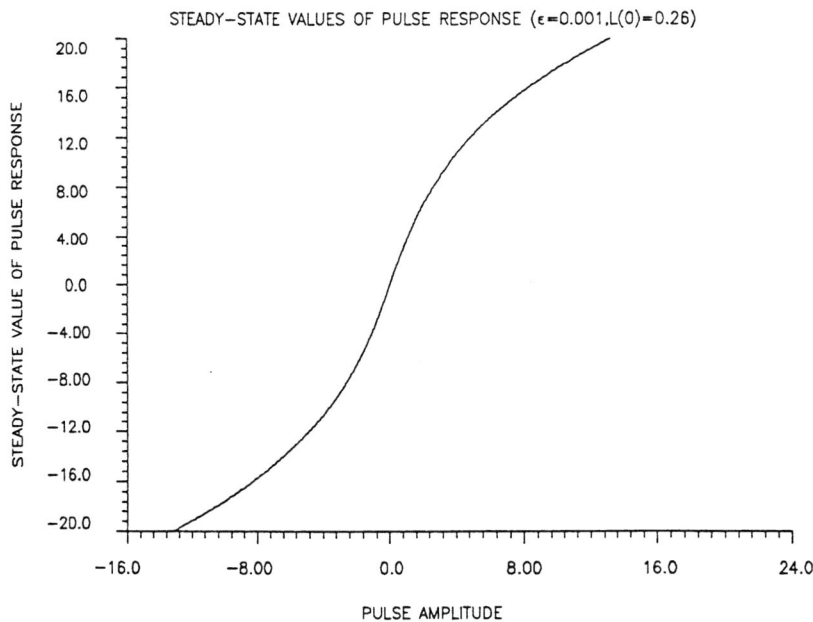

Fig. 19.   The steady–state values of the pulse responses as a function of input pulse amplitude for the cubic feedback system ($\varepsilon = 0.001$).

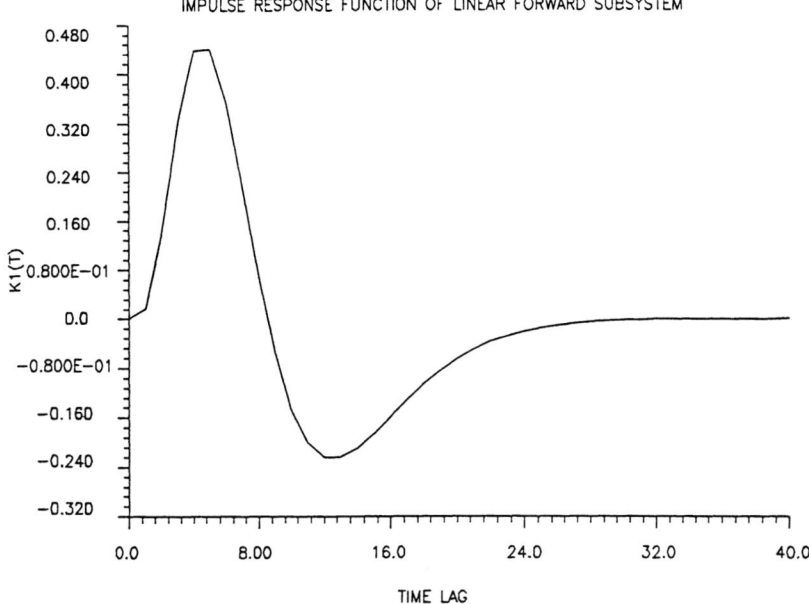

Fig. 20. Impulse response function of the underdamped linear forward subsystem.

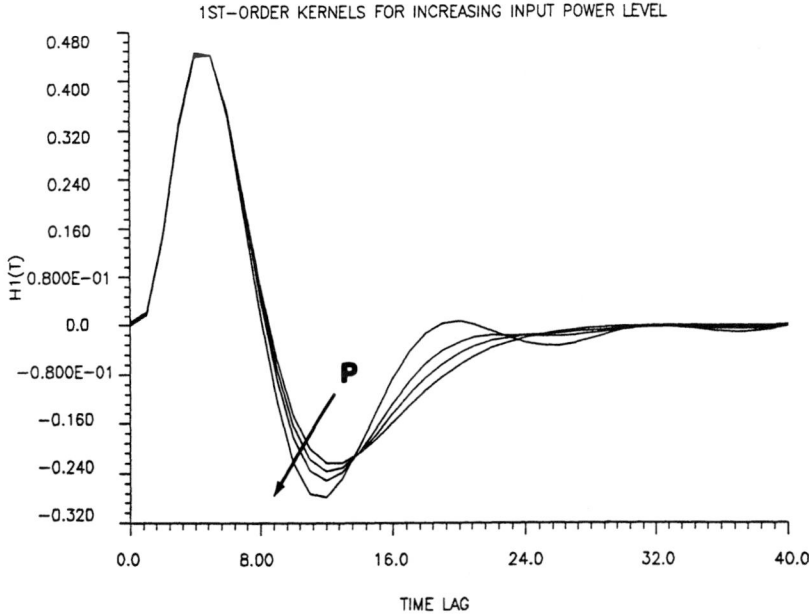

Fig. 21. First–order Wiener kernels of cubic feedback system ($\varepsilon = 0.008$) with the underdamped linear forward of Fig. 20, for $P = 0.1$, 1, 2 and 4.

of the undershoot portion of the kernel associated with a gradual shortening of its duration as P increases (i.e., we see a gradual broadening of the system bandwidth and upward shift of the resonance frequency as P increases). This is demonstrated in Fig. 22 where the FFT magnitudes of the kernels of Fig. 21 are shown. The changes in the waveform of these kernels with increasing P are consistent with our analysis in the previous example (i.e., Eqs. (56) and (62)). Next we compute the first–order Wiener kernels for $\varepsilon = 0.008$, $P = 1$ and increasing GWN input mean level. The resulting kernels for $\mu = 0, 1, 2$ and 4 are shown in Fig. 23, and they demonstrate insignificant effect of the nonzero mean levels due to the fact that γ (i.e., the area under k_1) is extremely small in this case, as predicted by Eq. (63). Finally, the system response to pulses of increasing amplitude ($A = 1, 2$ and 4) are shown in Fig. 24, demonstrating increasing resonance frequency and decreasing damping in the pulse response as A increases. Note also that the steady–state values of the pulse responses are extremely small, and the on–set/off–set response waveforms are similar (with reversed polarity), due to the very small value of $\gamma = K_1(0) = 1/L(0)$ (cf. Eq. (67)).

Sigmoid feedback systems

The next round of simulations deals with a different type of feedback nonlinearity. A sigmoid nonlinearity is chosen because, unlike the cubic one, it is bounded for any response signal amplitude and allows stable behavior of the feedback system regardless of input power. The following normalized arctangent function, shown in Fig. 25, for $\varepsilon = 1$ and $\alpha = 0.25$ was used in the simulations:

$$f(y) = \varepsilon \frac{2}{\pi} arctan(\alpha y) \tag{69}$$

For the previous low–pass forward subsystem, the resulting first–order Wiener kernels for $P = 1$ and $\varepsilon = 0, 0.1, 0.25$ and 0.5 are shown in Fig. 26 ($\alpha = 0.25$). The qualitative changes in waveform are similar to the cubic feedback case, for increasing feedback strength. However, for fixed sigmoid feedback strength (ε) the kernels resulting from increasing GWN input power level follow the reverse transition in waveform. This is demonstrated in Fig. 27, where the kernels obtained for $P = 1, 4, 16$ and 64 are shown ($\varepsilon = 1$ for all cases). The changes in kernel waveform follow the previously presented analysis, bearing in mind that the first–order Volterra kernel of this system is not the same as the impulse response function of the forward subsystem, but the impulse response function of the overall linear feedback system when the linear term of the sigmoid nonlinearity (i.e., its slope at zero) is incorporated in the (negative) feedback loop. Thus, the kernel waveform changes gradually from the impulse response function of the "linear" feedback system to that of the linear forward subsystem as P increases from very small to very large values. These two "limit" waveforms are shown in Fig. 28 for the case of $\varepsilon = 1$. The kernel waveform changes gradually from underdamped to overdamped as P increases (i.e., the gain of the "equivalent linearized feedback" decreases). This is demonstrated by the kernel FFT magnitudes shown in Fig. 29 for the kernels in Fig. 27.

Because of the bounded nature of the sigmoid nonlinearity, large values of ε and/or P do not lead to system instabilities as in the case of cubic feedback. In-

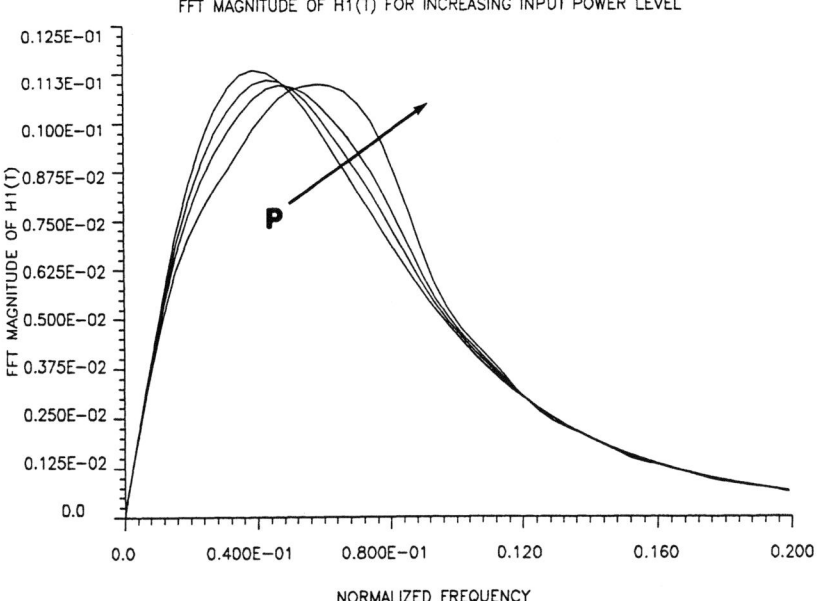

Fig. 22. FFT magnitudes of the first–order Wiener kernels shown in Fig. 21. Observe the gradual increase of bandwidth.

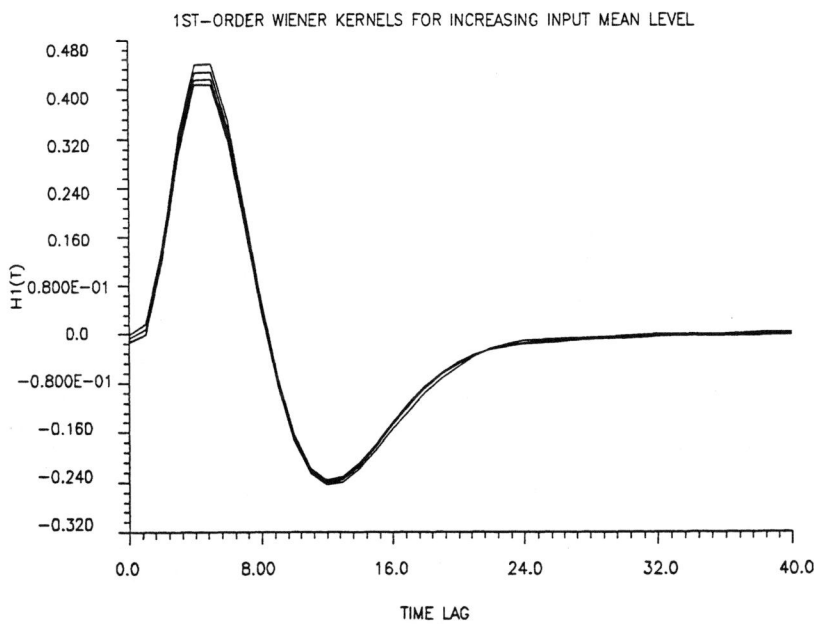

Fig. 23. First–order Wiener kernels of cubic feedback system ($\varepsilon = 0.008$) with underdamped forward for different GWN input mean levels $\mu = 0, 1, 2$ and 4 ($P = 1$).

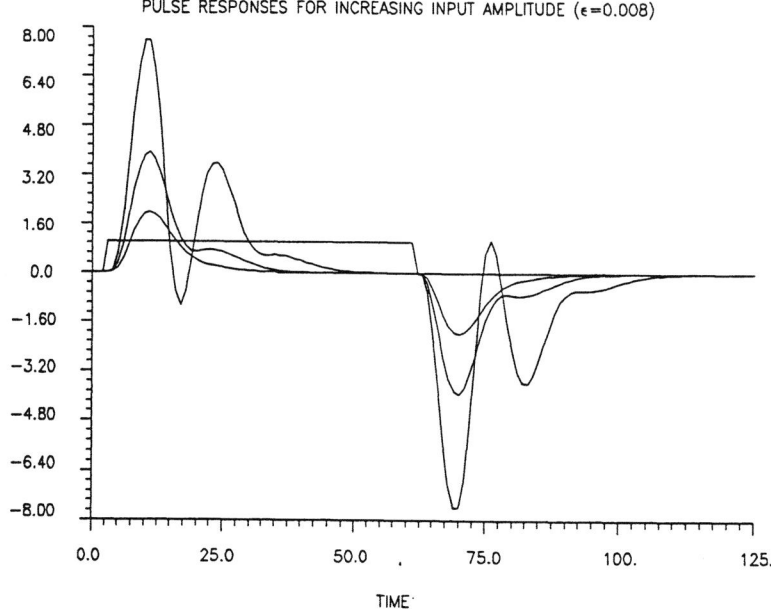

Fig. 24. Response of cubic feedback system ($\varepsilon = 0.008$) with underdamped forward to pulse inputs of different amplitudes $A = 1$, 2 and 4.

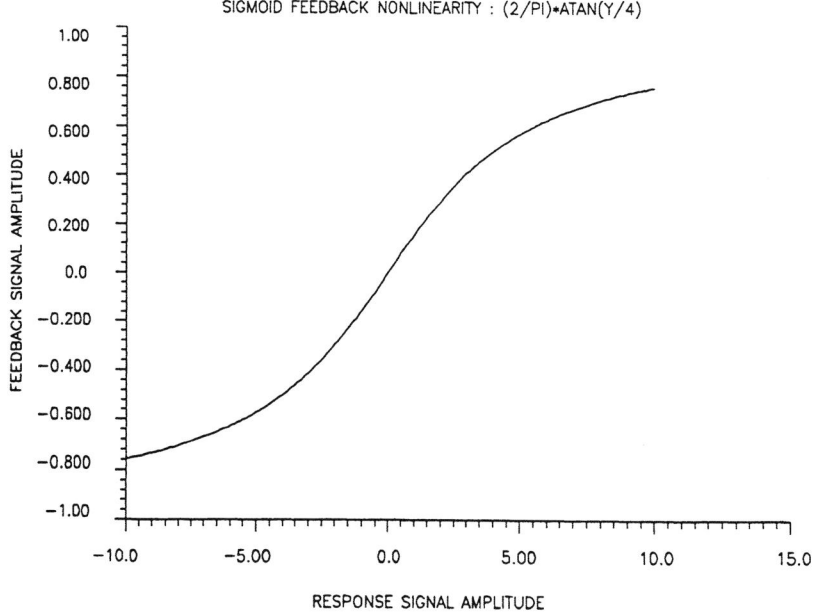

Fig. 25. Sigmoid feedback nonlinearity $f(y) = (2 / \pi) \, arctan \, (y/4)$.

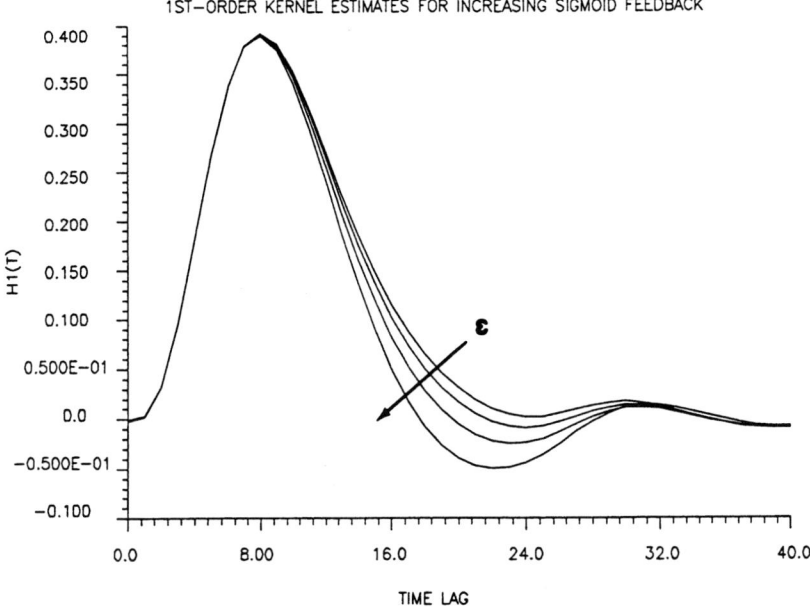

Fig. 26. First–order Wiener kernels of sigmoid feedback system with low–pass forward, for $\varepsilon = 0$, 0.1, 0.25 and 0.5 ($P = 1$).

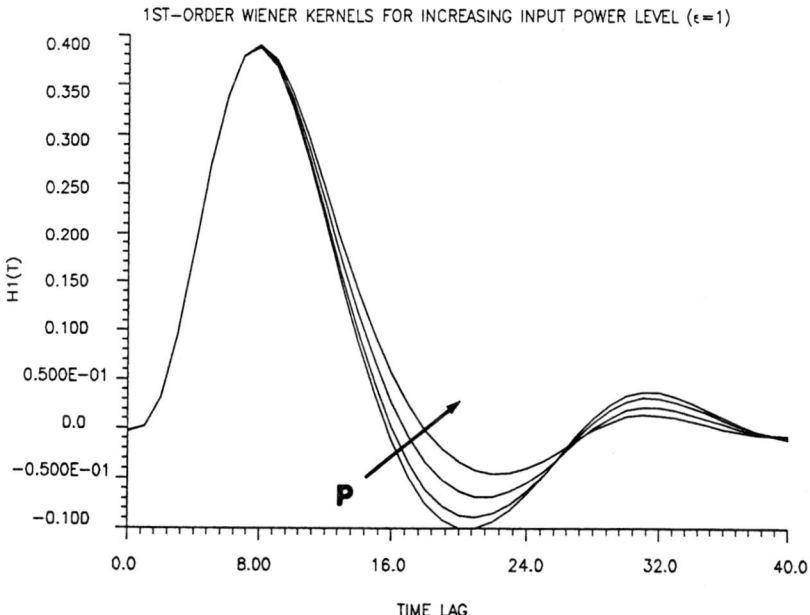

Fig. 27. First–order Wiener kernels of sigmoid feedback system ($\varepsilon = 1$) for $P = 1$, 4, 16 and 64.

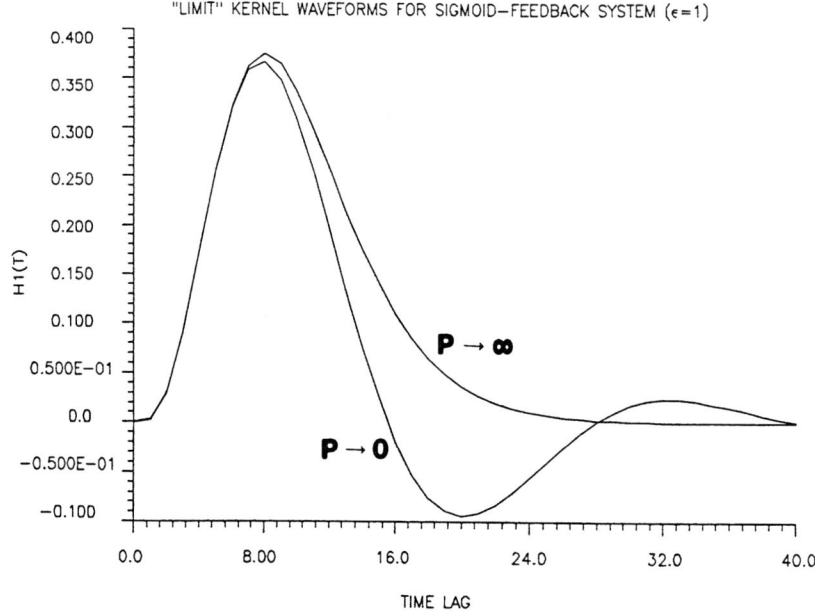

Fig. 28. The two "limit" kernel waveforms of sigmoid feedback system ($\varepsilon = 1$) with low-pass forward, for $P \to 0$ and $P \to \infty$.

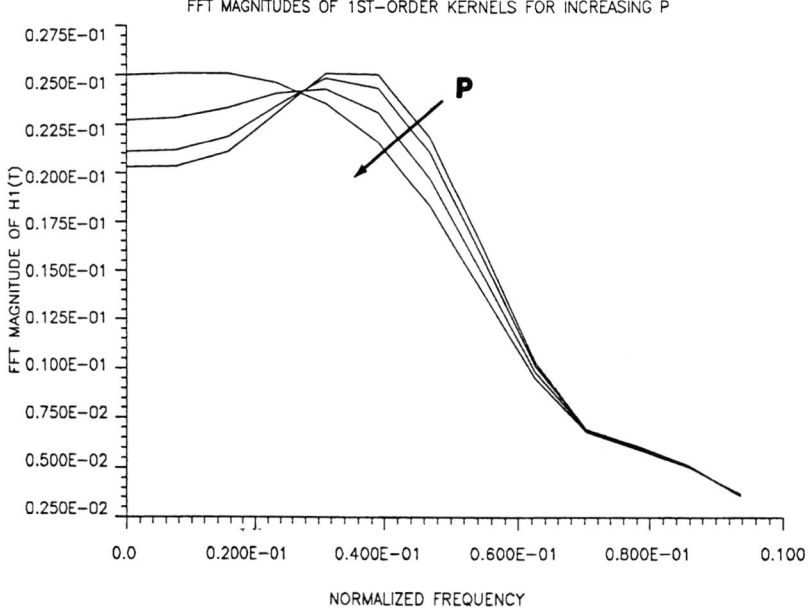

Fig. 29. FFT magnitudes of the first-order Wiener kernels shown in Fig. 27. Observe the gradual reduction of bandwidth as P increases.

creasing values of ε result in decreasing damping, eventually leading to oscillatory behavior. This is demonstrated in Fig. 30, where the kernels for $\varepsilon = 1$, 2 and 4 are shown ($P = 1$). The oscillatory behavior of the system, for very large values of ε, is more dramatically demonstrated in Fig. 31 where the first–order Wiener kernels for $\varepsilon = 100$ and 1000 are shown ($P = 1$). The actual system response $y(t)$ in these two cases is shown in Fig. 32. Clearly the system goes into perfect oscillation regardless of the GWN input, due to the overwhelming action of the nonlinear negative feedback that is both bounded and symmetric about the origin. The amplitude of this oscillation is proportional to ε, but it is independent of the input power level. In fact, the oscillatory response remains the same in amplitude and frequency for any input signal (regardless of its amplitude and waveform) as long as the value of ε is much larger than the maximum value of the input. The initial transient and the phase of the oscillation, however, may vary according to the input power and waveform. The frequency and amplitude of the oscillation depend on the linear forward subsystem. For instance, a low–pass subsystem with shorter memory (i.e., shorter time support) leads to higher frequency of oscillation, and so does an underdamped system with the same memory extent. The amplitude of the oscillation depends also on the strength of the negative sigmoid feedback as illustrated in Figs. 31 and 32.

The case of oscillatory behavior due to large sigmoid negative feedback is not covered by the Volterra–Wiener analysis presented in the previous section. It is, however, of great interest in physiology because of the numerous and functionally important physiological oscillators. It is a subject worthy of further exploration, albeit outside the scope of this article.

We return to the study of the sigmoid feedback for small values of ε. The effect of varying the slope of the sigmoid nonlinearity is demonstrated in Fig. 33, where the first–order kernels for $\varepsilon = 1$, $P = 1$ and slopes $\alpha = 0.125, 0.25, 0.5$ and 1 are shown. We observe gradually decreasing damping with increasing slope. This transition reaches asymptotically a limit in both directions of changing α values, as expected. For $\alpha \to \infty$, the sigmoid nonlinearity becomes the signum function and leads to perfect oscillations; and for $\alpha \to 0$ the gain of the feedback loop diminishes leading to a kernel identical to the impulse response function of the forward linear subsystem.

The effect of nonzero input mean level is demonstrated in Fig. 34, where the kernels for $\mu = 0, 1, 2$ and 4 are shown ($\varepsilon = 1$, $P = 1$, $\alpha = 0.25$). The kernels become more damped as the GWN input mean level increases, following the transition pattern of increasing input power level (i.e., decreasing gain of the "equivalent linearized feedback"). The system response to pulses of increasing amplitude are shown in Fig. 35 for $\varepsilon = 1$, $\alpha = 0.25$ and amplitudes $A = 1, 2$ and 4. Note the progressive difference between on–set and off–set responses.

In the case of the underdamped forward subsystem and sigmoid (negative) feedback the results are qualitatively similar to the previous case. The changes in the kernel waveform undergo a gradual transition from the linearized feedback system to the forward linear subsystem as the GWN input power level increases from very small to very large values. The two "limit" kernel waveforms are shown in Fig. 36 for this case. The effect of the sigmoid feedback is less dramatic in this case, since the kernel retains its underdamped mode for all values of P. There is, however, a downward shift of resonance frequency and increase of damping when P increases, as indicated

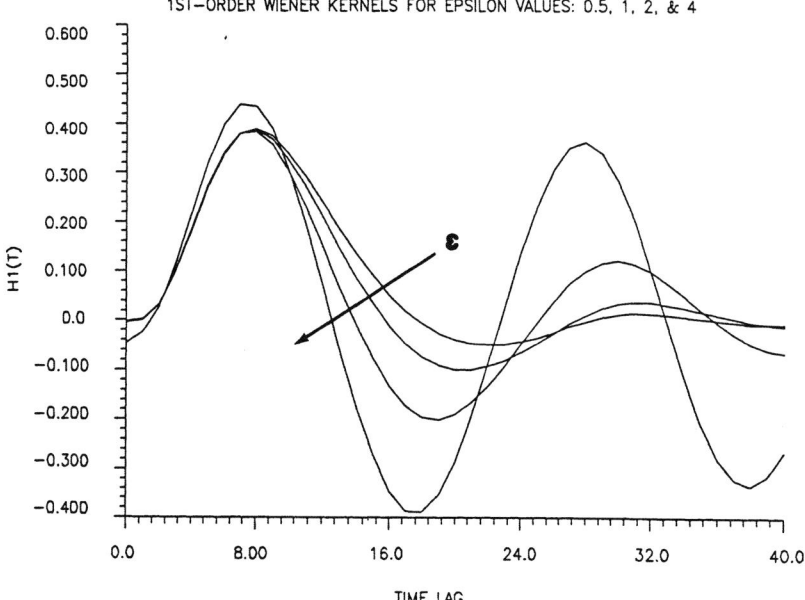

Fig. 30.   First–order Wiener kernels of sigmoid feedback system with low–pass forward, for $\varepsilon = 0.5$, 1, 2 and 4. Observe transition to oscillatory behavior as ε increases.

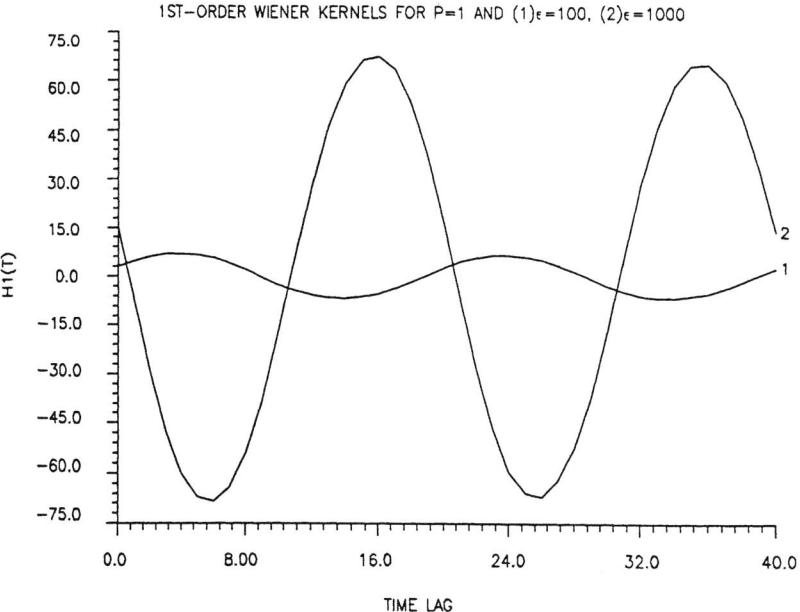

Fig. 31.   First–order Wiener kernels of sigmoid feedback system for $\varepsilon = 100$ and 1000. Oscillatory behavior has been established with amplitude proportional to ε.

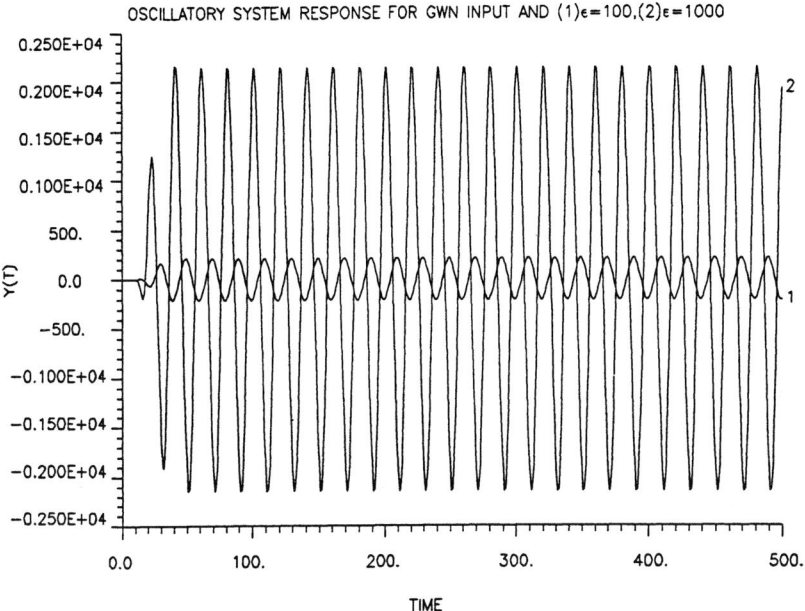

Fig. 32. Oscillatory system response for GWN input ($P = 1$) and $\varepsilon = 100$ and 1000.

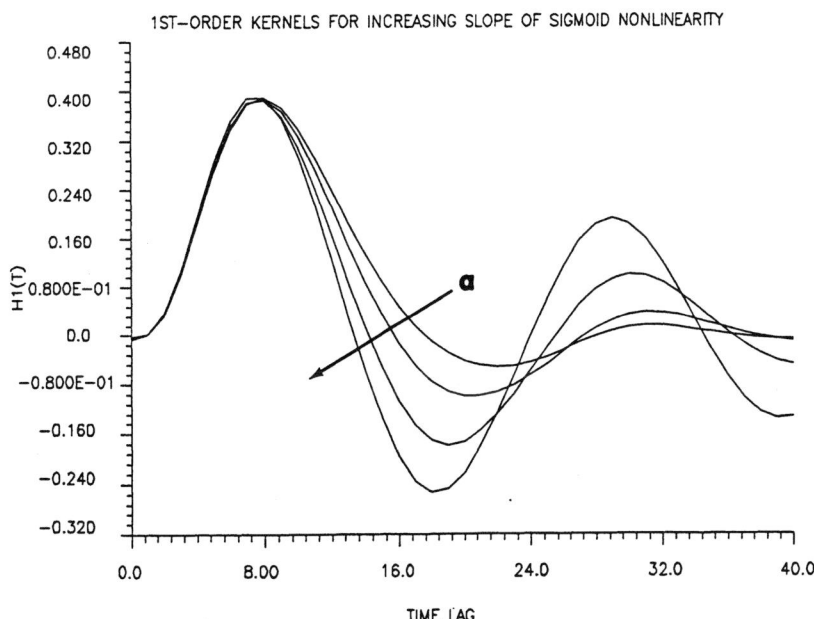

Fig. 33. First–order Wiener kernels of sigmoid feedback system ($\varepsilon = 1$, $P = 1$) for increasing slope of sigmoid curve: $\alpha = 0.125$, 0.25, 0.5 and 1.

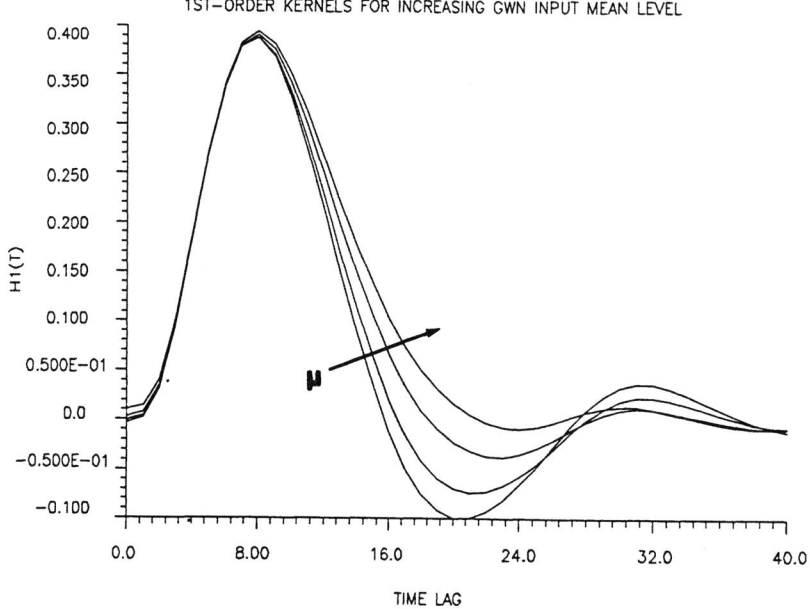

Fig. 34. First–order Wiener kernels of sigmoid feedback system ($\varepsilon = 1$, $P = 1$, $\alpha = 0.25$) for different GWN input mean levels $\mu = 0$, 1, 2 and 4.

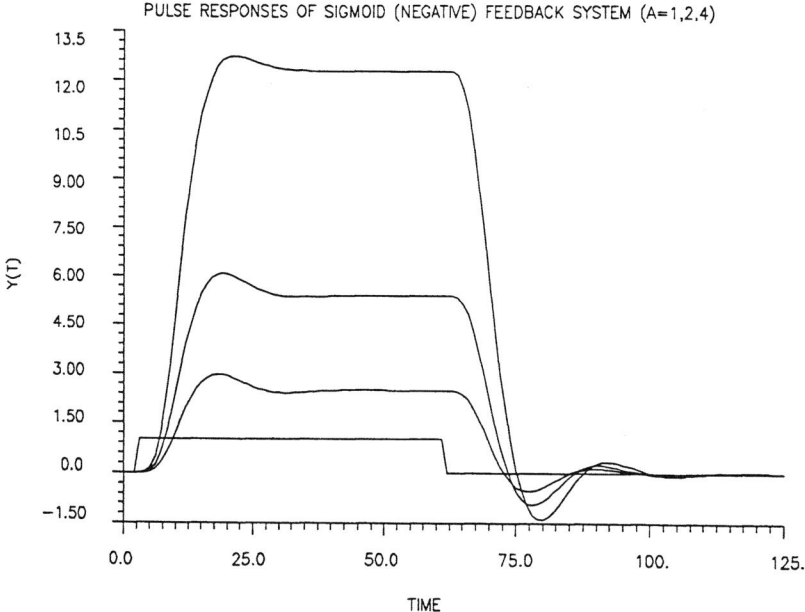

Fig. 35. Response of sigmoid feedback system ($\varepsilon = 1$, $\alpha = 0.25$) with low–pass forward, for pulse inputs of amplitudes $A = 1$, 2 and 4. Observe slightly faster than linear increase of steady–state response for increasing A.

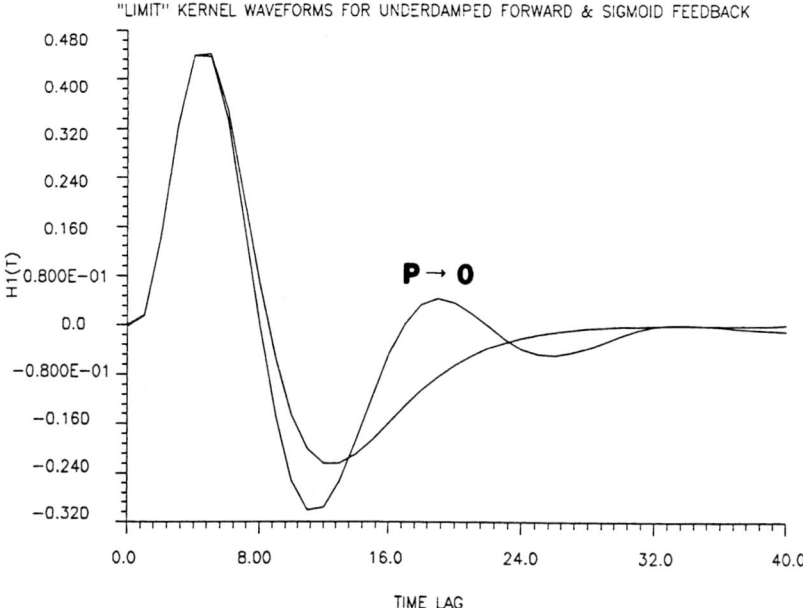

Fig. 36. The two "limit" Wiener kernel waveforms of (negative) sigmoid feedback system ($\varepsilon = 1$, $\alpha = 0.25$) with underdamped forward, obtained for $P \to 0$ and $P \to \infty$.

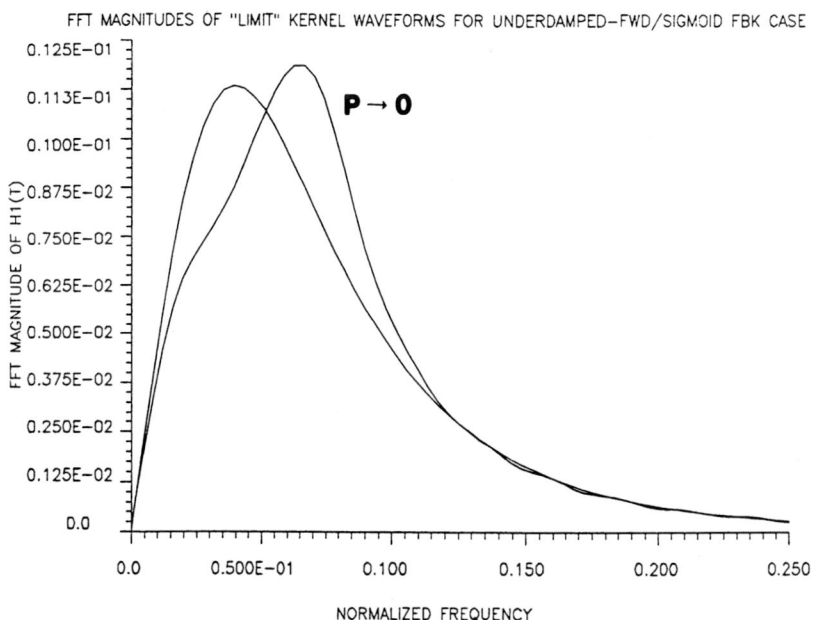

Fig. 37. FFT magnitudes of the two "limit" kernel waveforms shown in Fig. 36. Observe the lower resonance frequency for $P \to \infty$.

by the FFT magnitudes of the "limit" kernel waveforms (of Fig. 36) shown in Fig. 37. Note also that the reverse transition in kernel waveform occurs (i.e., upward shift of resonance frequency and decrease of damping with increasing P values) when the sigmoid feedback becomes positive. The one "limit" kernel waveform (for $P \to 0$) in the latter case is, of course, different than in the case of negative sigmoid feedback. This kernel has narrower bandwidth and is shown in Fig. 38 along with the other "limit" kernel waveform (for $P \to \infty$) which is the same as in the negative feedback case. The FFT magnitudes of these two "limit" kernel waveforms are shown in Fig. 39.

The reverse transition in kernel waveform is also observed when the polarity of cubic feedback is changed, as expected (see Eq. (63)). An example of the effect of positive cubic feedback is shown in Figs. 40 and 41, where the first–order Wiener kernel estimates for the previously discussed example of underdamped forward subsystem are shown for $\varepsilon = 0.008$ and $\varepsilon = -0.008$ ($P = 1$) in the time domain (Fig. 40) and in the frequency domain (Fig. 41). The positive cubic feedback leads to a decrease in resonance frequency and higher gain values in the resonant region. To retain stability in the case of negative cubic feedback for large values of P, a compressive (sigmoid–type) nonlinearity may be included in the forward loop following the linear subsystem. The qualitative behavior of sigmoid and cubic feedbacks follow the reverse pattern of transition in kernel waveforms as P increases. This pattern is also reversed when the feedback becomes positive instead of negative. The great advantage of sigmoid versus cubic feedback is that stability of the system behavior is retained regardless of the value of the GWN input power level. For this reason, sigmoid feedback is an appealing candidate for models of physiological feedback systems. For those systems that exhibit transitions to broader bandwidth and decreased damping as P increases (such as the retinal cells discussed above), candidate models may include either negative cubic feedback or positive sigmoid feedback. For those systems that exhibit the reverse transition patterns (i.e., to narrower bandwidth and increased damping) as P increases, candidate models may include either positive cubic feedback or negative sigmoid feedback.

An interesting example of a band–pass sensory system whose first–order Wiener kernel undergoes a transition to lower resonance frequencies as the input power level increases is found in auditory nerve fibers that have center (resonance) frequencies between 1.5 and 6 KHz (Moller, 1983).

To explore whether positive cubic feedback may constitute a plausible model in this case, we consider a band–pass linear forward subsystem (whose impulse response function is shown in Fig. 42) and positive cubic feedback with $\varepsilon = 0.01$. The obtained first–order Wiener kernel estimates of the overall feedback system are shown in Fig. 43 for GWN input power level $P = 1$ and 4. The FFT magnitudes of these kernels are shown in Fig. 44. We observe a gradual decrease of the resonance frequency as P increases, consistent with our previous analysis, and the experimental observations in auditory nerve fibers. However, these simulation results do not duplicate the experimentally observed broadening of the tuning curve (i.e., the resonance region) in auditory nerve fibers as P increases. Note also that this feedback system becomes unstable for larger values of the GWN input power level.

For these reasons, we explore also the possibility of negative sigmoid feedback for which system stability is guaranteed for all values of P. We consider a band–

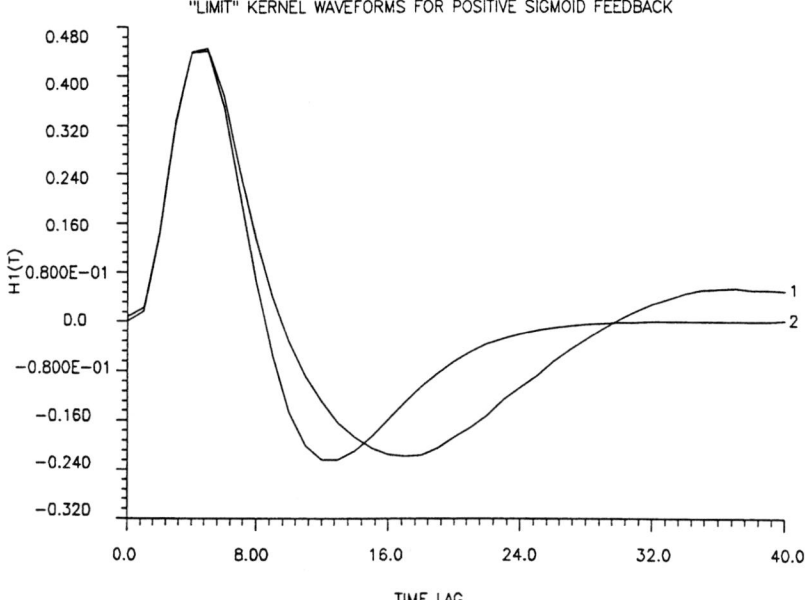

Fig. 38. The two "limit" Wiener kernel waveforms of positive sigmoid feedback system ($\varepsilon = -1$, $\alpha = 0.25$) with underdamped forward, obtained for $P \to 0$ (trace 1) and $P \to \infty$ (trace 2). Note that the kernel for $P \to \infty$ is the same as in Fig. 36 for $P \to \infty$.

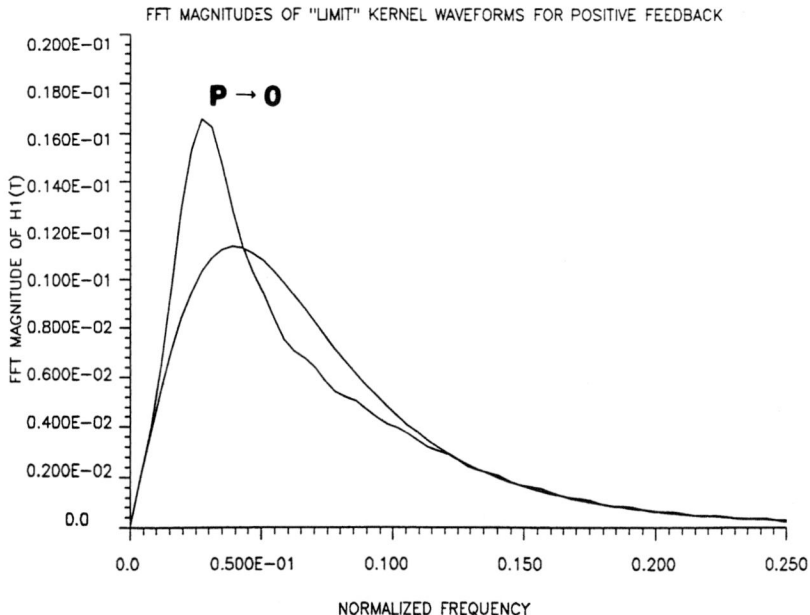

Fig. 39. FFT magnitudes of the two "limit" kernel waveforms shown in Fig. 38. Observe the higher resonance frequency for $P \to \infty$.

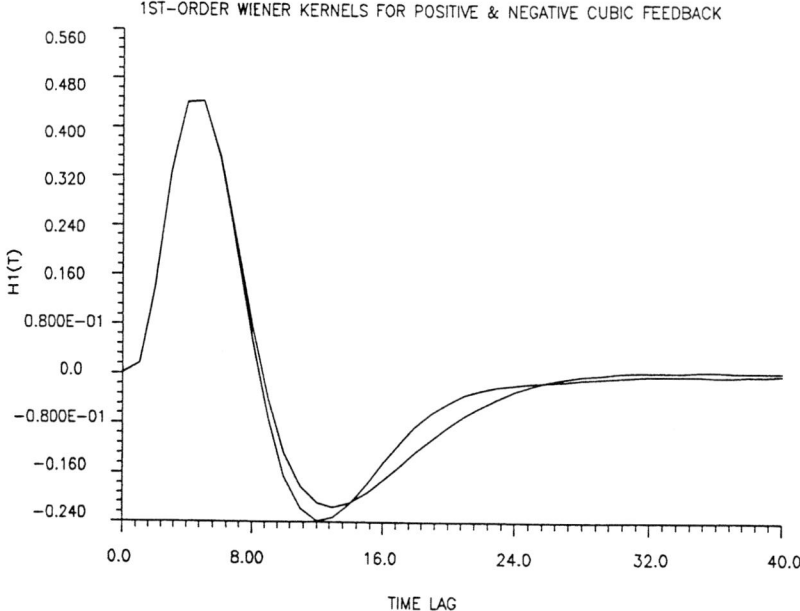

Fig. 40. First–order Wiener kernels for positive and negative cubic feedback ($\varepsilon = \pm 0.008$, $P = 1$) and the forward subsystem shown in Fig. 20. Positive feedback decreases the undershoot.

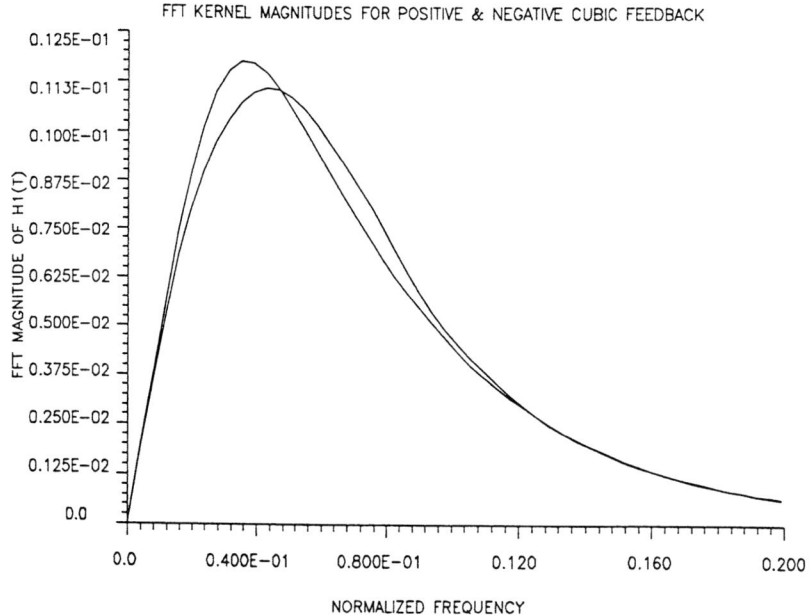

Fig. 41. FFT magnitudes of the kernels shown in Fig. 40. Positive cubic feedback leads to lower resonance frequency with higher gain.

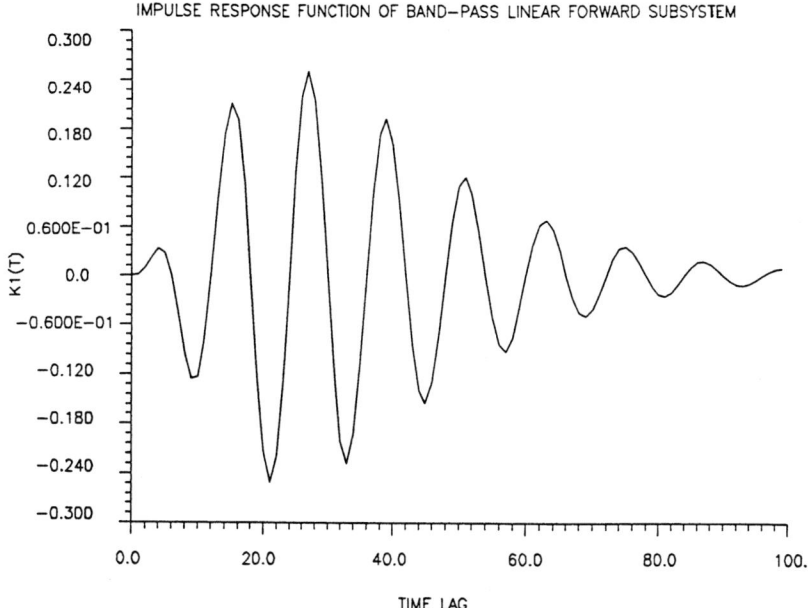

Fig. 42. Impulse response function of band–pass linear forward subsystem.

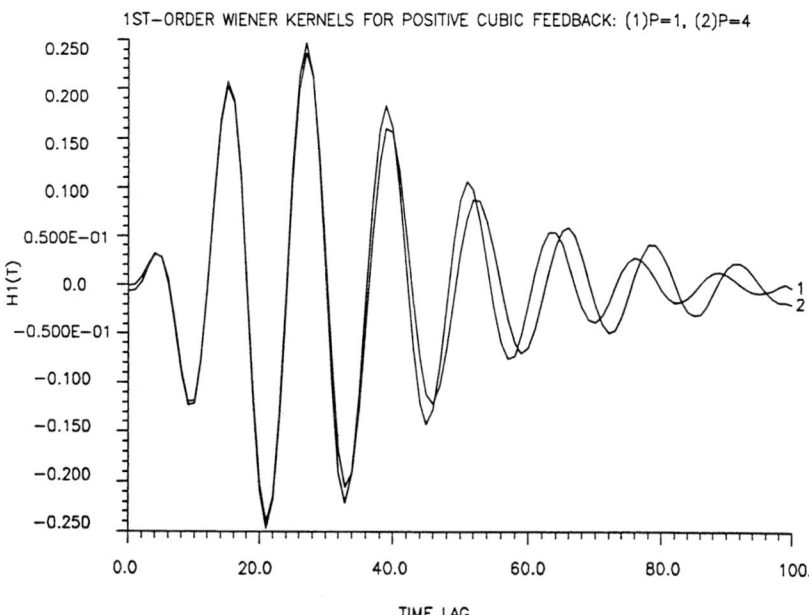

Fig. 43. First–order Wiener kernels of positive cubic feedback system ($\varepsilon = -0.01$) with the band–pass forward of Fig. 42, for $P = 1$ and 4. Note the lower resonance frequency for $P = 4$.

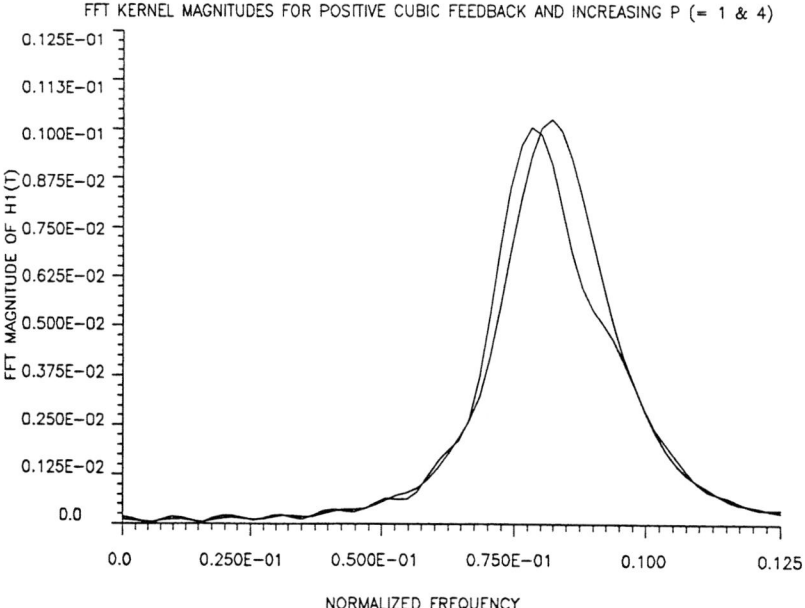

Fig. 44. FFT magnitudes of the kernels shown in Fig. 43, demonstrating lower resonance frequency for larger P.

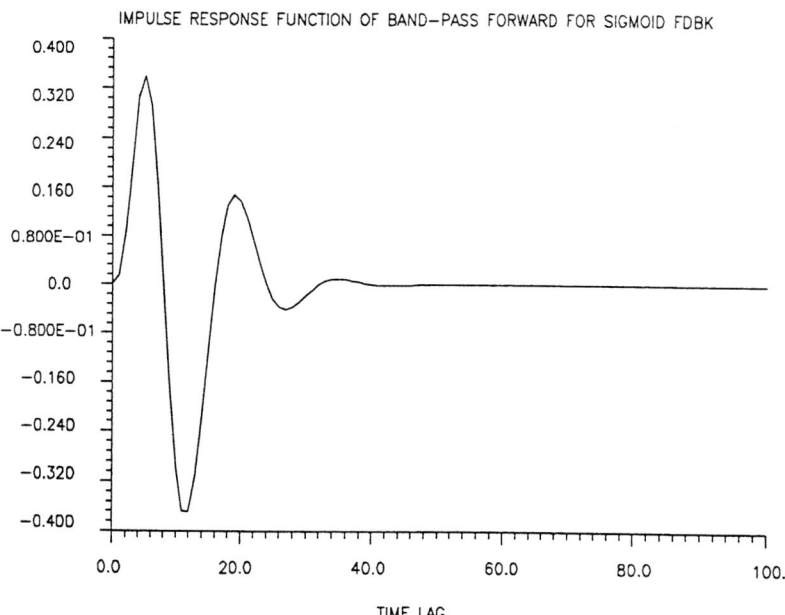

Fig. 45. Impulse response function of band–pass linear forward used in simulations of negative sigmoid feedback system.

pass linear forward subsystem, whose impulse response function is shown in Fig. 45, and the sigmoid (negative) feedback shown in Fig. 25. The obtained first–order Wiener kernel estimates for GWN input power level $P = 1$, 16, 256 and 4096 are shown in Fig. 46 (with appropriate offsets to allow easier visual inspection). The FFT magnitudes of these kernels are shown in Fig. 47, and they exhibit decreasing resonance frequency and broadening of the "tuning curve" as P increases. This transition pattern is similar to the one observed in auditory nerve fibers.

Second–order kernels of some nonlinear feedback systems

Our examples so far have employed nonlinear feedback with odd symmetry (cubic and sigmoid) and our attention has focused on first–order Wiener kernels of the resulting systems. These systems do not have even order kernels. However, if the feedback nonlinearity is not odd symmetric, then even order kernels exist. An example of this is given for quadratic feedback when the underdamped linear forward subsystem is the one shown in Fig. 20. Simulations were made for negative quadratic feedback of the form: εy^2, for $\varepsilon = 0.04$, 0.08 and 0.16 ($P = 1$), and the resulting second–order Wiener kernel estimates are shown in Figs. 48, 49 and 50, respectively. They have the approximate form and size predicted by our analytical derivations (cf. Eq. (50)) as indicated by the exact second–order kernel shown in Fig. 51. The first–order Wiener kernels were not affected by the quadratic feedback, as expected for small values of ε (there are, however, effects of order ε^2 and higher).

It is also important to note that, although the cubic or sigmoid feedback systems have no even–order Volterra kernels, Wiener analysis of these systems when nonzero mean is used for the GWN input yields even–order Wiener kernels dependent on the nonzero input mean (cf. Eq. (52)). This point was discussed in the previous section and general expressions were derived that relate the Wiener kernels for GWN input mean μ to "reference" Volterra kernels corresponding to a "reference" input mean μ_0 (see Eqs. (52) and (53)). This is illustrated by considering the cubic feedback system where only K_1 and K_3 are assumed to be significant for very small values of ε. The theoretically derived Eq. (53) becomes in this case (for $n = 2$):

$$
\begin{aligned}
H_2^\mu(\omega_1, \omega_2) &= 3\mu K_3(\omega_1, \omega_2, 0) \\
&= -3\varepsilon\mu\gamma K_1(\omega_1)K_1(\omega_2)K_1(\omega_1 + \omega_2)
\end{aligned}
\tag{70}
$$

and in the time domain:

$$
h_2^\mu(\tau_1, \tau_2) = -3\varepsilon\mu\gamma \int_0^{min(\tau_1,\tau_2)} k_1(\lambda)k_1(\tau_1 - \lambda)k_1(\tau_2 - \lambda)d\lambda
\tag{71}
$$

This result implies that the second–order kernel will retain its shape but increase linearly in absolute size with increasing μ (provided of course that ε is very small). This is illustrated in Figs. 52 and 53, where the second–order kernel estimates for $P = 1$ and $\mu = 0.5$ and 1 ($\varepsilon = 0.001$) are shown in the case of the overdamped forward subsystem of Fig. 2. We observe good agreement with the exact second–order kernel shown in Fig. 54. However, as μ increases a point may be reached when the fifth–order Volterra kernel K_5 begins to have an effect on H_2^μ. This is seen from the analytical expression that gives the contribution of K_5 to H_2^μ for this system:

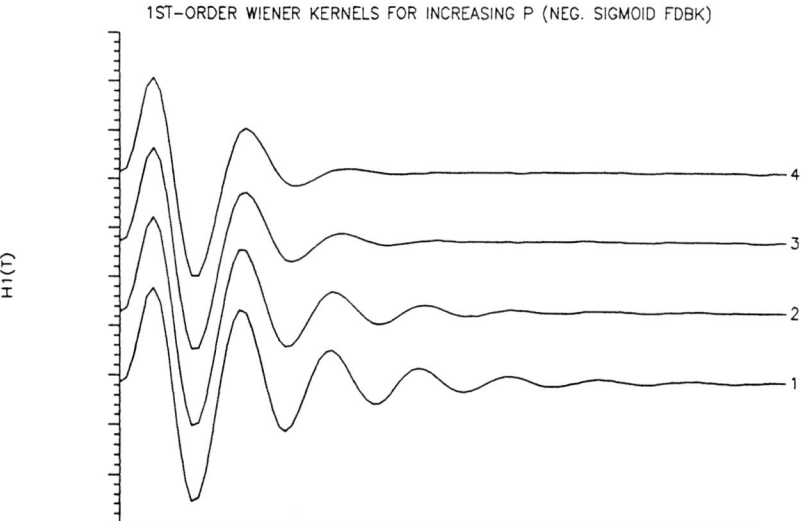

Fig. 46.   First–order Wiener kernels of negative sigmoid feedback system ($\varepsilon = 1$, $\alpha = 0.25$) with the band–pass forward of Fig. 45, for $P = 1$ (trace 1), 16 (trace 2), 256 (trace 3) and 4096 (trace 4).

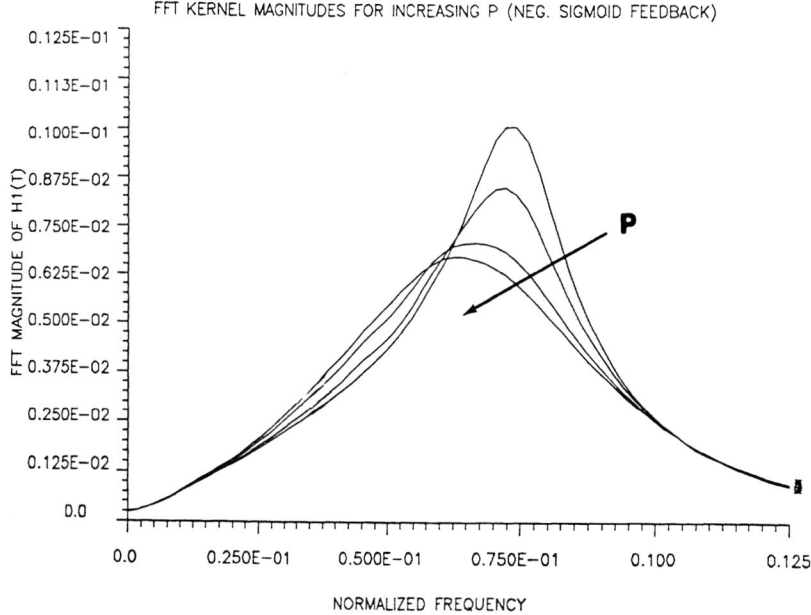

Fig. 47.   FFT magnitudes of the kernels shown in Fig. 46. We observe decreasing resonance frequency and gain as P increases, as well as broadening of the "tuning curve".

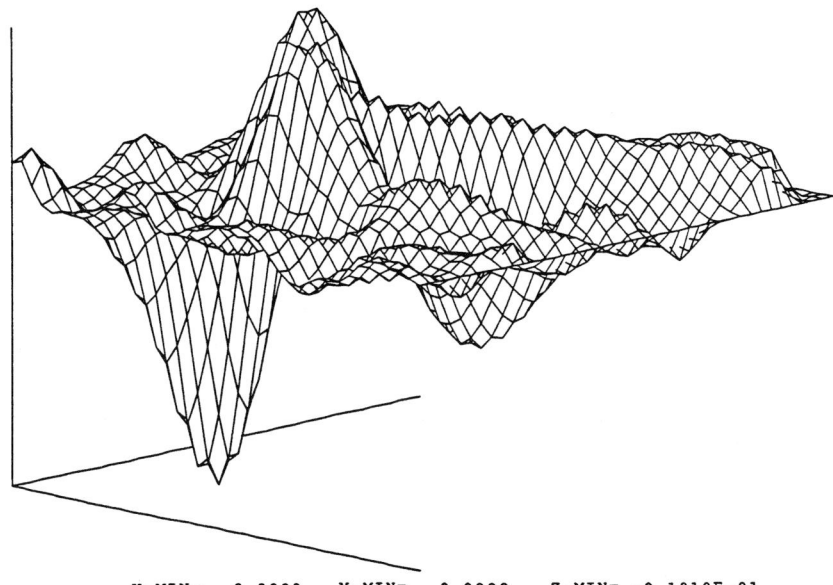

X-MIN=   0.0000   Y-MIN=   0.0000   Z-MIN= -0.1010E-01
X-MAX= 40.0000   Y-MAX= 40.0000   Z-MAX= 0.6185E-02

Fig. 48.   Second–order Wiener kernel of negative quadratic feedback system with the underdamped forward shown in Fig. 20 for $\varepsilon = 0.04$ and $P = 1$.

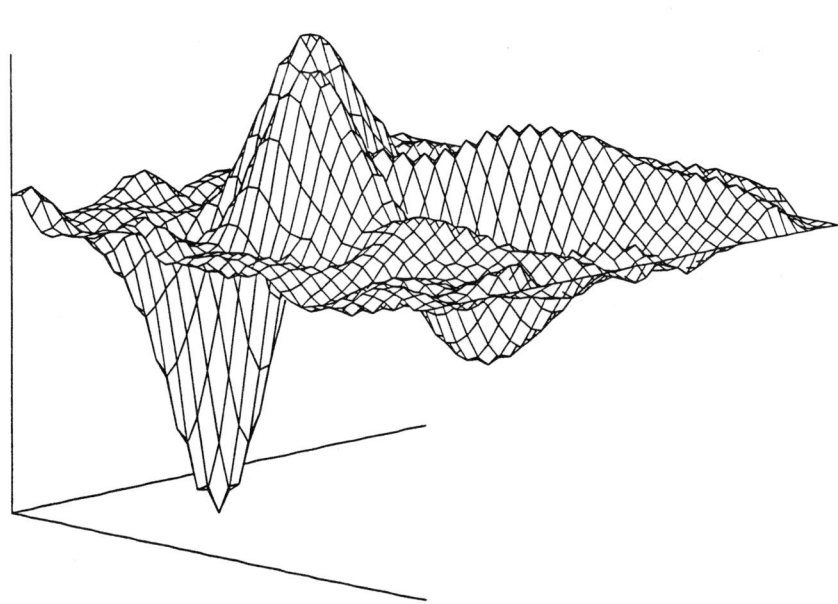

X-MIN=   0.0000   Y-MIN=   0.0000   Z-MIN= -0.1929E-01
X-MAX= 40.0000   Y-MAX= 40.0000   Z-MAX= 0.1155E-01

Fig. 49.   Second–order Wiener kernel of system described in the caption of Fig. 48, for $\varepsilon = 0.08$ and $P = 1$.

45

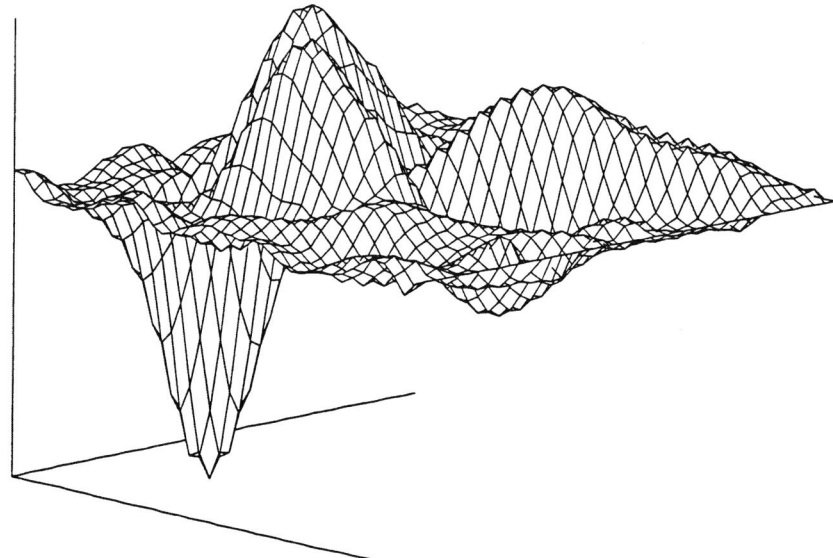

2ND-ORDER KERNEL OF QUADRATIC FEEDBACK SYSTEM (ϵ=0.16)

```
X-MIN=   0.0000   Y-MIN=   0.0000   Z-MIN= -0.3805E-01
X-MAX= 40.0000   Y-MAX= 40.0000   Z-MAX=  0.2343E-01
```

Fig. 50.   Second–order Wiener kernel of system described in the caption of Fig. 48, for $\varepsilon = 0.16$ and $P = 1$.

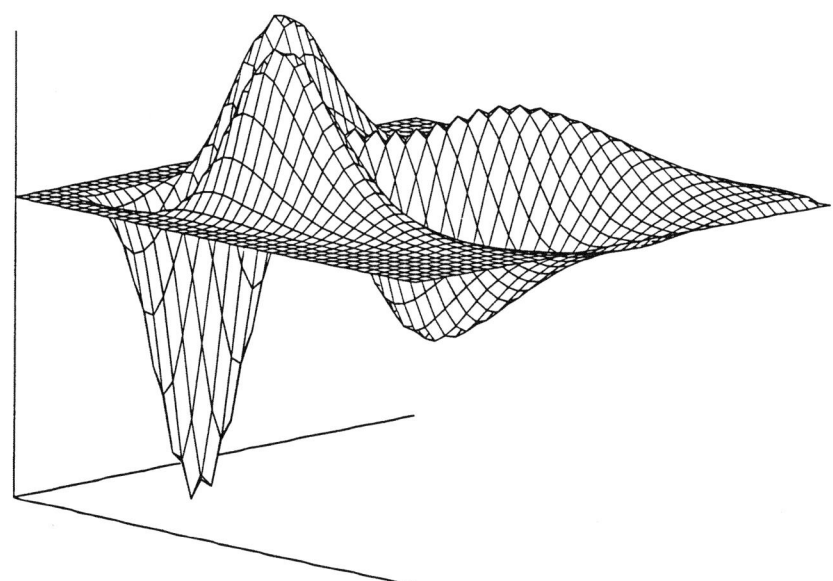

EXACT 2ND-ORDER KERNEL OF QUADRATIC FEEDBACK SYSTEM (ϵ=1)

```
X-MIN=   0.0000   Y-MIN=   0.0000   Z-MIN= -0.2574E+00
X-MAX= 40.0000   Y-MAX= 40.0000   Z-MAX=  0.1432E+00
```

Fig. 51.   Exact second–order Wiener kernel of the system described in the caption of Fig. 48, as given by Eq. (50) for nominal values: $\varepsilon = 1$, $P = 1$.

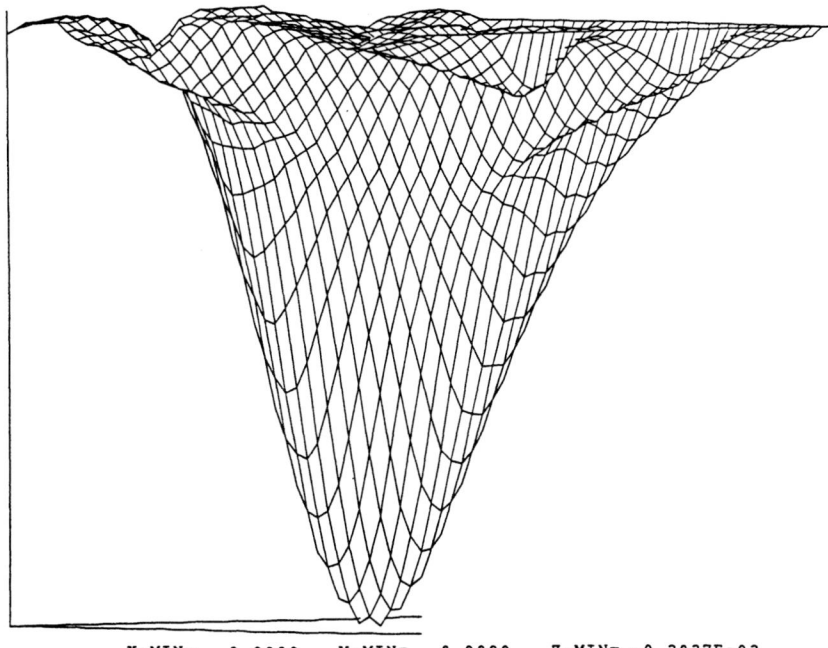

X-MIN=   0.0000   Y-MIN=   0.0000   Z-MIN= -0.2027E-02
X-MAX= 40.0000   Y-MAX= 40.0000   Z-MAX= 0.5985E-04

Fig. 52.   Second–order Wiener kernel of negative cubic feedback system ($\varepsilon = 0.001$) with the overdamped forward of Fig. 2, obtained for GWN input mean level $\mu = 0.5$ ($P = 1$).

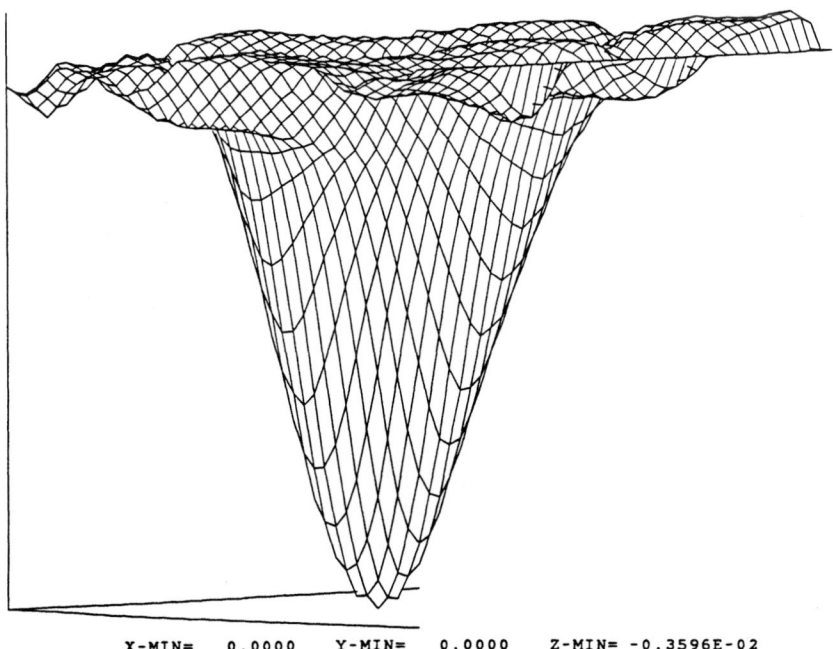

X-MIN=   0.0000   Y-MIN=   0.0000   Z-MIN= -0.3596E-02
X-MAX= 40.0000   Y-MAX= 40.0000   Z-MAX= 0.2536E-03

Fig. 53.   Second–order Wiener kernel of system described in the caption of Fig. 52, for $\mu = 1$.

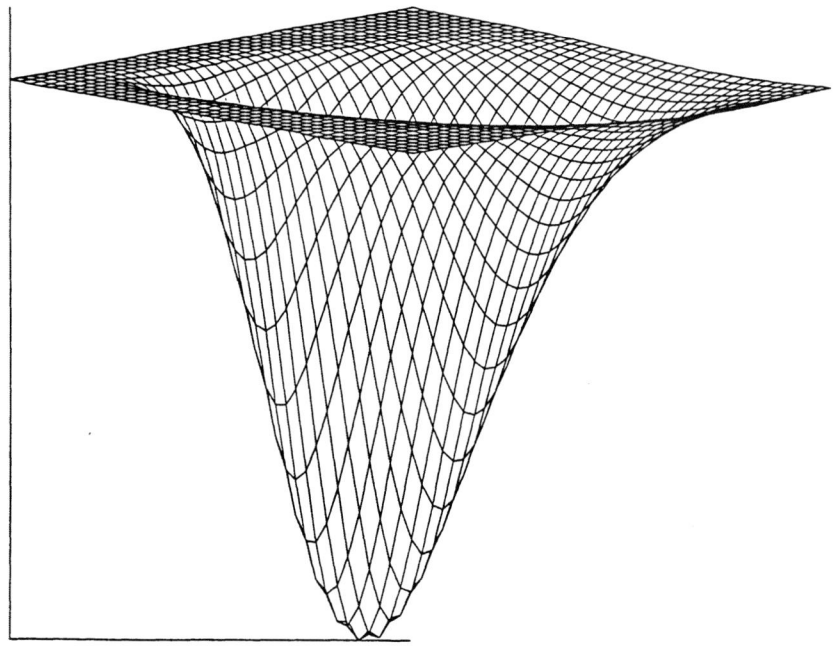

EXACT 2ND-ORDER KERNEL OF CUBIC FEEDBACK SYSTEM FOR μ≠0

Fig. 54.   Exact shape of second–order Wiener kernel of system described in the caption of Fig. 52 and given analytically by Eq. (71).

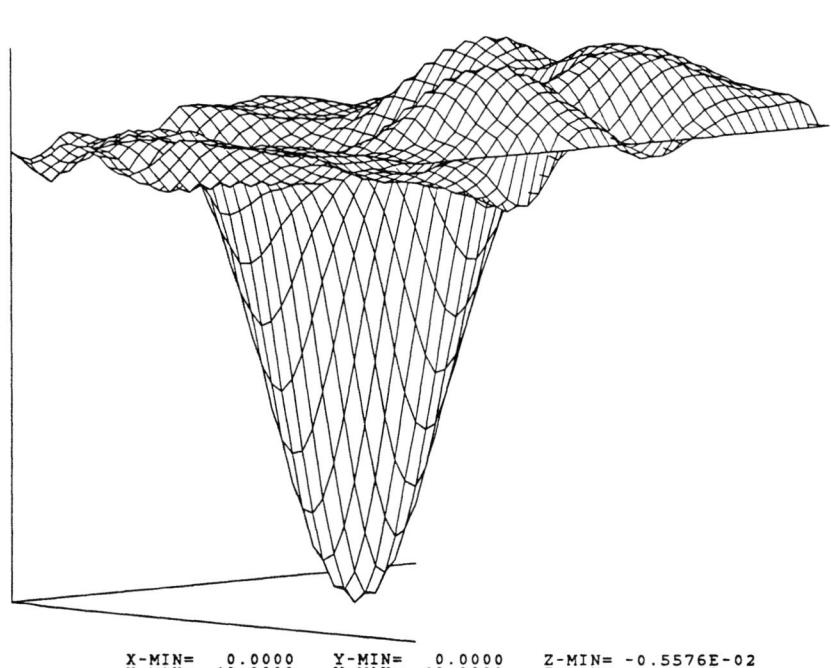

2ND-ORDER KERNEL OF CUBIC FEEDBACK SYSTEM FOR μ=2

```
X-MIN=    0.0000    Y-MIN=    0.0000    Z-MIN= -0.5576E-02
X-MAX=   40.0000    Y-MAX=   40.0000    Z-MAX=  0.9102E-03
```

Fig. 55.   Second–order Wiener kernel of system described in the caption of Fig. 52,
for $\mu = 2$.

$$H_{2,5}^{\mu}(\omega_1, \omega_2) = 9\varepsilon^2 \mu \gamma K_1(\omega_1) K_1(\omega_2) K_1(\omega_1 + \omega_2)$$

$$\left\{ (\mu\gamma)^2 [K_1(\omega_1) + K_1(\omega_2) + K_1(\omega_1 + \omega_2) + \gamma/3] \right.$$

$$\left. + \frac{P}{2\pi} \int_{-\infty}^{\infty} |K_1(u)|^2 F(\omega_1, \omega_2, u) du \right\} \tag{72}$$

where,

$$F(\omega_1, \omega_2, u) = K_1(\omega_1) + K_1(\omega_2) + K_1(\omega_1 + \omega_2) + 2Re\{K_1(u)\}$$

$$+ K_1(\omega_1 + u) + K_1(\omega_1 - u) + K_1(\omega_2 + u) + K_1(\omega_2 - u) + \gamma \tag{73}$$

For $\varepsilon \ll 1$, this contribution will become of comparable magnitude to the contribution of K_3 when the product $(\mu\gamma)$ becomes of order $1/\sqrt{\varepsilon}$. In that case, the first term in the brackets of (Eq. (72) will dominate the second term, if P does not take very large values. This, in fact, occurs in the aforementioned simulation example when $P = 1$ and μ becomes greater than 1. The second–order Wiener kernel estimates of this system for $\mu = 2$ and 4 ($P = 1$) are shown in Figs. 55 and 56 respectively, and they exhibit gradually different shape due to the increasing contribution of $H_{2,5}^{\mu}$.

2ND-ORDER KERNEL OF CUBIC FEEDBACK SYSTEM FOR μ=4

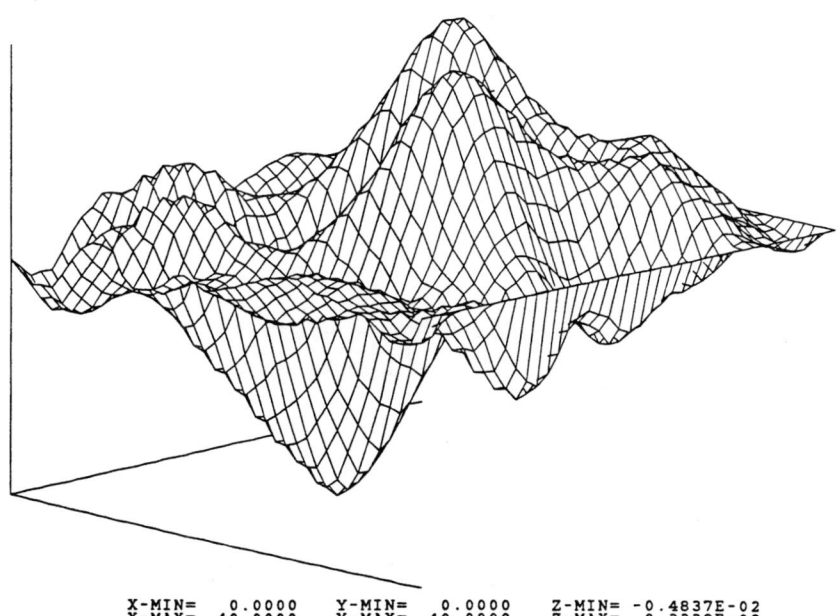

```
X-MIN=    0.0000    Y-MIN=    0.0000    Z-MIN=  -0.4837E-02
X-MAX=   40.0000    Y-MAX=   40.0000    Z-MAX=   0.3339E-02
```

Fig. 56.   Second–order Wiener kernel of system described in the caption of Fig. 52, for $\mu = 4$.

CONCLUSIONS

The study of Volterra–Wiener expansions of nonlinear differential equations has led to some results that begin to shed light into the important question of Wiener analysis of nonlinear feedback systems. Nonlinear feedback has been long thought to exist in many important physiological systems but its systematic and rigorous study has been hindered by the complexity (and often inadequacy) of the analytical methods used. This paper presents some relatively simple results, obtained for a class of nonlinear feedback systems, that relate Wiener kernel measurements with the effects of nonlinear feedback under various experimental conditions.

Numerous Wiener kernels of physiological systems have been measured experimentally in recent years, using GWN stimuli of various power levels and mean levels. For instance, in the study of retinal cells, GWN light stimuli have been used for a variety of mean luminance levels and modulation depths (power levels). The obtained first–order Wiener kernels have often exhibited changes in waveform and size for different mean and power levels of the stimulus. Simple cascade models (comprised of linear and static nonlinear subsystems) cannot explain the observed dependence of kernel waveform on stimulus mean and power level, since they can only account for (nonlinear) scaling changes of the kernels. However, a model that employs nonlinear feedback can account for such changes in kernel waveform.

What kind of changes in kernel waveform (as a result of varying the stimulus conditions) can a nonlinear feedback model explain? This is the question addressed by this paper for a certain class of nonlinear feedback systems. The guide to this study is provided by the theoretical Volterra–Wiener analysis of a class of nonlinear differential equations, that are equivalent representations of a class of nonlinear feedback systems (as discussed in Introduction). Explicit mathematical expressions were derived that relate Wiener kernel measurements to the characteristics of the feedback system and the stimulus parameters (cf. Eq. (29) or Eq. (31) for the first–order Wiener kernel). It was shown that the effects of varying stimulus parameters on the Wiener kernel measurements can be discerned easily for a broad class of nonlinear feedback systems. The theoretical results were tested with simulations and their validity was demonstrated in a variety of cases (cubic and sigmoid feedback with overdamped, underdamped, or band–pass linear forward subsystem). These test cases were chosen as to suggest possible interpretation of experimental results that have been published in recent years for two types of sensory systems: retinal horizontal cells and auditory nerve fibers. It was shown that relatively simple nonlinear feedback models can be constructed that reproduce the major changes in kernel waveforms observed experimentally in these sensory systems.

Specifically, it was shown that negative decompressive feedback (e.g., cubic) or positive compressive feedback (e.g., sigmoid) result in gradually decreasing damping (increasing bandwidth) of the first–order Wiener kernel as the GWN input power level and/or mean level increase. Conversely, positive decompressive or negative compressive feedback result in the reverse pattern of changes. The extent of these effects depends, of course, on the exact type of feedback nonlinearity and/or linear forward subsystem. There are also companion effects on the kernel size and system sensitivity (i.e., zero–frequency or resonance frequency gain).

It was demonstrated through analysis and computer simulations that the experimental observations of first–order Wiener kernel measurements for retinal horizontal cells can be qualitatively explained with the use of negative cubic feedback and a low–pass linear forward subsystem (viz., the gradual transition from an overdamped to an underdamped mode as the GWN stimulus power and/or mean level increase). Although the quantitative agreement was not exact, the approach to developing a closer approximation (possibly by use of cascaded feedback systems or different non-linearities) was illustrated. In the case of auditory nerve fibers, it was shown that the use of negative sigmoid feedback and a band–pass linear forward subsystem can reproduce qualitatively the effects observed experimentally on their "tuning curves" for increasing stimulus intensity (viz., a gradual downward shift of the center frequency and broadening of the tuning curve with increasing stimulus power level).

The effect of quadratic feedback on the second–order Wiener kernels (the first–order kernels are not affected) was also discussed and demonstrated through computer simulations. Likewise, the emergence of second–order Wiener kernels when GWN inputs with nonzero mean are used in probing cubic feedback systems was discussed and agreement with our theoretical derivations was demonstrated through simulations.

It is hoped that this work will inseminate an interest among systems neuro-physiologists in exploring the possibility of nonlinear feedback models in order to explain changes in Wiener kernel waveforms when the experimental stimulus conditions (i.e., power and mean level) vary. These changes in kernel waveform cannot be explained by simple cascade models of linear and static nonlinear components, that are currently popular in efforts to construct equivalent block–structured models from Wiener kernel measurements. The nonlinear feedback models may offer compact representations of the stimulus–response nonlinear dynamic relationship in these cases and may lead to physiologically meaningful interpretations of the system function. For instance, in the case of the auditory nerve fibers, the suggested model of negative sigmoid feedback may signify the gradually decreasing stiffness of hair cells as the sound stimulus intensity increases. Therefore, a host of new possibilities is made available by the presented analysis in interpreting Wiener kernels measurements in the context of physiologically meaningful attributes of nonlinear system function.

Acknowledgements

This work was supported by Grant No. RR–01861 from the Division of Research Resources of the National Institutes of Health. The author wishes to thank Dr. Moller, Dr. Naka and Dr. Sakai, whose Wiener kernel measurements in the auditory and visual system have provided the motivation for looking into the properties of nonlinear feedback systems vis–a–vis Wiener analysis. Thanks are also due to my brother Panos, who was the first to suggest (about 15 years ago) the possibility of nonlinear feedback in explaining the experimentally observed changes in horizontal cell kernel waveforms.

REFERENCES

Barrett, J.F., 1963, The use of functionals in the analysis of nonlinear physical systems, J. Electron. Control, 15:567–615.

Barrett, J.F., 1965, The use of Volterra series to find region of stability of a nonlinear differential equation, Int. J. Contr., 1:209–216.

Bedrosian, E. and Rice, S.O., 1971, The output properties of Volterra systems (nonlinear systems with memory) driven by harmonic and Gaussian inputs. Proc. IEEE, 59:1688–1707.

Brillinger, D.R., 1970, The identification of polynomial systems by means of higher order spectra, J. Sound and Vibration, 12:301–313.

French, A.S. and Butz, E.G., 1973, Measuring the Wiener kernels of a nonlinear system using the fast Fourier transform algorithm, Int. J. Control, 17:529–539.

Korenberg, M.J., 1973, Identification of nonlinear differential systems, Proc. of the Joint Autom. Contr. Conf., pp. 597–603.

Lee, Y.W. and Schetzen, M., 1965, Measurement of the Wiener kernels of a nonlinear system by cross–correlation. Int. J. Control, 2:237–254.

Marmarelis, P.Z. and Marmarelis, V.Z., 1978, "Analysis of Physiological Systems: The White–Noise Approach," Plenum, New York. Russian translation: Mir Press, Moscow, 1982.

Marmarelis, P.Z. and Naka, K–I., 1973, Nonlinear analysis and synthesis of receptive-field responses in the catfish retina, J. Neurophysiol., 36:605–548.

Marmarelis, V.Z. 1982, Nonparametric validation of parametric models, Math. Mod., 3:305–309.

Marmarelis, V.Z. (ed.), 1987, "Advanced methods of physiological system modeling, Volume I", USC Biomedical Simulations Resource, Los Angeles.

Marmarelis, V.Z., 1989, Linearized models of a class of nonlinear dynamic systems, Appl. Math. Modelling, 13:21–26.

Møller, A.R., 1983, Frequency selectivity of phase–locking of complex sounds in the auditory nerve of the rat, Hear. Res., 11:267–284.

Rugh, W.M., 1981, "Nonlinear System Theory: The Volterra/Wiener Approach," John Hopkins Univ. Press, Baltimore.

Sakai, H.M. and Naka, K–I, 1987, Signal transmission in the catfish retinal (V), J. Neurophysiol., 58:1329–1350.

Schetzen, M., 1980, "The Volterra and Wiener Theories of Nonlinear Systems," Wiley, New York.

Volterra, V., 1930, "Theory of Functionals and of Integral and Integro–Differential Equations," Dover, New York.

Wiener, N., 1958, "Nonlinear Problems in Random Theory," Wiley, New York.

PARAMETER ESTIMATION FOR NONGAUSSIAN PROCESSES VIA SECOND AND THIRD ORDER SPECTRA WITH AN APPLICATION TO SOME ENDOCRINE DATA

David R. Brillinger

Statistics Department
University of California
Berkeley, California 94720

INTRODUCTION

Quite a variety of processes that are observed in biomedicine have a pulse-like character and with bursts of activity occurring every so often with the principal variation corresponding to the location and size of the activity. Figure 1 provides examples of the temporal variation of the level of concentration of a particular hormone in the blood stream of a cow. One analytic representation for such processes is provided by

$$Y(t) = \sum_j A_j \, a(t-\sigma_j) \tag{1}$$

with the σ_j the times of initiation of pulses, with the A_j the respective amplitudes of the pulses and with $a(t)$ providing the pulse shape.

Concern in this work is with estimating the parameters of a particular process of the form (1). The intention is to demonstrate that for nongaussian processes one can get improved estimates by employing both second- and third-order statistics in the estimation procedure, as important information is not contained in second-order moments in the nongaussian case. A chance model is set down for such processes and the second and third order moments derived. A maximum likelihood type estimate based on power and bispectra is presented. The technique is quite flexible; all that is required are functional forms for the two spectra.

SOME PHYSIOLOGICAL BACKGROUND

The levels of lutenizing hormone (LH) were measured in 4 cows. This hormone is important in ovulation. The researchers, Rahe et al. (1980), desired to characterize the pattern of LH in systematic circulation during three periods (day 3, day 10-11, day 18-19) of the estrous cycle. This would help characterize the mechanism of release. Blood samples were collected every 10 minutes for a continuous 24-hour time interval. LH was seen to fluctuate in a pulse-like manner during each period. Figure 1 shows the fluctuations for day 10-11 in ng/ml. The pulses for the day 3 and day 18-19 series showed a periodic behaviour and are not studied in this work. The middle (midluteal) period was selected for out analyses because it could be modelled more simply, the release times evidencing no inherent pattern. Simple examination of the series of Figure 1 shows that the series are not gaussian, they are not symmetric about the mean level nor are they time reversable. Specific details of the experiment may be found in Rahe et al. (1980). One of the conclusions of the work following their analyses was that the pulse times were probably modulated by ovarian steroids. For more detail on the biological mechanisms involved see O'Sullivan (1986), O'Sullivan

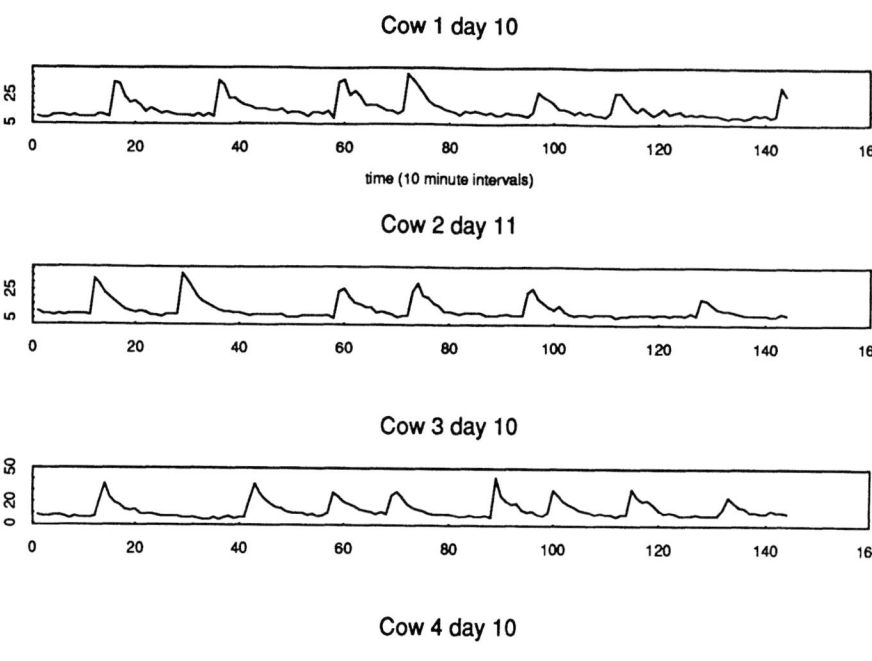

Fig.1. Examples of hormonal fluctuations.

steroids. For more detail on the biological mechanisms involved see O'Sullivan (1986), O'Sullivan and O'Sullivan (1988).

SOME STATISTICAL BACKGROUND

Chance Models

In the statistical approach to scientific problems it is usual to set down a chance model, that is a probabilistic description of the phenomenon of concern, and then to proceed deductively. The model will typically include some parameters of interest whose values are unknown, and most importantly, biologically interpretable.

The way that statistical history generally proceeded was for the parameters in some new circumstance first to be estimated by the method of moments. (Karl Pearson is one historical figure associated with this approach in the late 1800's and early 1900's.) This was followed by a switch to the method of maximum likelihood (introduced into statistical practice by R. A. Fisher in the 1920's and 1920's), where one sets down the probability density of the variates and considers it as a function of the unknown parameter. In the case that the variates involved are Gaussian, it and the method of moments are equivalent when one sticks to linear analyses. For nongaussian variates the linear and method of moment analyses are generally inefficient.

The analysis of random process data has been proceeding in similar fashion with initial concern for linear and quadratic statistics, gaussian models and the method of moments, followed more recently by a later switch to nongaussian assumptions and full likelihood analysis.

Statistics of the Empirical Fourier Transform

The time-side properties of processes, particularly nongaussian processes, are typically complicated. However in the case of stationary processes frequency-side properties are often simple and intuitive. Specifically, suppose that one has data $Y(t)$, $t=0,...,T-1$ available for analysis. The empirical Fourier transform of the data is given by

$$d^T(\lambda) = \sum_{t=0}^{T-1} Y(t)e^{-i\lambda t}$$

for λ real-valued. These variates have, approximately, mean zero, variance proportional to the power spectrum, covariance zero, joint third moment proportional to the bispectrum and higher moments corresponding to higher-order spectra. Further, statistics of different orders, second, third, fourth, ... are approximately independent, see Brillinger and Rosenblatt (1967). In some sense therefore bringing in a statistic of a higher-order brings in new information. These properties allow one to construct useful inference procedures in a variety of cases. In particular the problem of estimating a finite dimensional parameter will be considered.

Gaussian Fitting

Suppose that the process $Y(t)$ has power spectrum $f(\lambda) = f(\lambda|\theta)$ depending on an unknown parameter θ. (The formal definition is given in the Appendix). Then in many cases the periodogram values $I_s^T = (2\pi T)^{-1}|d^T(2\pi s/T)|^2$ are approximately independent exponential variates with means $f_s = f(2\pi s/T)$, $s = 1,...,T/2$. The minus log likelihood corresponding to such exponentials is given by

$$\sum_s (log\ f_s + I_s^T/f_s) \tag{2}$$

A parameter θ, describing the process $Y(t)$, and appearing in the power spectrum, may be estimated by minimizing expression (3). Differentiating and setting the result to 0 leads to the following estimating equations

$$\sum_s (I_s^T - f_s)\frac{\partial f_s}{\partial\theta}/f_s^2 = 0 \tag{3}$$

for θ. This estimation procedure was essentially set down by Whittle (1953). It is investigated in detail in the book Dzhaparidze (1986) where many references are given. Since it makes use only of second-order information, it is called Gaussian estimation.

Bispectral Fitting

In Brillinger (1985) a maximum likelihood type analysis based on second- and third-order moments was introduced. Let $I_{r,s}^T$ and $f_{r,s}$ denote the third-order periodogram and bispectrum at bifrequency $(2\pi r/T, 2\pi s/T)$, see Appendix. Setting down an approximate likelihood, i.e. neglecting terms of lower magnitude, based on both the second- and the third-order periodogram leads to the estimating equations

$$\sum_s (I_s^T - f_s)\frac{\partial f_s}{\partial\theta}/f_s^2 + \frac{2\pi}{T}\sum_r\sum_s (I_{r,s}^T - f_{r,s})\frac{\partial f_{r,s}}{\partial\theta}/f_r f_s f_{r+s} = 0$$

The first term here is the second-order one (3). The weights occurring in the second term correspond to the variance of the third-order periodogram, those in the first term correspond to the variance of the second-order periodogram.

To apply the technique one needs analytic expressions for the power spectrum and the bispectrum. For the example to be presented, the estimating equations were solved by posing the problem as one of iteratively reweighted least squares. A byeproduct of the procedure is the estimated standard errors.

THE PARTICULAR CASE OF CONCERN

The Chance Model

Suppose that pulses of LH are released at times σ_j with amplitudes A_j. Suppose that the time decay of a unit pulse is given by

$$a(t) = \exp\{-\alpha t\} \tag{4}$$

for $t \geq 0$ and by 0 otherwise. Then the level of hormone present at time t may be represented by

$$Y(t) = \sum_j A_j \, a(t-\sigma_j) \tag{5}$$

Writing

$$dM(t) = \sum_j A_j \, \delta(t-\sigma_j) \, dt$$

with $\delta(.)$ the Dirac delta, expression (5) may be written

$$Y(t) = \int a(t-u) \, dM(t)$$

Supposing the A_j and σ_j random, $M(t)$ is called a marked point process. When the σ_j come from a Poisson process and the A_j are constant, the process has long been called shot noise, eg. Weiss (1977).

One could build a full stochastic model for the data by assuming some specific distribution for the A_j and σ_j, eg. Poisson and gamma, and then seek maximum likelihood estimates of the unknown parameters. These estimates would be efficient in some sense.

In what follows it will be assumed that the process M has increments satisfying

$$cov\{dM(t+u),dM(t)\} = \sigma^2 \delta(u) dt du$$

and

$$cum\{dM(t+u),dM(t+v),dM(t)\} = \gamma \sigma^3 \delta(u) \delta(v) dt du dv$$

as would follow if the process M had independent increments and in particular for the A_j independent and $\{\sigma_j\}$ a Poisson process.

For the process (5), the full likelihood seems unwieldy to set down, so we turn to bispectral fitting.

The Power Spectrum and Bispectrum

The moments of the process (5) may be evaluated by elementary computations. The autoco-

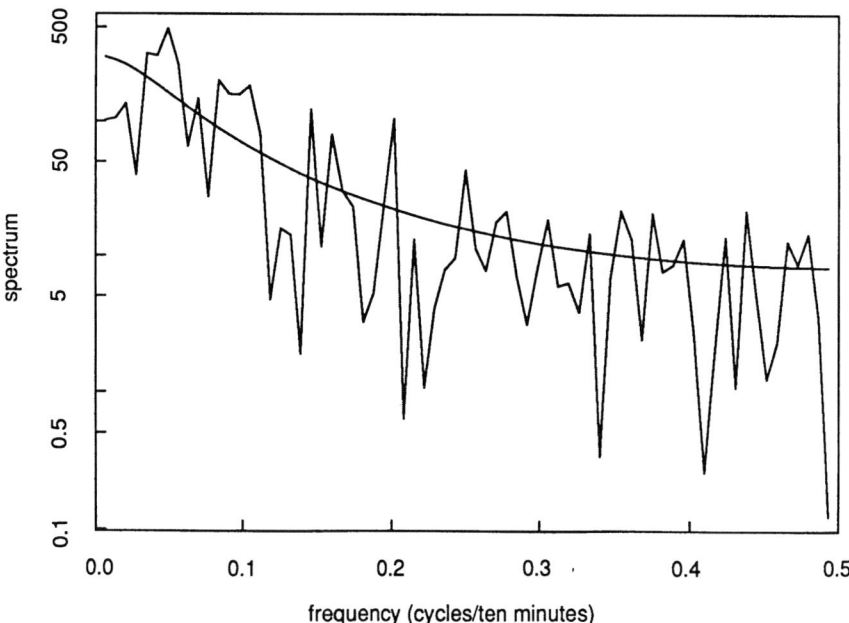

Fig. 2. Periodogram and fitted power spectrum.

covariance function and power spectrum are given by $c(u) = \exp\{-\alpha|u|\}/(2\alpha)$ and

$$f(\lambda) = \frac{\sigma^2}{2\pi} \frac{1 - |\zeta|^2}{|1 - \zeta|^2} \frac{1}{2\alpha}$$

respectively, where $\zeta = \exp\{-(\alpha+i\lambda)\}$. The third order moment function and bispectrum are given by $c(u,v) = \exp\{-\alpha(u+v)+3\alpha min(0,u,v)\}/(3\alpha)$ and

$$\frac{\gamma\sigma^3}{(2\pi)^2} \frac{1-\zeta_1\zeta_2\zeta_3}{(1-\zeta_1)(1-\zeta_2)(1-\zeta_3)} \frac{1}{3\alpha}$$

respectively, where $\zeta_1 = \exp\{-(\alpha+i\lambda)\}$, $\zeta_2 = \exp\{-(\alpha+i\mu)\}$, $\zeta_3 = \exp\{-(\alpha-i\lambda-i\mu)\}$.

It is to be remarked that the chance model has been set down in continuous time, but the spectra here have had to be derived for discrete time as the data is sampled. The parameter to be estimated is $\theta = (\alpha,\sigma,\gamma)$.

RESULTS

Both Gaussian and bispectral fitting were applied to the hormone series graphed in Figure 1 employing the chance model above. Table 1 provides the results of the Gaussian fitting for Cow 3. The parameter α provides the rate of fall-off of the pulse of Eq. (4). The value obtained fits with the principal pulse of the figure. It possibly relates to the rate of clearing of LH from the blood stream. The parameter σ corresponds to the variability of the pulse times and their amplitudes.

Table 1. Second-Order Fit

Parameter	α	σ
Estimate	.385	6.53
s.e.	.094	.47

Table 2 provides the results of the bispectral fitting. Now a further parameter γ can be estimated. It results from the nongaussianity of the process. The estimate of γ is five times its estimated standard error, providing evidence of nongaussianity. The estimate of α has changed from .385 to .461 and, notably, its standard error (estimate) has dropped from .094 to .038. The third-order information provided by the bispectrum, has played an important role in improving the estimate.

Table 2. Second- and Third-Order Fit

Parameter	α	σ	γ
Estimate	.462	6.71	3.30
s.e.	.038	.41	.62

Figures 2, 3 and 4 provide the estimated spectra and their fitted values. The construction of the estimates is described in the Appendix. The estimates obtained by the bispectral fit have been employed in these figures. As Marmarelis (1982) notes such figures may be employed to examine the reasonableness of a model. The crude, three parameter, model is seen to fit to a reasonable extent.

The computations were carried out on a Sun 3 workstation and took a matter of minites for these series of 144 observations.

DISCUSSION AND FUTURE WORK

The study reported is preliminary and involved a small number,144, of observations. It went beyond method of moment procedures, which simply equate empirical moments to their expected values, and was of maximum likelihood type.

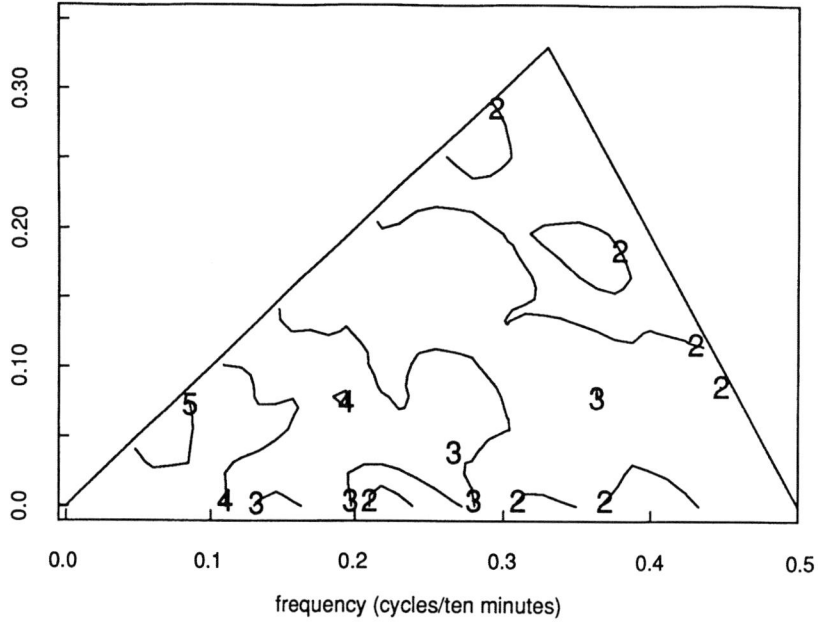

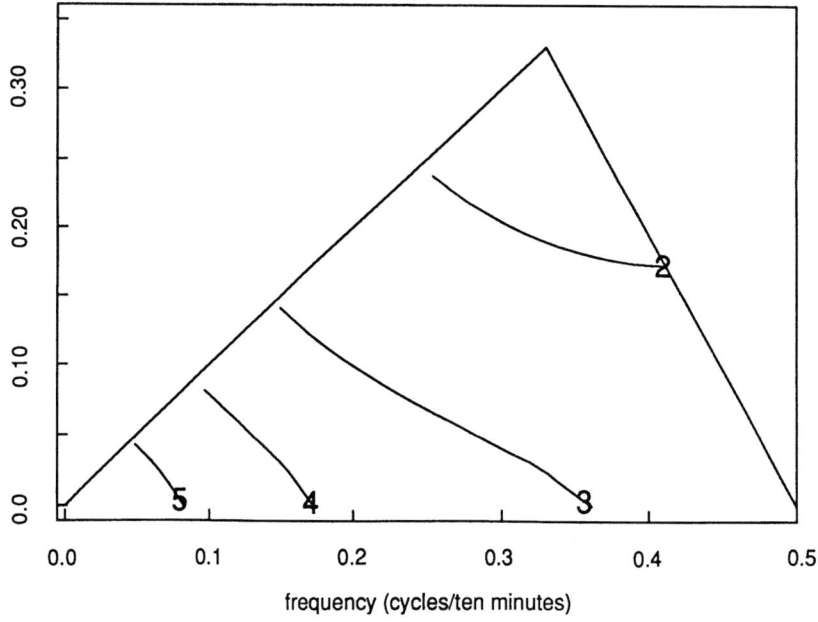

Fig.3. Estimated and fitted log modulus bispectrum

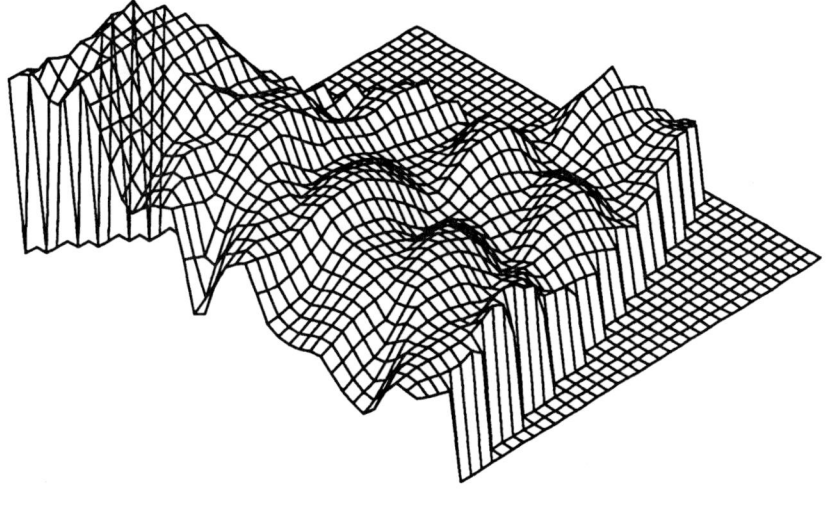

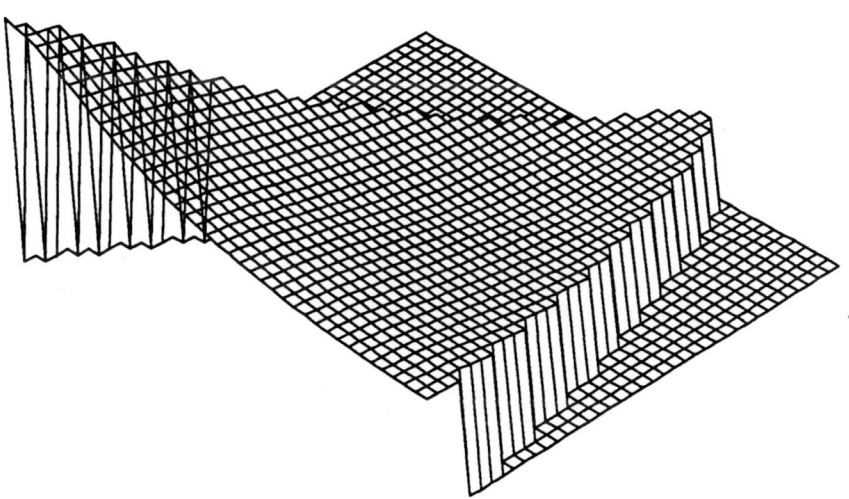

Fig.4. Estimated and fitted log modulus bispectrum

It was here assumed that the σ_j were stochastic. In the approach of O'Sullivan and O'Sullivan (1988) they are assumed fixed and estimated as well. Clearly, that is appropriate for some circumstances. Other studies of data of the present type, eg. the paper just referred to, have been comparative in character comparing the estimates obtained under differing experimental conditions. Work of the present sort is a necessary preliminary to those studies.

Future work would involve: better handling of the bias of the spectral estimates, more refined modelling of the marked point process, $M(t)$, other pulse shapes, trispectral fitting and full maximum likelihood. Alternate models for $M(t)$ will lead to alternate expressions for the spectra.

In conclusion one can say: an estimate of a biologically important parameter has been substantially improved through the incorporation of third-order moment information. into the estimation process.

ACKNOWLEDGEMENTS

The author is grateful to Professor H. J. Newton for providing the plasma LH series analysed. The research was supported by the NSF Grant MCS-8316634.

REFERENCES

Brillinger, D.R., 1985, Fourier inference: some methods for the analysis of array and nongaussian series data. Water Resources Bull. , 21: 743-756.

Brillinger, D.R. and Rosenblatt, M., 1967, Asymptotic theory of estimates of k-th order spectra, in "Spectral Analysis of Time Series," B. Harris, ed., J. Wiley, New York.

Dzhaparidze, K., 1986, "Parameter Estimation and Hypothesis Testing in Spectral Analysis of Stationary Time Series," Springer, New York.

Marmarelis, V.Z., 1982, Non-parametric validation of parametric models, Mathematical Modelling, 3: 305-309.

O'Sullivan, J., 1986, Statistical modelling of episodic hormone release. M. A. Thesis, University of California, Berkeley.

O'Sullivan, F. and O'Sullivan, J., 1988, Deconvolution of episodic hormone data: an analysis of the role of season on the onset of puberty in cows. Biometrics , 44:339-353.

Rahe, C.H., Ownes, R.E.,Fleeger, J.L., Newton H.J. and Harms, P.G., 1980, Pattern of plasma lutenizing hormone in the cyclic cow: dependence upon the period of the cycle, Endocrinology , 107:498-503.

Weiss, G., 1977, Shot noise models for the generation of synthetic streanflow data, Water Resources Res. , 13:101-108.

Whittle, P., 1953, The analysis of multiple stationary time-series, J. Royal Statist. Soc. B 15:125-139.

APPENDIX

A.1 Parameters

Let $Y(t)$, $t=0,\pm1,\pm2,...$ denote a stationary time series. Let it have mean c, covariance function $c(u) = cov\{Y(t+u),Y(t)\}$ and third moment function $c(u,v) = E\{[Y(t+u) - c][Y(t+v) - c][Y(t) - c]\}$. The power spectrum of Y at frequency λ is defined by

$$f(\lambda) = \frac{1}{2\pi}\sum c(u)e^{-iu\lambda}$$

The bispectrum at bifrequency (λ,μ) is defined by

$$f(\lambda,\mu) = \frac{1}{(2\pi)^2}\sum\sum c(u,v)e^{-i(u\lambda+v\mu)}$$

The fundamental domain of bifrequencies for the bispectrum is $0 \leq \mu \leq \lambda \leq \pi$ and $0 \leq 2\lambda+\mu \leq 2\pi$. This is the region of Figure 3.

60

A.2 Estimates

There are a variety of fashions by which a bispectrum may be estimated. In this paper the technique of segmenting and averaging is employed because of the ease with which it may be programmed. Specifically let the data be broken into L, possibly overlapping, stretches of length V. Next compute the tapered Fourier transform of the l-th stretch,

$$d^V(\lambda;l) = \sum_{v=0}^{V-1} h\left(\frac{v+1}{V+1}\right)Y(lV+v)e^{-iv\lambda}$$

for $l = 0,...,L-1$. (The taper employed in the work of this paper is Tukey's namely $h(u) = (1 - \cos2\pi u)/2$.) Next form the third order periodogram of the l-th stretch

$$I^V(\lambda,\mu;l) = \frac{1}{(2\pi)^2 h_3}d^V(\lambda;l)d^V(\mu;l)\overline{d^V(\lambda+\mu;l)}$$

where $h_3 = \sum h((v+1)/(V+1))^3$. The estimate of the bispectrum is now

$$f^T(\lambda,\mu) = \frac{1}{L}\sum_{l=0}^{L-1}I^V(\lambda,\mu;l)$$

The power spectrum is estimated by similarly averaging the second-order periodograms of the L stretches. In the computations of the paper $L = 13$, $V = 36$ with the stretches overlapping 50%.

ANALYSIS OF THE WHOLE-NERVE RESPONSES FROM THE EXPOSED

AUDITORY NERVE IN MAN TO PSEUDORANDOM NOISE

Aage R. Møller

Department of Neurological Surgery
University of Pittsburgh School of Medicine
Pittsburgh, PA   15213

INTRODUCTION

Evoked potentials that originate in the auditory nervous system are now widely used for diagnostic purposes (see, e.g., Jacobson, 1985), as well as for intraoperative monitoring to detect injuries to the auditory system that may occur during neurosurgical operations (see, e.g., Møller, 1987a). The most commonly used stimuli for such studies are transient sounds, such as click sounds or short tonebursts. However, natural sounds are usually more or less continuous in nature and are time varying. The electrical potentials that such sounds evoke are of much lower amplitude than those evoked by transient sounds, and complex signal processing is necessary to extract useful features from the responses recorded to continuous sounds. Transient sounds produce evoked potentials of higher amplitude than those produced by continuous sounds, because transient sound activates a greater number of neural elements at nearly the same time. The responses that can be recorded from the auditory nerve, the nerve tracts in the brain, and the brainstem nuclei to transient sounds, therefore, have a higher amplitude than that of the response to continuous sounds. When auditory evoked potentials are recorded from electrodes placed at a great distance from their source, such as on the scalp (farfield responses), it is important that the response has a higher amplitude.

Although, for the reasons just discussed, it is easier to study the auditory system by evaluating responses to transient (click) sounds than to continuous (natural) sounds, the way in which the auditory system processes click sounds is different from the way it processes natural sounds in several respects. Thus, when click sounds are used for diagnostic purposes, abnormalities in the "event-related potentials" that are recorded may not properly reflect pathological changes in the function of the auditory system. For example, the responses to click sounds may be affected only to a limited degree by injuries that cause major auditory deficits, such as a reduced ability to perceive speech. There is, therefore, a need to develop electrophysiological methods to test sensory systems such as the auditory system by using more natural stimuli. For the auditory system, continuous broadband sounds seems to be a suitable choice because of their similarity to natural sounds. A recent study (Møller and Sekiya, 1988) showed that when the auditory nerve is injured the waveform of the transient response may not be similar to the waveform of the cross-correlograms of the response to pseudorandom noise. The use of noise stimuli may thus offer advantages over the traditional use of transient stimuli in studies of injuries to the auditory nerve. The use of such broadband sounds as noise also makes it possible to study a system's nonlinearities (Møller, 1983b, 1987b).

Studies have shown that cross-correlograms of the responses from the ear, auditory nerve, and auditory nuclei -- when recorded with gross electrodes (whole-nerve potentials) to stimulation of tones and noise that are amplitude modulated with pseudorandom noise -- have a waveform similar to that of event-related potentials that can be obtained in response to transient sounds (Møller, 1981a,b, 1987b). However, there are some important differences between the cross-correlograms and the responses evoked by transient sound. Notably, while the latency of the response to transient sounds decreases with increasing sound intensity, the various components of the cross-correlograms appear at nearly the same delays over a wide range of stimulus intensities.

When sounds that are amplitude modulated with pseudorandom noise are used as stimuli, the response is cross-correlated with the pseudorandom noise. Thus, it is the modulation waveform that is regarded as the input to the system, and the correlograms are estimates of the first-order Wiener kernels of the system's properties to (small) changes in the amplitude of the input. A similar technique has been used to study the responses from single nerve cells in the cochlear nucleus to stimulation by tones and noise that were amplitude modulated with pseudorandom noise (see Møller 1976a,b) (for details about methods see Møller, 1974, 1983a).

For low-frequency sounds the probability of the occurrence of a nerve impulse is greater during certain phases of the stimulus. This is known as "phase-locking," and it has been studied to a great extent in animal experiments involving recording from single auditory nerve fibers. The ability of a single auditory nerve fiber to phase-lock to the waveform of a sound decreases above 2 to 3 kHz, and it is usually not detectable for stimulus frequencies above 5 to 6 kHz. When the stimulus is broadband noise the neural discharges in a single auditory nerve fiber phase-lock to the halfwave rectified bandpass-filtered version of the stimulus sound. The bandpass filter is the cochlear spectral filter, and the unidirectionality of the sensory cells (hair cells) causes the halfwave rectification.

These animal experiments have shown that the time pattern of discharges in single fibers of the auditory nerve is related to the time pattern of the vibration of the basilar membrane in the cochlea, at the location of the sensory cell that innervates the particular nerve fiber. Our own animal experiments using pseudorandom noise as the stimulus have shown that the time pattern of the spike train in single auditory nerve fibers can be modeled by a linear bandpass filter followed by a non-memory nonlinearity representing the unidirectional sensitivity of the sensory cells in the cochlea (hair cells) and the spike generator (Møller, 1977, 1983a,b). However, the characteristics of the cochlear bandpass filter are different at different sound intensities (Møller, 1977, 1983,a,b; Harrison and Evans, 1982). This nonlinearity seems to be a result of a slow process, and it does not seem to create any "hard" nonlinearities.

Several studies of the discharge patterns of single auditory nerve fibers have shown that some nerve fibers respond to one phase of displacement of the basilar membrane while others respond to displacement in the opposite direction (Konishi and Nielsen, 1978; Zwislocki and Sokolich, 1973); other fibers respond to the velocity of the motion of the basilar membrane in either or both directions of displacement. Each individual nerve fiber mainly responds to displacement or velocity in only one direction of motion of the basilar membrane, because each hair cell acts in a way similar to a single halfwave rectifier (Zwislocki and Sokolich, 1973; Konishi and Nielsen, 1978). However, when the whole-nerve potential is considered the effect of this nonlinearity is more complex, because the whole nerve receives similar contributions from the responses to both phases of a sound, as well as components that are related to the derivative of the sound. Since derivation is a linear process, it may be included in the linear filter that precedes the rectifier.

The whole-nerve response and the responses from single nerve fibers or nerve cells are related to each other in the way that the whole-nerve response is the convolution between a single-nerve-fiber action potential and the sum of the distribution of the discharges over time of all contributing nerve fibers, assuming that all nerve fibers generate action potentials that have the same waveshape (Goldstein, 1960).

The whole-nerve potential that can be recorded from the auditory nerve in response to continuous sound at a certain intensity may thus be assumed to be generated by a system that consists of a bank of linear bandpass filters representing the frequency selectivity of the basilar membrane, followed by non-memory nonlinearities representing the unidirectional sensitivity of the sensory cells (hair cells) and the spike generator, followed by a linear filter that is a result of the convolution between the spike distribution and the action potential. It seems that the only "hard" nonlinear element in this system is the neural transduction process.

A model of the periphery of the auditory system (Fig. 1) would therefore consist of a bank of bandpass filters representing the functions of the cochlea, and which can be regarded to be linear at a certain sound level, followed by a halfwave rectifier that represents the unidirectional sensitivity of hair cells. In some of the channels, this rectifier is preceded by an inverter that represents a certain proportion of cells that are sensitive to the opposite phase of the sound. The velocity sensitivity of some hair cells is represented by a differentiator. All channels are assumed to have identical spike generators, and we assume that the activity in all nerve fibers is convolved with a unit

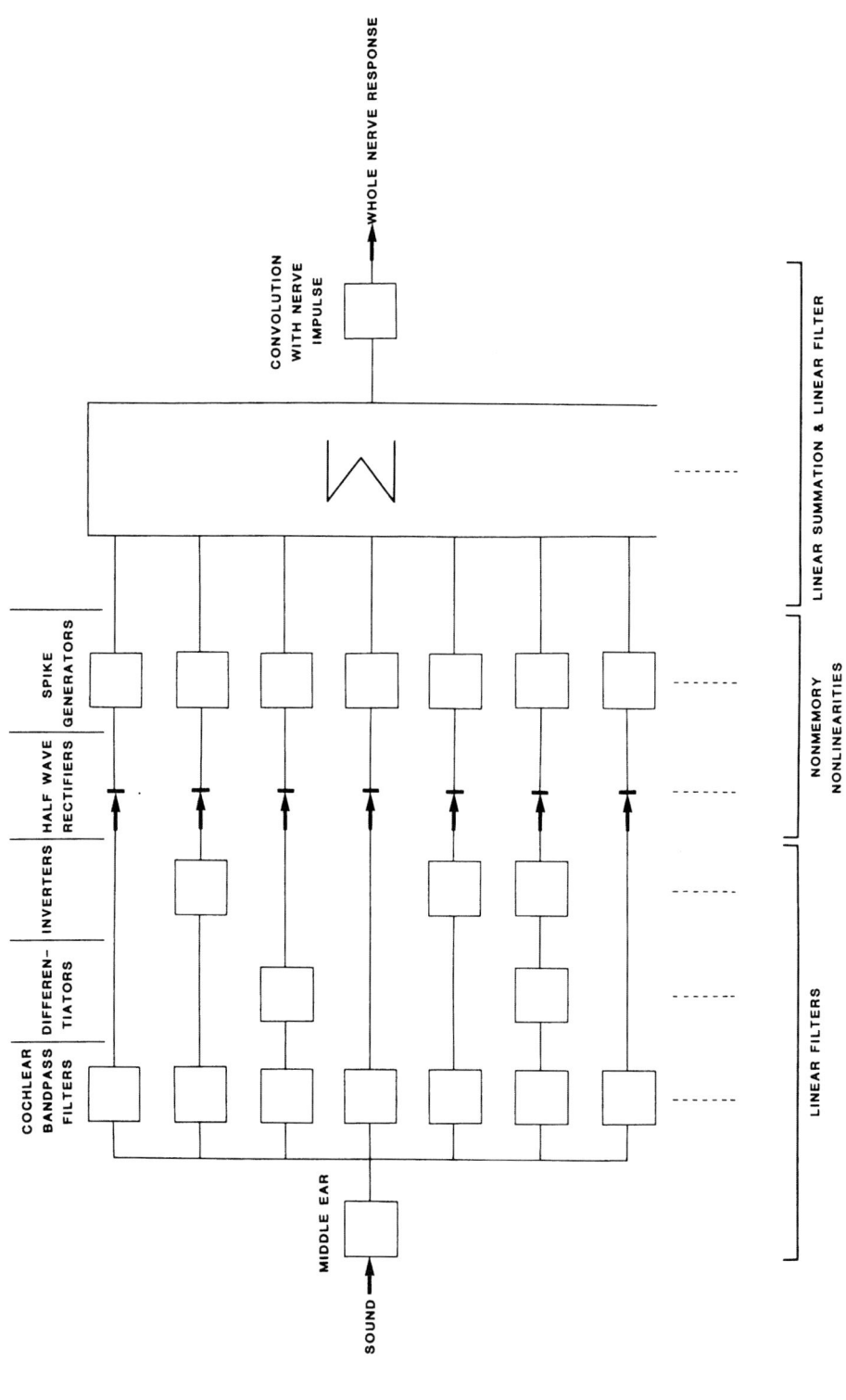

Fig. 1. Model of main functions of the ear involved transduction of sound into the potentials that can be recorded from the auditory nerve.

discharge to form the recorded compound action potential and summed linearity. The cochlear filters are preceded by a linear filter that represents the middle ear and sound transmission to the tympanic membrane.

The part of the model shown in Fig. 1, that represents the function of the cochlea is similar to the model proposed by Weiss (1966), with the main difference being that Weiss's model does not show that deflection of the basilar membrane in both directions can cause excitation in auditory nerve fibers. The spike generator is regarded to be a zero-memory nonlinearity, which is an approximation because the upper limit of phase-locking, in fact, represents a lowpass filter with a cutoff frequency of about 5 kHz.

We will now describe the methods and some of the results of using pseudorandom noise in studies of the responses from the exposed intracranial portion of the auditory nerve in patients undergoing neurosurgical operations.

METHODS

Responses were recorded from the exposed intracranial portion of the auditory-vestibular nerve (cranial nerve VIII) in patients who were operated upon to relieve vascular compression of cranial nerves (CN V to relieve face pain; CN VII to relieve hemifacial spasm; and CN VIII to relieve disabling positional vertigo) using the microvascular decompression technique described by Jannetta (1977, 1981a,b). The potentials were recorded using a monopolar electrode consisting of a Teflon-insulated, multistrand, silver wire, to the uninsulated tipe of which a small cotton ball was sutured (Møller and Jannetta, 1983; Møller, 1987a). The recording electrode was placed on the cochlear portion of the nerve in the middle of its intracranial course, and the reference electrode was placed either on the opposite earlobe` or on the shoulder (noncephalic reference). The recorded potentials were amplified 20,000 times, highpass filtered (cutoff frequency of 3 Hz, rolloff of 6 dB/octave), and lowpass filtered (cutoff of 3 kHz and rolloff of 18 or 24 dB/octave).

Pseudorandom noise was presented through miniature stereo earphones to the ear on the side from which the recordings were made (see Møller, 1987a for details). The noise was generated by lowpass filtering ternary m-sequences with a length of 19,682 steps. After digital lowpass filtering (cutoff frequency of 2.5 kHz), linear interpolation was used to reduce the number of samples to 2,048. The generation of the noise and tests of its properties are described in detail in earlier publications and will not be repeated here (Møller, 1982, 1983a,b). Two of the sixteen noise sequences that were described earlier (Møller, 1982) were used in the experiments to be discussed here.

The noise was presented as periods of a continuous sound, and the end of each noise period was directly followed by the beginning of an identical period of noise. The responses to 50 to 400 periods of the noise were averaged during the actual recording and stored on computer disks for off-line processing. These averaged responses thus represent the response to one period of the pseudorandom noise, with the uncorrelated noise being reduced by the averaging process (ideally by a factor of $1/\sqrt{N}$, where N is the number of responses that are being averaged). The recorded averaged responses were highpass filtered by subtracting a lowpass-filtered version of the response (running rectangular average, duration 4.0 ms) from the response. This zero-phase, highpass filtering eliminated the slow potentials that resulted mainly from variations in how much of the recording electrode was covered by fluid.

Circular cross-correlations between these filtered, averaged responses and one period of the pseudorandom noise were computed (Møller 1974, 1983a). The output of a linear model that has the computed cross-correlogram as its impulse response was computed by convolving the cross-correlograms with one period of the pseudorandom noise that was used to drive the earphone.

This same recording arrangement was also used for intraoperative monitoring, the purpose of which was to detect changes indicative of injury to the auditory nerve from surgical manipulation. For intraoperative monitoring, click sounds were generated by presenting a 100-µs square wave to the earphone at a rate of about 19 pps. The peak sound pressure was 115 dB SPL. Changes in the potentials evoked by the click sounds indicated possible nerve injury. The procedure was approved by the Institutional Review Board of the University of Pittsburgh, and the patients gave informed consent.

RESULTS

A typical cross-correlogram of the response from a patient with normal hearing, recorded from the middle part of the intracranial portion of the eighth nerve to pseudorandom noise, is shown in Fig. 2, together with the compound action potential to click stimulation recorded from the same location on the nerve and obtained at about the same time. It is seen that the waveforms of the compound action potentials and the cross-correlogram are similar. However, there is a considerable individual variation regarding the waveform of the first-order correlograms, and in many cases additional waves are seen at longer delays.

If it is assumed that the system under test can be regarded as a cascade of linear filters and non-memory nonlinearities and that the noise used as the test signal has a negligible amount of anomalies, then the first-order cross-correlation is a valid estimate of the impulse response of the linear portion of such a nonlinear system (Marmarelis and Marmarelis, 1978; Møller, 1983a,b, 1986, 1987b). The difference between the response and the model output is a measure of the nonlinearities of a system that has no inherent noise. We have made use of this in determining the degree of nonlinearity in the auditory system (Møller, 1983b, 1986, 1987b).

When this method is used in the analysis of the response from the human auditory nerve (Fig. 3), a large difference becomes apparent between the response and the output of the linear model. The existence of this difference indicates that the system has a non-negligible non-linearity.

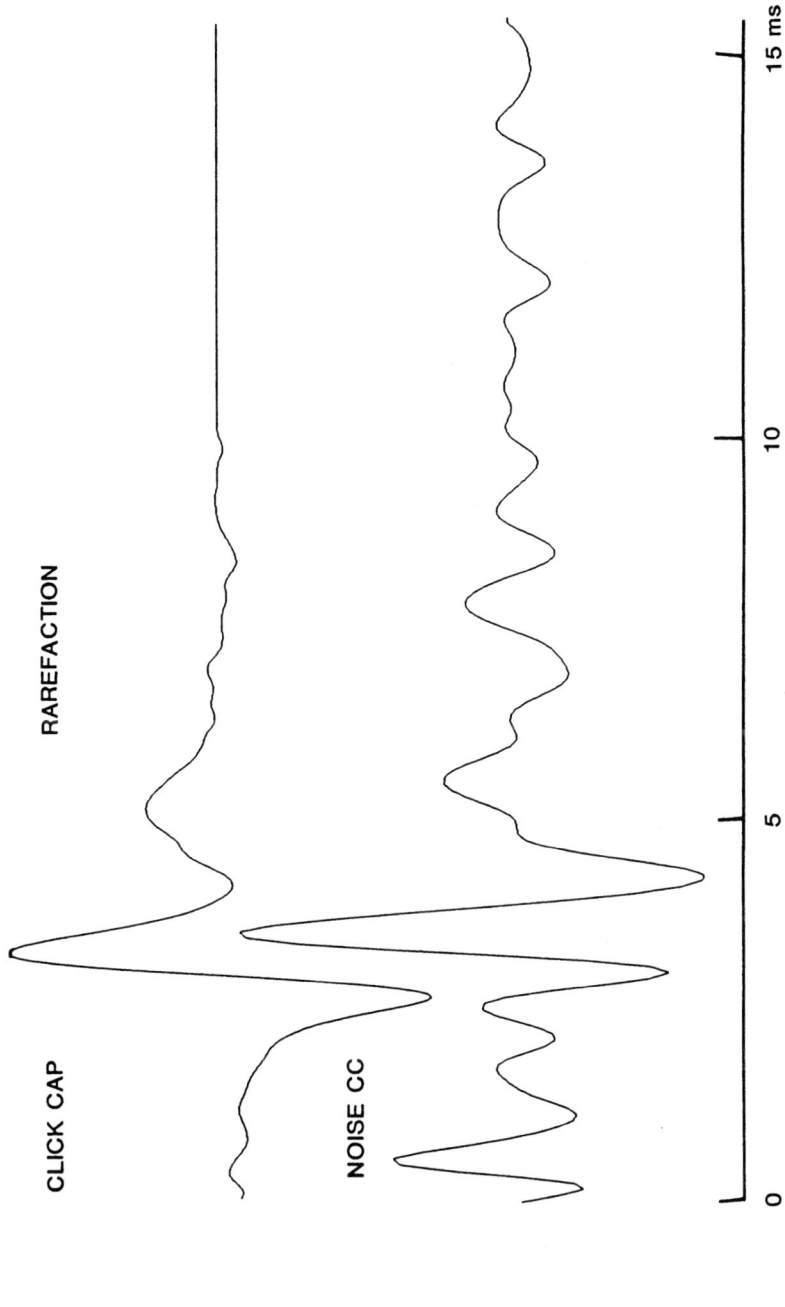

Fig. 2. Compound action potentials in response to click sounds presented at a peak intensity of 115 dB SPL (top tracing), and a cross-correlogram of the responses to pseudorandom noise (bottom tracing). The responses to 200 periods of pseudorandom noise presented at a sound intensity of about 90 dB SPL were averaged. An upward deflection in the cross-correlogram indicates a positive correlation between the rarefaction phase of the sound and a negative nerve potential. The compound action potential to click stimulation shows negativity as an upward deflection. The same conventions are used in all subsequent figures.

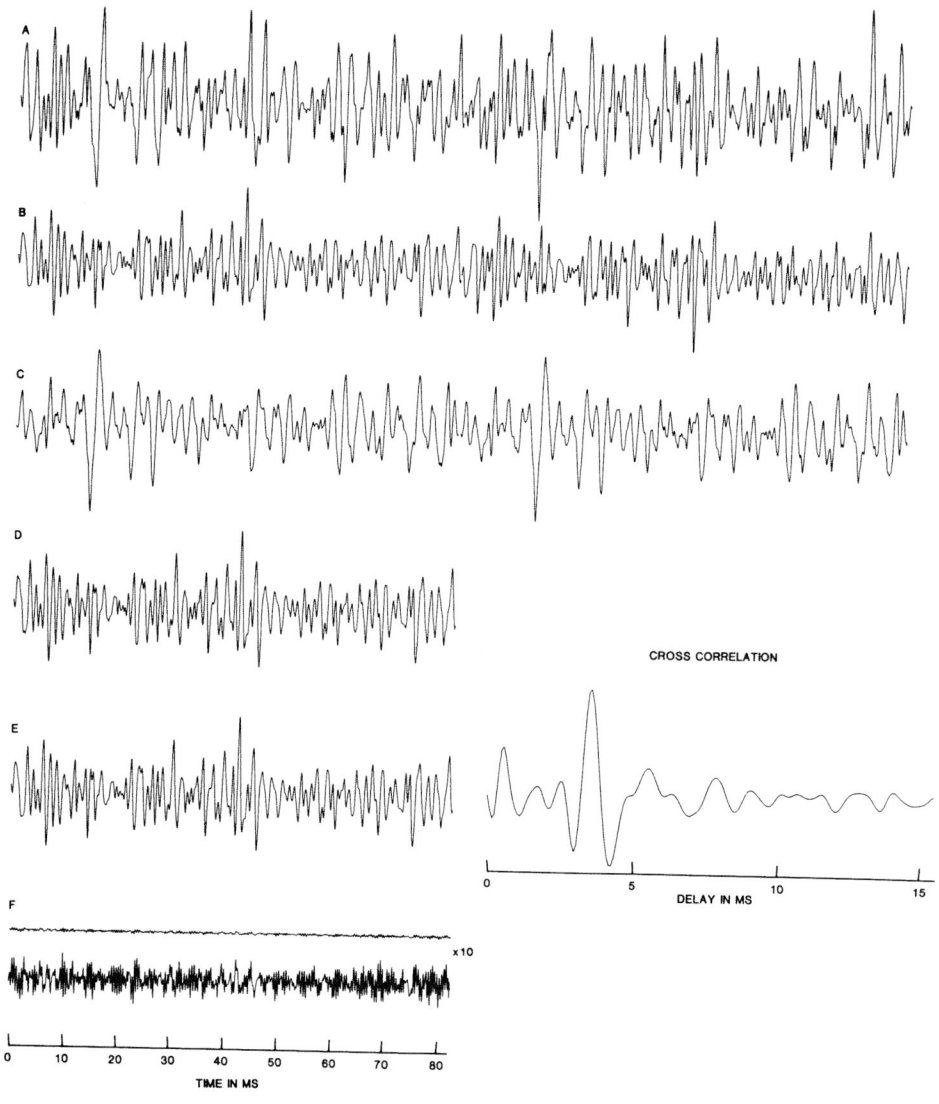

Fig. 3. A: The average of the responses to 200 periods of pseudorandom noise presented at an intensity of 90 dB SPL. B: Output of a linear model that has the cross-correlation of the response as its impulse response and the pseudorandom noise as its input. C: Difference between nerve response (A) and model output (B). D: The response shown in A after the latter half was subtracted from the former half (folding). E: Similarly folded output of the model. F: Difference between the folded response (D) and the folded output (E) (also shown on a 10-times expanded amplitude scale). Insert: Cross-correlogram of the noise used to drive the earphone and the response.

We have made use of the fact that the pseudorandom noise that we use is of the inverse-repeat type (the latter half of the noise sequence is identical to the first half, but inverted) in obtaining information about the nature of the nonlinearity (Møller, 1983a,b, 1986, 1987b). While subtracting the latter half of the response to one noise period from the former half will not change the linear portion of the response, all even-order nonlinearities will be cancelled. The difference between such 'folded' responses and the similarly folded output of the model is a measure of the odd-order nonlinearities of the system under test. When the recorded potentials and model output shown in Fig. 3,A and B are folded (D,E, and F), we find that the amount of odd-order nonlinearities in the recorded potentials is small compared to the amount of even-order nonlinearities, as indicated by the difference between the response and the output of the model.

The difference between the response and the model output, as seen in Fig. 3, could also be a result of uncorrelated noise in the recorded potentials. The recorded potentials contain a considerable amount of such uncorrelated noise in the form of various electrical interferences, unrelated nerve signals, and amplifier noise. The purpose of averaging responses to many periods of the noise is to decrease the amount of uncorrelated noise, but averaging will not totally eliminate uncorrelated noise from these recordings.

The amount of such uncorrelated noise that remains in the averaged recordings from which the cross-correlograms were computed may be estimated by comparing two consecutive averages of responses to the same noise (Fig. 4). The difference between such consecutive recordings represents the inherent noise level of the recording as well as the change in the (correlated) response of the auditory nerve and in the recording conditions during the time of recording. This difference also represents how the properties of the system that is tested vary between the times when the two consecutive responses are obtained. It is seen that this difference is small (Fig. 4,A), but when the response amplitude decreases as a result of lowering the stimulus intensity, the relative amount of such uncorrelated noise increases (Fig. 4,B).

A comparison of the folded response and the folded output of a linear model that has the cross-correlogram as its impulse response is shown in Fig. 5 for another patient, together with the sum of the two halves of the response and of the output of the model ("folded added"). The substantial amplitude of the "folded added" response represents the even-order nonlinearities. Since the model is linear the amplitude of the equally "folded added" model output is zero.

Although the pseudorandom noise that is used in the present study is known to have only a small amount of anomalies, it is important to ascertain that the components seen in the cross-correlograms can be attributed to the system under test and are not a result of anomalies in the noise that is used. We therefore compared the cross-correlograms computed from the responses to noise based on two different ternary sequences. Comparison of these two cross-correlograms (Fig. 6,A,B) shows that the difference between the cross-correlograms at short lags is no greater than the difference between the cross-correlograms of consecutive responses to the

71

same noise. However, at long lags the small peaks which reproduce in consecutive correlograms when the same noise is used do not reproduce when the correlograms are based on the responses to different noises. This indicates that the components that are seen in the correlograms at long lags most likely result from anomalies in the noise.

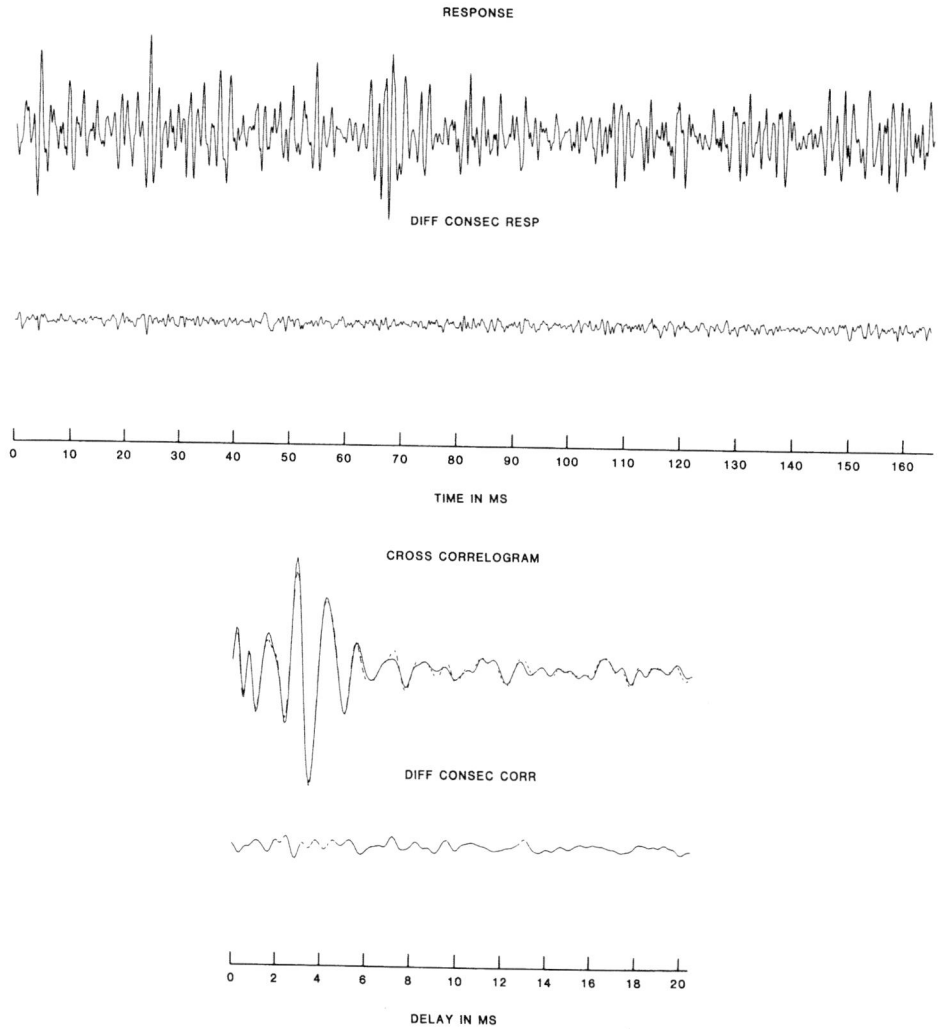

Fig. 4,A: Top: The responses to 100 presentations of pseudorandom noise and the difference between two consecutive recordings in response to the same noise. Bottom: Cross-correlograms of the responses to the two recordings shown on top and the difference between the two correlograms. The sound intensity was 90 dB SPL.

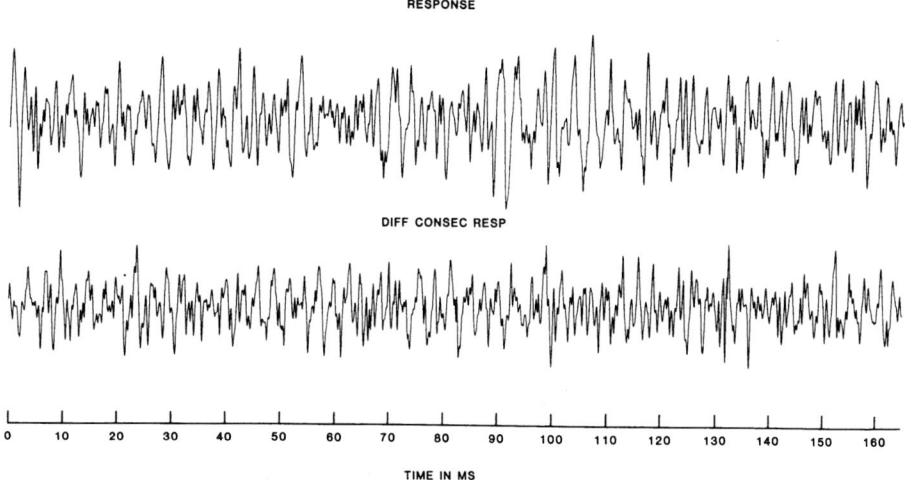

RESPONSE

DIFF CONSEC RESP

| | | | | | | | | | | | | | | | | |
0 10 20 30 40 50 60 70 80 90 100 110 120 130 140 150 160

TIME IN MS

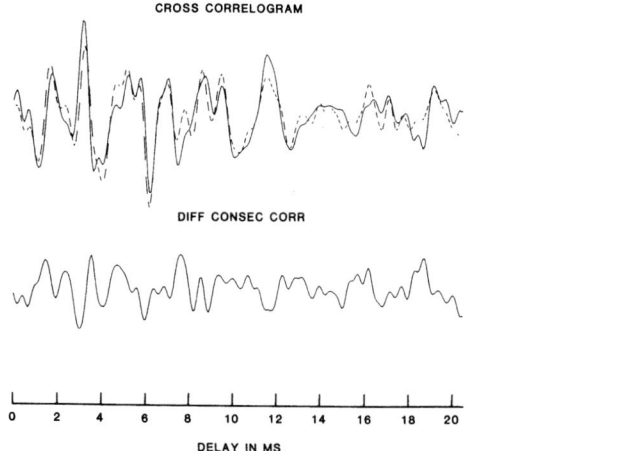

CROSS CORRELOGRAM

DIFF CONSEC CORR

| | | | | | | | | | |
0 2 4 6 8 10 12 14 16 18 20

DELAY IN MS

Fig. 4,B: Recordings similar to those in A, but showing the responses to 100 presentations of the same noise at a lower intensity (70 dB SPL).

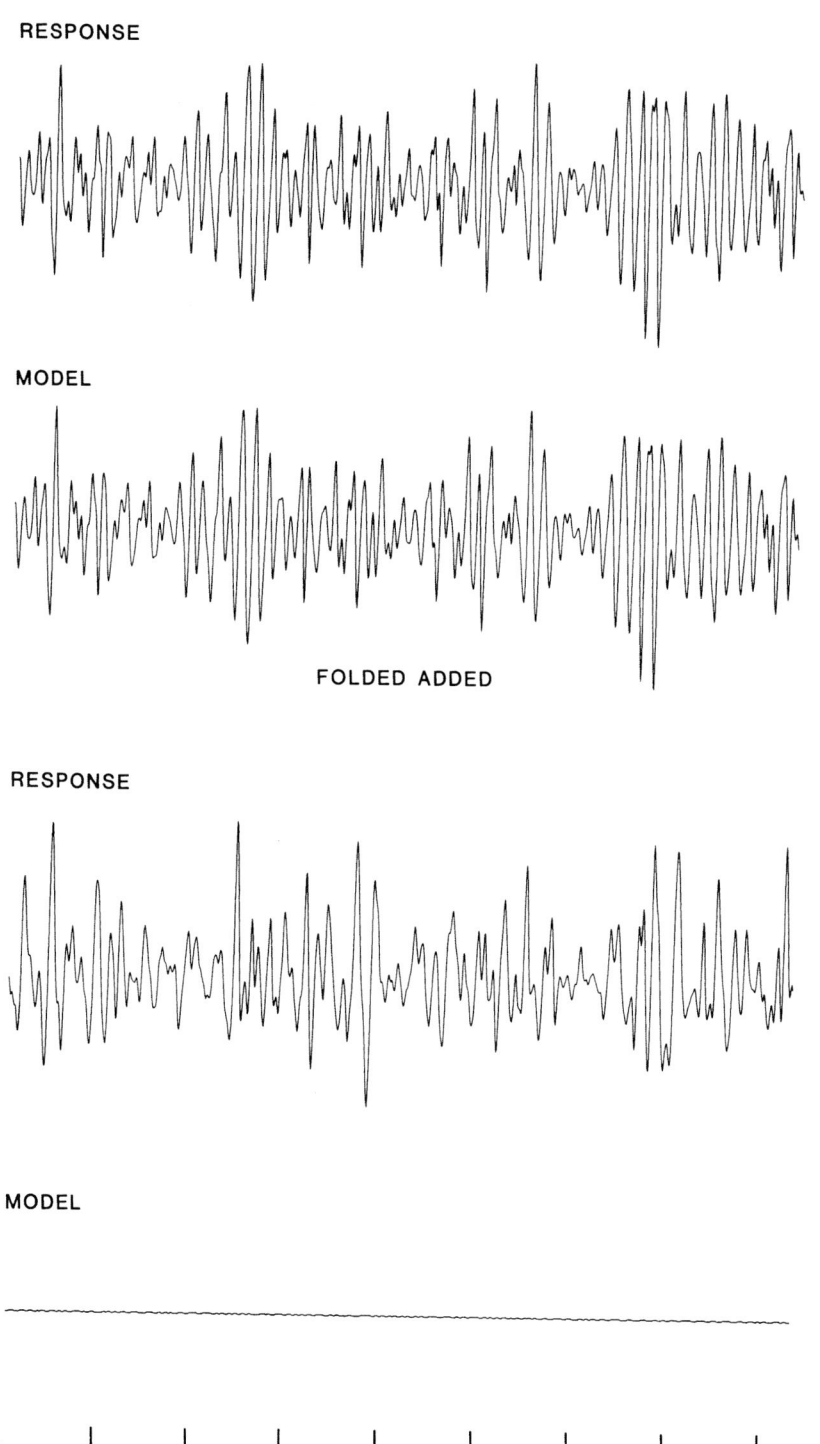

Fig. 5,A: The results of folding the responses from the eighth nerve in two ways. The stimulus intensity was 90 dB SPL in this patient.

FOLDED SUBTRACTED

RESPONSE

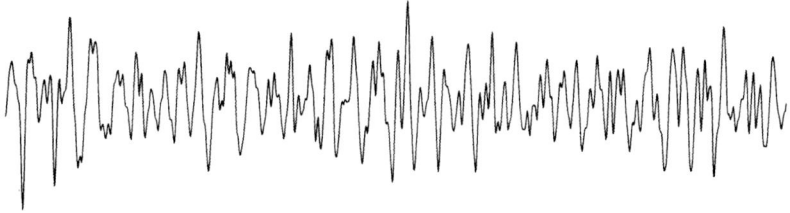

MODEL

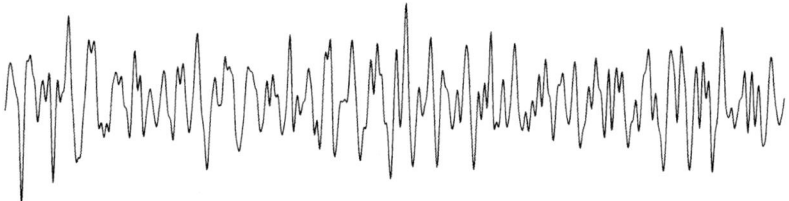

FOLDED ADDED

RESPONSE

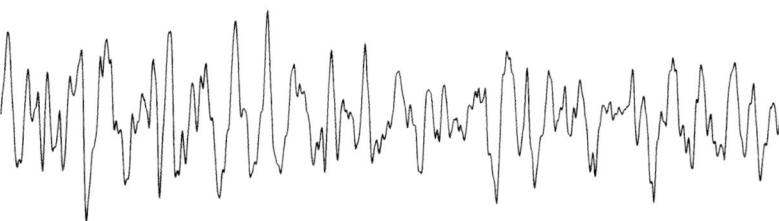

MODEL

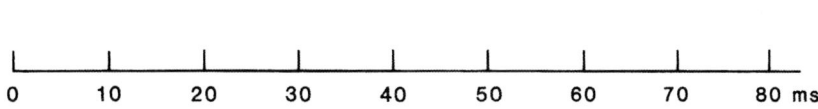

```
L___|___|___|___|___|___|___|___|___|
0   10  20  30  40  50  60  70  80 ms
```

Fig. 5,B: Records similar to those shown in A, but obtained at a lower stimulus intensity (70 dB SPL). Each record is the average of the responses to 100 periods of the noise.

75

SAME NOISES

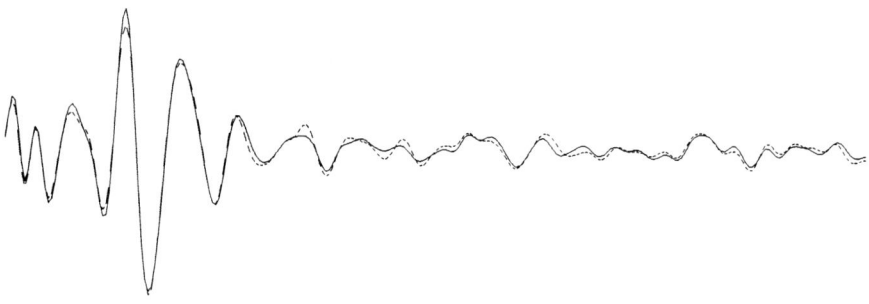

DIFFERENCE

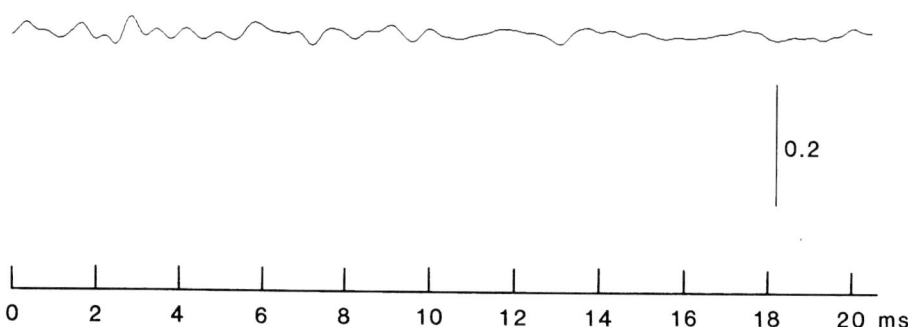

0.2

| | | | | | | | | | | |
0   2   4   6   8   10   12   14   16   18   20 ms

Fig. 6,A: Two cross-correlograms of the response recorded
from the auditory nerve in response to the same noise, and the
difference between the cross-correlograms.

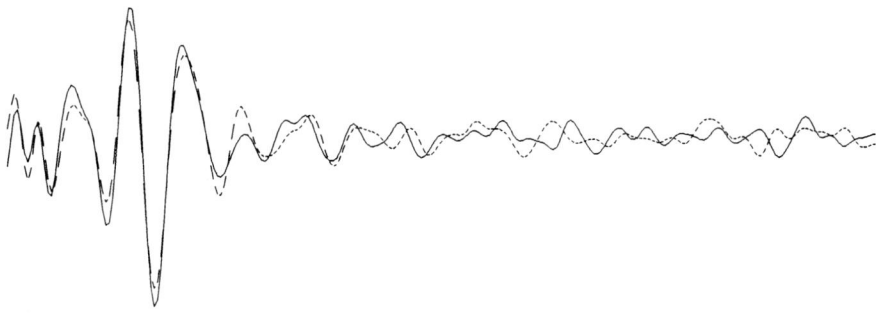

DIFFERENCE

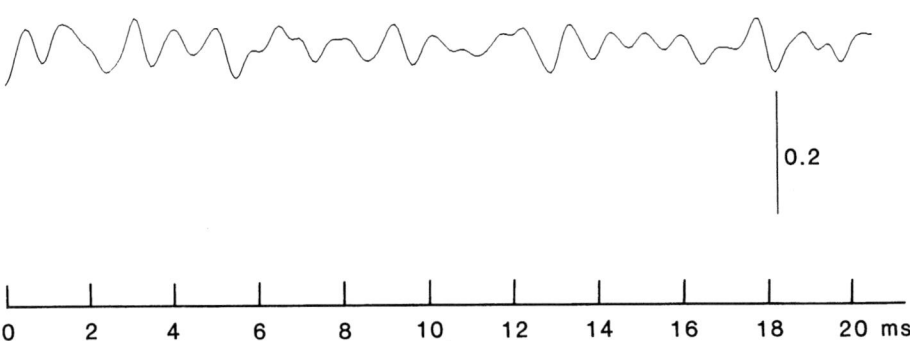

0.2

```
L____|____|____|____|____|____|____|____|____|____|
0    2    4    6    8   10   12   14   16   18   20 ms
```

Fig. 6,B: Same as the top, but using two different noises. The stimulus intensity was 90 dB, and each correlogram was computed from the average of 200 responses.

DISCUSSION

The results of the work discussed here show that the cross-correlograms of the responses from the exposed intracranial portion of the auditory nerve generally resemble those of the responses to transient sounds (click sounds). The cross-correlograms of the responses to pseudorandom noise may be assumed to reflect the ability of many auditory nerve fibers to discharge nearly synchronously and in such a way that the discharges are delivered in preferred phases of the sound wave.

The whole-nerve responses from the human auditory nerve are highly nonlinear, and the nonlinearity is mainly of an even order. Contrary to what is the case for recordings from single auditory nerve fibers in which the nonlinearity can be explained by the halfwave rectification by the sensory cells, the nonlinearity in the gross response is more complex. Since convolution is a linear process, the transformation of neural discharges in many nerve fibers into a whole-nerve action potential can be regarded as a linear filter, and this process in itself thus does not contribute to the nonlinearity. The fact that the whole-nerve response contains nonlinearities that are similar (but not identical) to a fullwave rectification may, however, be explained by the fact that both the rarefaction and condensation phases of the sound contribute to the whole-nerve response, because the whole-nerve response is made up of responses from single-nerve fibers, some of which respond to the rarefaction phase of the sound and some of which respond to the condensation phase of the sound.

Although the amount of even-order nonlinearities in the whole-nerve response of the auditory nerve to pseudorandom noise is greater than it is in the response to amplitude-modulated sounds (Møller, 1987b), the content of odd-order nonlinearities is no higher. The reason for this may be that nonlinearities caused by saturation of a system that only responds in one direction are mainly of even order.

In comparing compound action potentials to transient sounds and cross-correlograms of the responses to noise, it must be kept in mind that while the compound action potentials reflect a summed number of discharges that are delivered by all nerve fibers immediately after the transient sound, the cross-correlograms reflect the correlation between the sound and the phase-locked discharges.

The results of the work reported here show that when noise is used as the test sound in studies of the function of the auditory system, information can be gained that cannot be derived from studying the responses to transient sound stimuli. It remains to be seen to what extent this new approach can be useful in studies of the function of normal and pathological auditory systems.

REFERENCES

Goldstein, M.H., Jr., 1960, A statistical model for interpreting neuroelectric responses. Inf. Contr. 3:1.
Harrison, R.V. and Evans, E.F., 1982, Reverse correlation study of cochlear filtering in normal and pathological guinea-pig ears. Hear. Res. 6:303.
Jacobson, J.T., 1985, "The Auditory Brainstem Response," College Hill-Press, San Diego.
Jannetta, P.J., 1977, Observations on the etiology of trigeminal neuralgia, hemifacial spasm, acoustic nerve dysfunction and glossopharyngeal neuralgia. Definitive microsurgical treatment and results in 117 patients. Neurochirurgia 20:145.
Jannetta, P.J., 1981a, Vascular decompression in trigeminal neuralgia, in: "The Cranial Nerves," M. Samii and P.J. Jannetta, eds., Springer-Verlag, Heidelberg, West Germany.
Jannetta, P.J., 1981b, Hemifacial spasm, in: "The Cranial Nerves," M. Samii and P.J. Jannetta, eds., Springer-Verlag, Heidelberg, West Germany.

Konishi, T. and Nielsen, D.W., 1978, The temporal
    relationship between basilar membrane motion and nerve
    impulse imitation in auditory nerve fibers of guinea
    pigs. <u>Jap. J. Physiol.</u> 28:291.
Marmarelis, P.Z. and Marmarelis, V.Z., 1978, "Analysis of
    Physiological Systems. The White-Noise Approach."
    Plenum Press, New York.
Møller, A.R., 1974, Use of stochastic signals in evaluation of
    the dynamic properties of a neuronal system. <u>Scand. J.
    Rehab. Med. Suppl.</u> 3:37.
Møller, A.R., 1976a, Dynamic properties of the responses to
    single neurons in the cochlear nucleus. <u>J. Physiol.</u>
    259:63.
Møller, A.R., 1976b, Dynamic properties of primary auditory
    fibers compared with cells in the cochlear nucleus. <u>Acta
    Physiol. Scand.</u> (Stockh.) 98:157.
Møller, A.R., 1977, Frequency selectivity of single auditory
    nerve fibers in response to broadband noise stimuli. <u>J.
    Acoust. Soc. Am.</u> 62:135.
Møller, A.R., 1981a, Latency in the ascending auditory
    pathway determined using continuous sounds: Comparison
    between transient and envelope latency. <u>Brain Res.</u>
    207:184.
Møller, A.R., 1981b, Neural delay in the ascending auditory
    pathway. <u>Exp. Brain Res.</u> 43:93.
Møller, A.R., 1982, Use of pseudorandom noise in studies of
    the dynamic properties of the linear part of a sensory
    neural system. <u>in</u>: "Proceedings of the Fifteenth Hawaii
    International Conference on System Sciences" (held in
    Honolulu, Hawaii, 1982), Vol. 2. B.D. Shriver, R.R.
    Grams, T.H. Walker, and R.H. Sprague, eds., Western
    Periodicals Company, North Hollywood, California.
Møller, A.R., 1983a, Use of pseudorandom noise in studies of
    frequency selectivity: The periphery of the auditory
    system. <u>Biol. Cybern.</u> 47:95.
Møller, A.R., 1983b, Frequency selectivity of phase-locking of
    complex sounds in the auditory nerve of the rat. <u>Hear.
    Res.</u> 11:267.
Møller, A.R. and Jannetta, P.J., 1983, Monitoring auditory
    functions during cranial nerve microvascular
    decompression operations by direct recording from the
    eighth nerve. <u>J. Neurosurg.</u> 59:493.
Møller, A.R., 1986, Systems identification using pseudorandom
    noise applied to a sensorineural system. <u>in</u>: "Computer
    and Mathematics with Applications (C.M.A.)" (special
    issue in memory of Dr. Richard Bellman), Vol. 12. G.
    Adomian, ed., Pergamon Press, Elmsford, New York.
Møller, A.R., 1987a, "Evoked Potentials in Intraoperative
    Monitoring," Williams & Wilkins, Baltimore, Maryland.
Møller, A.R., 1987b, Auditory evoked potentials to continuous
    amplitude-modulated sounds: Can they be described by
    linear models? <u>Electroenceph. Clin. Neurophysiol.</u> 66:56.
Møller, A.R. and Sekiya, T., 1988, Injuries to the auditory
    nerve: A study in monkeys. <u>Electroenceph. Clin.
    Neurophysiol.</u> 70:248.
Weiss, T.F., 1966, A model of the peripheral auditory system.
    <u>Kybernetik</u> (Berlin) 3:153.
Zwislocki, J.J. and Sokolich, W.G., 1973, Velocity and
    displacement responses in auditory nerve fibers. <u>Science</u>
    182:64.

ACKNOWLEDGEMENT

This work was supported by the National Institutes of Health (Grant No. R01 NS21378-05). The author is grateful to Peter J. Jannetta, M.D., for making patients under his care available for this study, and to his associate and residents for their assistance during data collection. I also wish to thank Margareta B. Møller, M.D., Dr.Med.Sci., for her audiological and otological evaluation of the patients in this study.

NONLINEAR MODELS OF TRANSDUCTION AND ADAPTATION IN LOCUST PHOTORECEPTORS

A.S. French, A.E.C. Pece and M.J. Korenberg

Department of Physiology, University of Alberta, Edmonton
Canada T6G 2H7, and Department of Electrical Engineering
Queen's University, Kingston, Ontario, Canada K7L 3N6

INTRODUCTION

The absorption of light by the photoreceptor cells of vertebrates and invertebrates initiates a chain of events which eventually leads to a change in membrane current, and thus of membrane potential. When light is presented as a brief flash, approximating an impulse or delta function, the change in membrane potential is delayed by a period of up to 500 ms and has an asymmetric voltage-time profile, decaying more slowly than it rises (Fein & Szuts, 1982). In several animal species, the transduction of single photons in a dark-adapted photoreceptor cell can be observed as discrete fluctuations in membrane potential or "bumps" (Yeandle, 1958; Borsellino and Fuortes, 1968) which often exceed 1 mV per photon. The voltage-time profiles of bumps are similar to those of responses to larger flashes of light, suggesting that all light responses result from linear or nonlinear combinations of single photon bumps.

Several types of nonlinear behavior have been seen in photoreceptor responses. Increasing the intensity of a flash stimulus or the background light leads to a sublinear increase in response amplitude as well as a faster voltage-time profile (Fuortes and Hodgkin, 1964; Baylor and Hodgkin, 1974), while the initial phase of the response in some preparations may show a supralinear component with increasing intensity (Payne and Fein, 1986; Grzywacz et al., 1988). Early experiments with white noise stimulation of photoreceptors also demonstrated a significant nonlinear component in the response to a continuously varying input (Eckert and Bishop. 1975; Gemperlein and McCann, 1974). However, the major nonlinear feature of photoreceptors is the reduction in sensitivity with increasing light levels, reflecting the ability of photoreceptors to adapt to the wide range of mean light intensities encountered in the natural environment.

Initially, it was believed that nonlinear behavior in photoreceptors was only significant at light levels where many photons were transduced together within the response time of the photoreceptor. However, estimates of the number of simultaneous photon transductions required to produce nonlinear behavior have fallen progressively from about 30 in fly photoreceptors (Hamdorf and Kirshfeld, 1980), to 10 photons in the fly (Dubs, 1981), and 2 photons in locust photoreceptors (French and Kuster, 1985).

Since the absorption of a single photon causes many ion channels to open or close, there is significant amplification involved in light transduction. This amplification of the light response, as well as the delay between light absorption and the change in membrane current, was initially linked to a cascade of biochemical reactions by Fuortes and Hodgkin (1964) before any details of photoreceptor biochemistry were known. Modification of an early stage in the cascade by feedback from a later stage was suggested by Borsellino and Fuortes (1968) to account for changes in gain and time course. Cascade models have recently been used to account for all of the observed nonlinearities in *Limulus* photoreceptors using either feedback or feed-forward mechanisms (Grzywacz and Hillman, 1988).

In the present work we used locust photoreceptors, which have large photon bumps and no detectable spontaneous bumps in the absence of light (Lillywhite, 1977). Nonlinear behavior in response to white noise modulated light amplitude was examined by means of a very efficient and accurate method for estimating Wiener kernels (Korenberg, 1988; Korenberg et al., 1988). Different cascade models were tested by measuring nonlinear behavior at several different mean light intensities. The results of these experiments and simulations suggest that feedback control of gain is the most likely mechanism of light adaptation in these photoreceptors.

METHODS

Locusts. *Locusta migratoria.* were taken from a laboratory colony which was maintained in a photocycle of 12 hours dark and 12 hours light. Animals were used 1 to 2 weeks after their final moult to the adult stage and they were dark-adapted for at least one hour before the experiments. The animals were relatively intact, with only the wings, antennae and legs removed, and they were immobilized with dental wax, taking care not to obstruct any respiratory spiracles. All experiments were performed at room temperature (20 ± 2°C).

Light stimulation

The locust head was mounted at the center of a 5 cm diameter Cardan arm, which held a high intensity green light emitting diode (Hewlett-Packard HLMP 3950). A 33-bit binary shift register with feedback was used to generate a random binary sequence with a clock rate of 1 kHz (Marmarelis and Marmarelis, 1978). This signal was then filtered by a nine-pole active low pass filter with a corner frequency of 50 Hz and a single-pole high pass filter with a corner frequency of 0.5 Hz to yield an approximately Gaussian random noise signal. A DC component was added to the random noise signal and the amplitude of the random noise and the DC component were adjusted independently. The combined signal was then used to drive a constant voltage to current convertor, which in turn drove the light emitting diode. At each mean level of light intensity, the amplitude of the random component was adjusted to be as high as possible while maintaining a positive current through the diode at all times.

Light intensity was calibrated in effective photons per second (ep/s) for each cell by counting the individual bumps at low light levels (5-10 ep/s). This calibration was not exact because it required a human observer to discriminate individual bumps. However, a storage oscilloscope was used to simplify the task. Any error in calibration would affect all of the measurements on a particular cell, leaving relative levels unaffected. Extrapolation of the calibration to the entire working range of light intensities was performed by measuring the relative light output from the diode at different current levels with a phototransistor (Motorola MRD 3050).

Electrical recording

Glass microelectrodes filled with 3 M potassium acetate and having resistances of 50-100 MΩ were lowered through a hole in the cornea of the right eye to penetrate retinular cells. The reference silver chloride electrode was placed extracellularly in the left eye. Intracellular potentials were measured with a high impedance amplifier (Getting model 5), high pass filtered at 0.5 Hz by a single-pole filter, and then low pass filtered at 100 Hz by a single-pole filter. The LED current and the intracellular potential were sampled at 10 ms intervals by a 12-bit analog to digital convertor and stored by a digital computer. The signal from the photoreceptor did not contain frequencies above 50 Hz, the Nyquist frequency of the sampling rate, and the low pass filter was only employed to eliminate high frequency noise in the recording.

THEORY

Feed-forward gain control

Figure 1 shows a dynamic feed-forward gain control mechanism. The nonlinear element n performs the function:

$$z(t) = x(t)\{1-s(t)\} \tag{1}$$

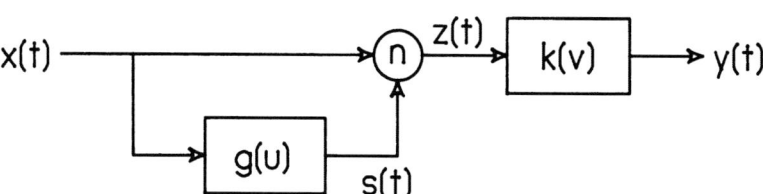

Fig. 1. A separable cascade model of feed-forward gain control adaptation. The elements $g(u)$ and $k(v)$ represent linear filters and the element n is a static nonlinearity.

Equation (1) may be expanded as follows:

$$z(t) = x(t) - \int g(u)x(t)x(t-u)du \tag{2}$$

and therefore:

$$y(t) = \int k(v)x(t-v)dv - \int\int k(v)g(u)x(t-v)x(t-u-v)dudv \qquad (3)$$

Substituting $w=u+v$, and assuming that $g(u)=0$ for $u<0$:

$$y(t) = \int k(v)x(t-v)dv - \int\int k(v)g(w-v)x(t-v)x(t-w)dwdv \qquad (4)$$

If the output of the system is represented as a Volterra series expansion:

$$y(t) = h_0 + \int h_1(u)x(t-u)du + \int\int h_2(u,v)x(t-u)x(t-v)dudv + \ldots \qquad (5)$$

then comparison of equations (4) and (5) yields the Volterra kernels:

$$h_0 = 0 \qquad (6a)$$

$$h_1(u) = k(u) \qquad (6b)$$

$$h_2(u,v) = -g(u-v)k(v) \qquad (6c)$$

with all higher order kernels equal to zero.

Feedback gain control

Figure 2 shows a feedback control system with similar elements to those of Fig. 1. The nonlinear element is again described by Eq. 1 so that its output, $z(t)$, may be expanded as follows:

$$z(t) = x(t) - \int g(u)x(t)z(t-u)du \qquad (7)$$

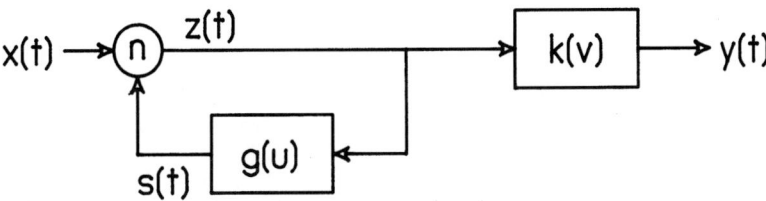

Fig. 2. A separable cascade model of feedback gain control adaptation. The elements are similar to those of Fig. 1.

Equation 7 is a Fredholm integral equation of the second kind (Davis, 1962) and cannot be solved analytically. A series approximation to Eq. 7 can be obtained by expanding the term in $z(t)$ within the convolution integral, introducing new time variables where necessary, and assuming that the linear functions $g(u)$ and $k(v)$ are physically realizable and therefore zero at negative times:

$$z(t) = x(t) - \int g(u)x(t)x(t-u)\{1-\int g(w)z(t-u-w)dw\}du \tag{8}$$

The output of the feedback system, $y(t)$, may then be obtained from convolution of this series with the final linear element, $k(v)$:

$$y(t) = \int k(v)x(t-v)dv - \int\int k(v)g(u)x(t-v)x(t-u-v)dudv + \ldots \tag{9}$$

If the system output is approximated by a Volterra series, as in Eq. 5, then the Volterra kernels are:

$$h_0 = 0 \tag{10a}$$

$$h_1(u) = k(u) \tag{10b}$$

$$h_2(u,v) = -g(u-v)k(v) \tag{10c}$$

$$h_3(u,v,w) = g(u-v)g(v-w)k(w) \tag{10d}$$

etc.

An alternative form of the nonlinear component n in Figs. 1 and 2 would compute the ratio of the two signals:

$$z(t) = x(t)/\{1+s(t)\} \tag{11}$$

In this case it is possible to follow similar derivations and arrive at Volterra kernels with identical first- and second-order terms as in Eqs. 6 and 10. However, in both the ratio models there are also third- and higher order kernels. These kernels are different in each of the two ratio models and also differ from the feedback product model.

Separability of the second-order kernels

Equations 6 and 10 both predict that the second-order Volterra kernel, $h_2(u,v)$ should be separable into two functions, $g(u)$ and $k(v)$, if it is first re-arranged to give the related second-order kernel:

$$p_2(u,v) = h_2(u+v,v) \qquad\qquad (12)$$

so that:

$$p_2(u,v) = -g(u)k(v) \qquad\qquad (13)$$

Both equations predict that the first-order kernel, $h_1(u)$, should be identical to the impulse response of the linear system which is in series with the nonlinearity.

Computational methods

The fast orthogonal algorithm (Korenberg, 1988; Korenberg et al., 1988) was used to estimate Wiener kernels of the phototransduction process. In order to obtain reliable estimates of the kernels, at least 40,000 pairs of input-output data were usually required. At low background light intensities the output data contained significant photon noise, so that longer data lengths were required, up to 150,000 data pairs.

The kernels were estimated with 25 lags, giving a total memory length of 240 ms. The second-order kernels were then re-arranged by Eq. 12 to obtain $p_2(u,v)$. In order to extract the function $g(u)$, the three rows of $p_2(u,v)$ containing the highest values of $h_1(v)$ were averaged, i.e. if the time to peak of $h_1(v)$ were t_p, we estimated:

$$g(u) = -\{p_2(u,t_p-1) + p_2(u,t_p) + p_2(u,t_p+1)\}/3 \qquad\qquad (14)$$

The function $k(v)$ was extracted by a similar procedure. Scaling factors for $h_1(u)$ and $g(u)$ were estimated by minimizing the mean square error between the experimental output data and the outputs of the models, using a section of the original input data record containing 1000 data pairs. Similar scaling factors were obtained for each of the models. Simulations of the model outputs were performed by discrete convolution and simple algebra using original input data values at the same 10 ms sample rate.

RESULTS

First- and second-order Wiener kernels were measured at a range of background light intensities from 450 ep/s to 12,500 ep/s. The modulation depth of the noise amplitude was adjusted to be as close as possible to 100% of the background level in each case. Figure 3 shows the first-order kernel, $h_1(u)$, at 450 ep/s and 2000 ep/s for the same cell. The first-order kernel had a similar general appearance to the flash response in these cells (Howard et al., 1984; French and Kuster, 1985). Increasing the background light intensity reduced the amplitude of the first-order kernel but accelerated its time course, with most of the temporal change occurring at the lowest light levels. First-order kernels at higher background intensities were very similar to the one at 2000 ep/s.

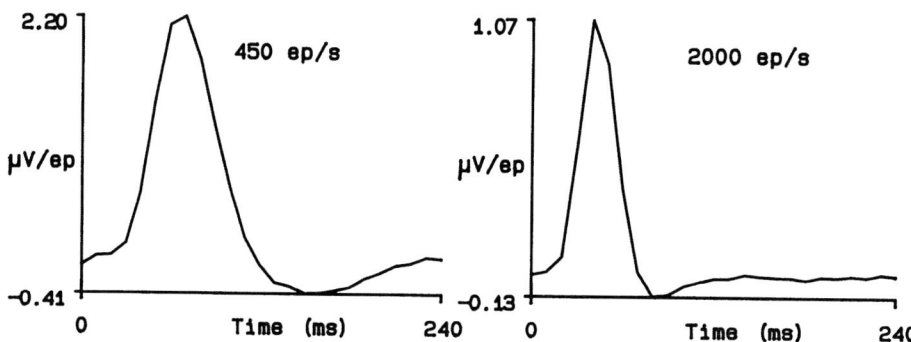

Fig. 3. First-order kernels, $h_1(u)$, at two background light intensities.

Figures 4 and 5 show the second-order Wiener kernel $h_2(u,v)$, obtained at the same two background intensities for the same cell as in Fig. 3. At 450 ep/s the major features were a positive peak on the diagonal, $u=v$, at approximately 60 ms, and a negative peak, slightly off the diagonal at approximately 75 ms. These two features correspond to the early facilitation and the later depression of response seen before in locust photoreceptors (French & Kuster, 1985). At a background level of 2000 ep/s the kernel was dominated by the negative peak, which now occurred earlier at about 50 ms. Again, a similar second-order kernel was obtained at higher background intensities.

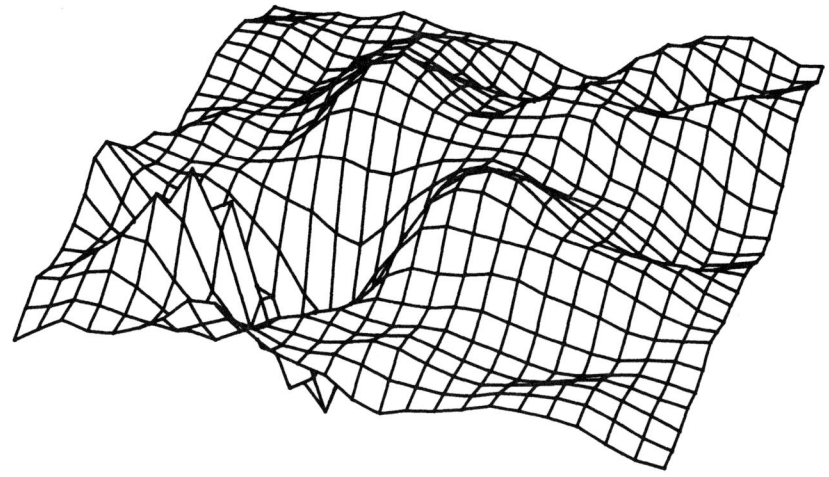

Fig. 4. Perspective plot of the second-order kernel, $h_2(u,v)$, at 450 ep/s. The kernel is symmetric with its origin at the extreme left. The axes, u and v, range from 0 to 240 ms. The peak amplitude of the kernel is 0.67 nV/ep^2, and the peak depression is -0.70 nV/ep^2.

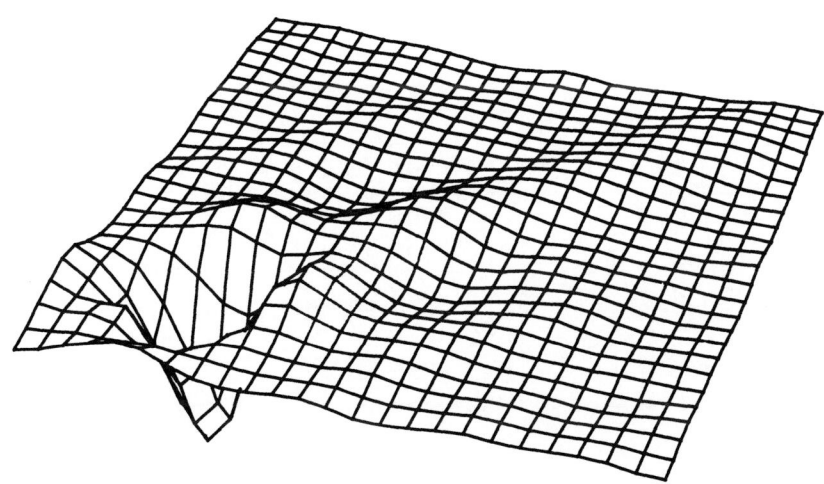

Fig. 5. Perspective plot of the second-order kernel $h_2(u,v)$, at 2000 ep/s. Plotting details are the same as Fig. 4. except for the peak amplitude and depression, which are 0.018 nV/ep^2 and -0.08 nV/ep^2 respectively.

In order to test the separability of the second-order kernels, they were transformed according to Eq. 12 to yield the kernel $p_2(u,v)$, then the functions $g(u)$ and $k(v)$ were obtained using Eq. 13. Figures 6 and 7 show the results of the transformations and Figs. 8 and 9 show $k(v)$ and $g(u)$ at background levels of 450 and 2000 ep/s respectively. The functions $k(v)$ were very similar to the corresponding first-order kernels, $h_1(u)$, in both amplitude and time course, as predicted by Eqs. 6b and 10b. However, a problem with the separability appears if the reverse operation is considered. Combining the first-order kernel, $h_1(u)$, and the function $g(u)$ should allow reconstruction of the transformed second order kernel, $p_2(u,v)$. Although this was true for the general shape of the kernel, including the inhibitory peak, it was not true for the early facilitation seen at 450 ep/s in Figs. 4 and 6. At 2000 ep/s this problem was much less apparent and at higher background levels the early facilitation was further reduced, making a separable model quite acceptable.

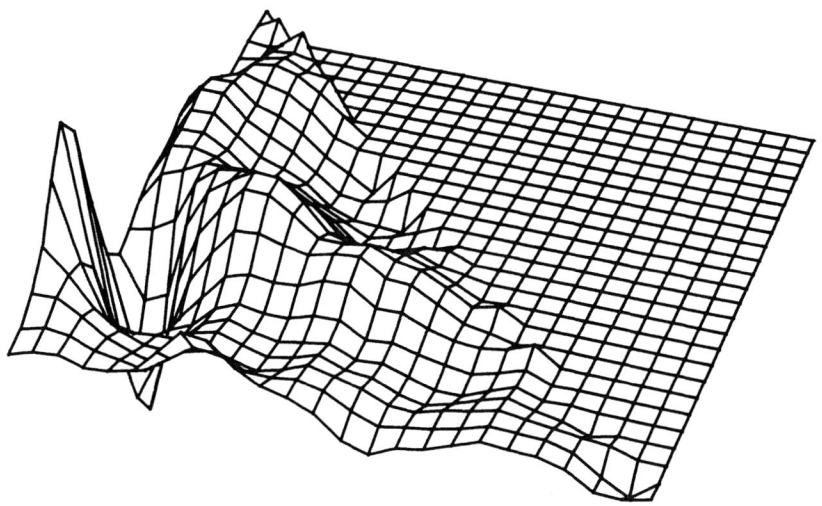

Fig. 6. Perspective plot of the transformed second-order kernel, $p_2(u,v)$, obtained from Fig. 4 via Eq. 12. The horizontal axes have the same values as in Figs. 4 and 5 with u along the axis closer to the observer.

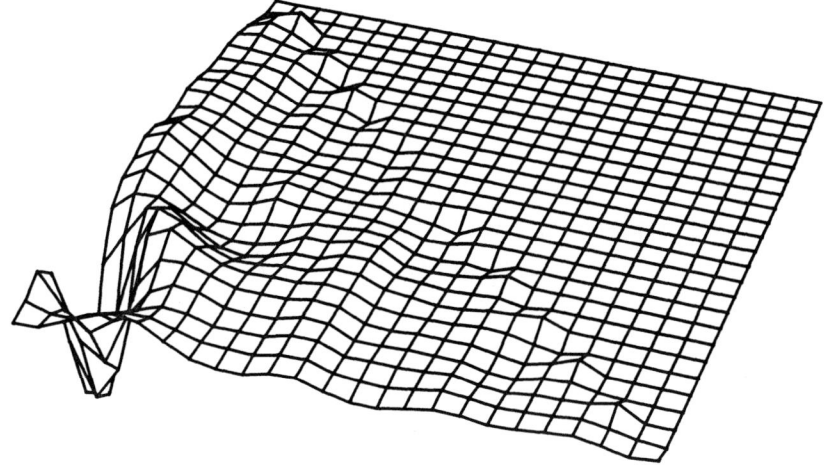

Fig. 7. Perspective plot of the transformed kernel, $p_2(u,v)$, obtained from Fig. 5 via Eq. 12. Plotting as in Fig. 6.

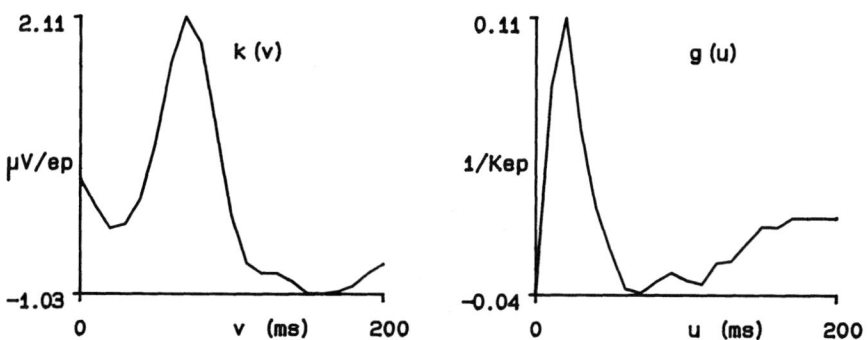

Fig. 8. The separated functions $k(v)$ and $g(u)$ obtained from the second-order kernel of Fig. 6.

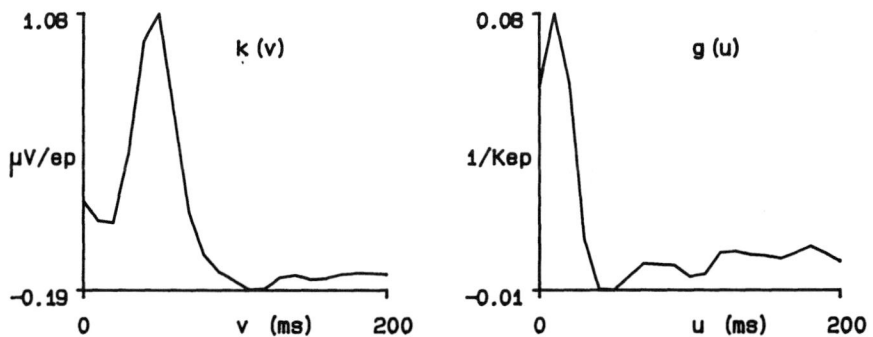

Fig. 9. The separated functions $k(v)$ and $g(u)$ obtained from the second-order kernel of Fig. 7.

The first- and second-order Wiener kernels of a system equal the Volterra kernels of corresponding order if all higher order kernels are negligible (Marmarelis and Marmarelis, 1978). We therefore used $g(u)$ and $h_1(u)$ derived from the experimental Wiener kernels, together with the four cascade models and portions of the experimental input, to compare simulated outputs from the cascades with the actual experimental output. At any given background level, $g(u)$ and $h_1(u)$ derived from the kernels measured at that level gave accurate predictions of the experimental output with any of the cascade models.

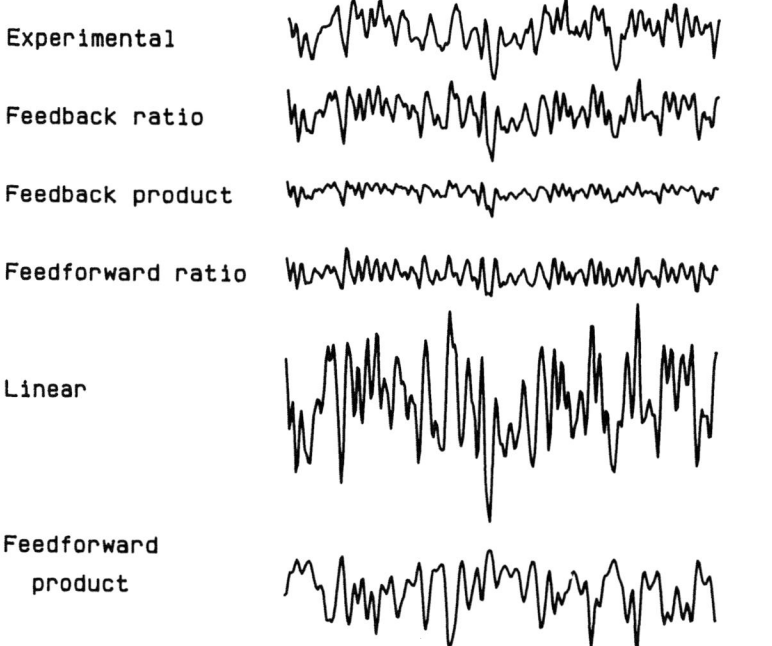

Fig. 10. The experimental output of a locust photoreceptor (top) in response to random light fluctuation at a background level of 12,500 ep/s, together with simulations from the same input data, the linear model, and the four nonlinear models, using the kernels obtained from stimulation at 2000 ep/s. Time calibration bar is 1 s, upper vertical calibration bar is 10 mV, and lower calibration bar is 2 mV (lowest trace only).

Any complete model of a nonlinear system should be capable of predicting the experimental output under any conditions. For the present experiments this requires that a single pair of functions, $g(u)$ and $k(v)$, should be able to predict the output at all background light intensities. Figure 10 shows the results of simulating the output at a background level of 12,500 ep/s with the kernels obtained at 2000 ep/s. Although several models reproduced some of the features of the actual output, the feedback ratio model, corresponding to Fig. 2 with the nonlinearity of Eq. 11, gave the best approximation to the experimental output. The linear model, which includes no gain control, produced a response with exaggerated gain, while the feed-forward product model produced an inverted response because the second-order component of the output exceeded the first-order component.

DISCUSSION

A feedback mechanism of gain adaptation was proposed in one of the earliest cascade models of phototransduction in *Limulus* by Fuortes and Hodgkin (1964). More recently, a cascade model with feedback has been proposed by Grzywacz and Hillman (1988) to account for steady-state responses in *Limulus*. A current model of phototransduction and adaptation in vertebrates also includes feedback in a biochemical cascade (Pugh and Altman, 1988). In the present work, we found that a feedback model with a ratio nonlinearity was not only able to account for a large part of the nonlinear behavior at each background light level, but could also predict the system output over a wide range of light intensities. If the reduction in gain is actually accomplished by inhibition of an enzyme within the phototransduction mechanism (Grzywacz and Hillman, 1988), then the ratio nonlinearity of Eq. 11 might be a reasonable mathematical approximation of the actual mechanism.

The feedback model with ratio nonlinearity gives a good approximation to phototransduction and adaptation at light levels of 2000 ep/s and above, but at lower light levels there are other nonlinear phenomena which it cannot explain. First, the change in time course of the response at low light levels cannot be explained by such a simple model with functions $g(u)$ and $k(v)$ which are invariant over different light levels. This clearly requires some nonlinear dependence of these elements on background light intensity. Second, the second-order kernel,

$h_2(u,v)$, is not perfectly separable at low background levels, because the early facilitation cannot be reconstructed from the first-order kernel, $h_1(u)$, and the function $g(u)$. Both these problems suggest that an additional nonlinear process is involved in changing the time course of the photon response, but this process must saturate at low light levels. One way to account for this would be to place an additional filter before the input of Fig. 2. Feedback from a later stage would then have to accelerate the response of the additional filter until it became a delta function at about 2000 ep/s, effectively disappearing from the cascade.

Although the present cascade models are imperfect in explaining the complete mechanisms of phototransduction and adaptation in locust photoreceptors, they can account for the observed behavior over a wide range of stimulus conditions. They can also be quite closely related to the underlying biochemical stages of the processes involved. Although more investigations and more complicated models will eventually be required, the nonlinear systems analysis approach presented here seems to be a useful tool in understanding these important phenomena.

Acknowlegements: We thank Dr. Stanley Klein, U.C. Berkeley, for suggesting separable models to us and for helpful discussion. Support for this work was provided by the Medical Research Council of Canada, the Alberta Heritage Foundation for Medical Research, and the Natural Sciences and Engineering Research Council of Canada.

REFERENCES

Baylor, D.A. and Hodgkin, A.L., 1974, Changes in time scale and sensitivity in turtle photoreceptors, *J. Physiol.*, 242:729.

Borsellino, A., Fuortes, M.G.F., 1968, Responses to single photons in visual cells of *Limulus*, *J. Physiol.*, 196:507.

Davis, H.T., 1962, "Introduction to Nonlinear Differential and Integral Equations," Dover Publications, New York.

Dubs, A., 1981, Nonlinearity and light adaptation in the fly photoreceptor, *J. Comp. Physiol.*, 144:53.

Eckert, H. and Bishop, L.G., 1975, Nonlinear dynamic transfer characteristics of cells in the peripheral visual pathway of flies. Part I: The retinula cells, *Biol. Cybern.*, 17:1.

Fein, A. and Szuts, E.Z., 1982, "Photoreceptors: Their Role in Vision," Cambridge University Press, Cambridge.

French, A.S. and Kuster, J.E., 1985, Nonlinearities in locust photoreceptors during transduction of small numbers of photons, *J. Comp. Physiol.*, 156:645.

Fuortes, M.F.G. and Hodgkin, A.L., 1964, Changes in time scale and sensitivity in the ommatidia of *Limulus*, *J. Physiol.*, 172:239.

Gemperlein, R. and McCann, G.D., 1975, A study of the response properties of retinula cells of flies using nonlinear identification theory, *Biol. Cybern.*, 19:147.

Grzywacz, N.M. and Hillman, P., 1988, Biophysical evidence that light adaptation in *Limulus* photoreceptors is due to negative feedback, *Biophys. J.*, 53:337.

Grzywacz, N.M., Hillman, P. and Knight, B.W., 1988, The quantal source of area supralinearity of flash responses in *Limulus* photoreceptors, *J. Gen. Physiol.*, 91:659.

Hamdorf, K. and Kirschfeld, K., 1980, Prebumps: evidence for double-hits at functional subunits in a rhabdomeric photoreceptor. *Z. Naturforsch.*, 35c:173.

Howard, J., Dubs, A. and Payne, R., 1984, The dynamics of phototransduction in insects, *J. Comp. Physiol.*, 154:707.

Korenberg, M.J., 1988, Identifying nonlinear difference equation and functional expansion representations: the fast orthogonal algorithm, *Ann. Biomed. Eng.*, 16:123.

Korenberg, M.J., French, A.S. and Voo, S., 1988, White noise analysis of nonlinear behavior in an insect sensory neuron: kernel and cascade approaches, *Biol. Cybern.*, 58:313.

Lillywhite, P.G., 1977, Single photon signals and transduction in an insect eye. *J. Comp. Physiol.*, 122:189.

Marmarelis, P.Z. and Marmarelis, V.Z., 1978, "Analysis of Physiological Systems. The White Noise Approach," Plenum Press, New York.

Payne, R. and Fein, A., 1986, The initial response of *Limulus* ventral photoreceptors to bright flashes. Released calcium as a synergist to excitation. *J. Gen. Physiol.*, 87:243.

Pugh, E. and Altman, J., 1988, A role for calcium in adaptation, *Nature*, 334:16.

Yeandle, S., 1958, Electrophysiology of the visual system - Discussion, *Am. J. Ophthal.*, 46:82.

IDENTIFICATION OF INTENSIVE NONLINEARITIES IN CASCADE MODELS

OF VISUAL CORTEX AND ITS RELATION TO CELL CLASSIFICATION

Robert C. Emerson, Michael J. Korenberg[*] and Mark C. Citron[**]

Department of Ophthalmology / Box 314 and Center for Visual Science, University of Rochester, Rochester, NY 14642; [*]Department of Electrical Engineering, Queen's University, Kingston, Ontario, Canada K7L 3N6; [**]Childrens Hospital of Los Angeles, Neurology Research, P.O. Box 54700, Terminal Annex, Los Angeles, CA 90054

INTRODUCTION

Nonlinear transformations in the nervous system seem to have the function of increasing the specificity of a neuron by reducing the class of stimuli to which that neuron responds, and often by enhancing responses to other, "preferred," stimuli. This specificity can confer the benefit that the large number of neurons in the brain can be apportioned to concentrate on different aspects of a given information-processing task.

Although this report is primarily about nonlinear coding of stimulus intensity, our initial interest in cortical nonlinearities has been motivated by our studies of the neural basis of motion processing in single units of the cat's visual cortex (Emerson et al, 1987b; Emerson & Gerstein, 1977b). We have found evidence for specific nonlinear spatiotemporal interactions that seem to underlie the particular form of directional selectivity (DS) that we see in all DS cortical cells. These neurons produce strong unmodulated responses to optimally oriented stimuli moving in one direction, and essentially no response for movement in the opposite, "null," direction (Emerson & Coleman, 1981). While other mechanisms limit the orientation range of acceptable stimuli, DS divides the population in half.

To understand the complex spatiotemporal consequences of moving stimuli, we have sought a model that mimics these neural properties. Such a model would allow quantitative testing of mechanistic hypotheses. The model that best fits our cortical data is one that was proposed by Adelson and Bergen (1985) on psychophysical grounds, and is well known for capturing the percept of smooth motion. This model not only mimics well the excitatory responses of "complex" cells (Hubel & Wiesel, 1962) to single light and dark bar presentations, but it also accounts completely for nonlinear interactions that occur between pairs of bars that represent samples of movement (Emerson et al, 1987a). Even DS "simple" cells (Hubel & Wiesel, 1962), show movement-related nonlinear interactions that are mimicked by the model (Emerson & Citron, 1988). The primary nonlinear transformation in this model is a pure squaring operation that occurs separately for each of two linear filters that differ in phase from each other by 90° in space and time, i.e., they are in quadrature. The sum of these two squared terms then produces an unmodulated signal for one movement-direction only. The signal represents the modulus or "energy" for movement in that direction, hence the name, the "energy" model.

As there is compelling evidence that the nonlinear transformation occurs in the cortex (Emerson et al, 1987b), we wanted to know whether the seemingly unnatural operation of squaring actually could be carried out by a cortical neuron. Because the shape of nonlinear transformations in the cortex is an important determinant of the computational operations a neuron can perform, we have set out to measure the exact shape of a cortical nonlinearity. In our initial attempt, reported here, we need to simplify the task, because mathematical identification of nonlinear cortical interactions between spatially separated inputs is too large a first step. Therefore, we have chosen to examine the most elementary form of nonlinearity exhibited by a simple cell, its output firing rate when the luminance at a single RF position is modulated over a large range, i.e., its intensive nonlinearity. Although many of these same neurons are DS, here we deliberately ignore responses to motion by concentrating on luminance inputs in time at only one position. The more localized mechanism may afford us the opportunity to assign a specific neural cause for the transform.

Unfortunately, it is not easy to measure directly a cortical transformation, because the relevant inputs to a given neuron are not as accessible as the action potentials that comprise its output. Therefore, we have chosen a modeling technique, system identification, that offers the possibility of generating mathematical descriptions for each of the blocks that make up the system structure. This approach, in turn, requires expressing the structure of the system under scrutiny as a network of interconnected blocks. The network reduces to a "cascade" model, when the sequence of operations can be assumed to occur serially.

Cascade modeling of nonlinear systems has a venerable history, which has been reviewed recently (Korenberg & Hunter, 1986). Briefly, workers in this area have found it useful to characterize transformations as being linear and possessing memory (dynamic linear: "L"), or nonlinear without memory (static nonlinear: "N"), or an alternating sequence of the two. Most neural elements can be considered in these terms. Our work has been partly motivated by Mancini's (1983/1984) success in demonstrating that responses of simple cells in the cat's visual cortex can be approximated by the sequence L followed by N, where N includes polynomial terms up to only second degree (see DISCUSSION). A slightly more general version that consists of sequential L, N, and L stages (the LNL, or sandwich model) has become an especially useful combination. Spekreijse (1969) and Spekreijse and Oosting (1970) were among the first not only to use the LNL model to represent a physiological system, but also to describe mathematically, or "identify," the functions within each block, directly from measured responses to a set of stimuli that was rich enough to test the system thoroughly. Shortly thereafter, Korenberg (1973a, b, c) showed that complete identification of the model could be obtained through the use of crosscorrelation in a single white-noise experiment. This approach has been improved gradually (Korenberg, 1982, 1985) to the point where it is now practical to identify an LNL model directly from data obtained in an insect sensory neuron (Korenberg et al, 1988), or in the cat's visual cortex, as shown here. Another notable example of sandwich modeling in the visual system is the meticulous work of Shapley and Victor (Shapley & Victor, 1978; Shapley & Victor, 1980; Shapley & Victor, 1981; Victor & Shapley, 1979), who have used the approach in the frequency domain to describe the control of contrast gain in the cat's retina. In addition, the LNL cascade has been used elegantly by Sakai and Naka (1987b) to reconstruct the second-order kernel of a type-N cell in the catfish retina.

Our successful application of the LNL model to the intensive nonlinearity in a cortical simple cell has suggested, in agreement with Mancini (1983/1984), that the nonlinearity occurs late enough to represent the firing-rate floor, or threshold, for the neuron being measured, and that the onset of firing is gradual enough to allow a low-degree polynomial to describe adequately the nonlinearity. We conclude that this gradual threshold-function, in combination with inverters already available in the visual cortex, could easily calculate the squaring function needed by complex cells for excitatory responses to light and dark stimuli, and by the Adelson and Bergen motion energy model. Some of this work has been presented briefly (Emerson et al, 1988).

METHODS

Surgical Preparation

Details of the preparation have been reported elsewhere (Citron & Emerson, 1983; Citron et al, 1981; Emerson et al, 1987b; Emerson & Coleman, 1981; Emerson & Gerstein, 1977a). Briefly, cats were surgically anesthetized as necessary with thiopental sodium for exposing area 17 and installing on the skull a fluid-tight cylindrical well. Contact and trial-case lenses focussed the eyes on a tangent screen or video screen at 115 cm. Before single-cell recording, we switched the animal to a continuous intravenous infusion of 1-2 mg/Kg/hr of thiopental mixed with 0.125 mg/Kg/hr of Alloferin (Roche) for paralysis. Heart rate and reflexes were monitored continuously to insure that anesthetic dosage was sufficient to keep the animal in a relaxed and unresponsive state.

Receptive Field Classification: Conventional Stimuli

We made extracellular single-unit recordings in area 17 of lightly anesthetized adult cats. Initially we hand-plotted on the tangent screen cortical RFs in the dominant eye to determine location, size, presence of DS, and approximate preferred orientation. We then inserted a mirror into the path to the tangent screen and centered a 8° by 11° video monitor on the RF. We presented moving edges, moving light bars, and flashing light and dark bars to assess the type of RF and the extent of DS. These stimuli are standard for classifying cortical RFs (Citron *et al*, 1981; Emerson, In Press; Sherman *et al*, 1976). Simple cells show separate regions of response to presentation (ON responses) and removal (OFF responses) of a light stimulus, whereas cells of the complex family give both ON and OFF responses to either light or dark stimuli (Emerson & Coleman, 1981; Hubel & Wiesel, 1962; Movshon *et al*, 1973a, b).

Random Stimuli

Having classified the RF, we then presented a video-generated random-bar "white-noise" stimulus that has been described previously (Citron & Emerson, 1983). It consisted of 16 contiguous 0.5° by 11° bars, each of which could change its luminance randomly and independently of the other bars every video frame (about every 16 msec). This random grating included only three intensities, a light level at 444 cd/m^2, a mean luminance at 222 cd/m^2, and a dark level at nearly 0 cd/m^2. Each of the three luminances occurred with equal probability, as determined by a 32 bit linear congruential random number generator (Knuth, 1969, Sect. 3.2.1 and 3.3.3.), whose values were decimated over the 16 bars in each video frame. Thus, the stimulus was a pseudorandom ternary spatiotemporal grating with a uniform probability distribution. The stimulus was generated offline and recorded through the frame buffer on a frame-by-frame video recorder (GYYR) for playback in real time during the experiment. As all possible combinations of dark and light bars could occur in a given frame, the grating included bars of both contrast signs, moving up and down in short sequences at random regions of the RF, and at speeds whose values were limited only by the 16-msec temporal quantization, and the 0.5° bar width.

Data presented here are based on responses to three repetitions of a 7-minute white-noise sequence (about 25,000 video frames), averaged over the three repetitions. We always presented bars at the optimal orientation for the RF, and bars were long enough to fill or extend beyond the RF. Therefore these experiments were two dimensional, only one dimension in space, and the other in time.

Data Analysis

The times of neural impulses were measured often with a resolution of 1 msec, but for the neuron shown here, 16 msec, and stored on disk for offline analysis. We have used crosscorrelation between the output of the neuron and the luminance at the stimulus position of interest (Lee & Schetzen, 1965) to calculate Wiener kernels in Fig. 1 for initial assessment of linear and especially nonlinear RF properties. Wiener-kernel analysis is a well established general method for analyzing nonlinear interactions (Citron *et al*, 1981; Marmarelis *et al*, 1986; Marmarelis & Marmarelis, 1978), and has been used in the cortex by us previously (Citron & Emerson, 1983; Emerson *et al*, 1987b). For other figures, as discussed in more detail below, we have used orthogonal techniques for estimating kernels, and least-square error fits for polynomial descriptions of static nonlinearities (Korenberg, 1988; Korenberg *et al*, 1988). While identifying cascade models, because of computational limitations, we considered averaged responses to only the first 15,000 stimulus frames out of the 25,000 in the 7-min white-noise sequence. Separate analysis on the remaining 10,000 frames showed similar results, which attests to the stationarity of our physiological experiment.

RESULTS

As mentioned above, responses to flashing stimuli differ between simple and complex cells. We chose a simple-cell example from a sample of 11 simple cells, five complex cells, four B (complex) cells (Henry *et al*, 1978), and two nonoriented cells recorded in five random-grating experiments in area 17 of the cat. The first task is to determine whether it is possible in a simple cell to elicit strong enough nonlinearities to study. Simple cells are justifiably regarded as being highly linear in comparison with complex cells (Emerson, In Press; Movshon *et al*, 1978a).

Selecting for the Nonlinearity Among Multiple Spatial Inputs

A straightforward requirement of a linear system is that inverse inputs cause inverse outputs. Therefore, examining the time-course of responses for a visual neuron to brief presentations of light and dark bars provides an initial check of its intensive linearity. Figure 1 shows results of such a test at positions 3-16 of our cortical simple cell. Around position 12, light-bar responses in B are positive, while dark-bar responses in A are equally strong, but negative; i.e., they are inverses of each other. At position 8, however, the dark-bar responses of A are so strong in the positive direction that the light-bar responses in B would have to be very large and negative to maintain the intensive linearity exhibited at position 12.

This linear expectation is not upheld in B, perhaps because the light bar drove the neuron so far negative that the firing rate reached the floor (0 spikes/sec), often called the neural "threshold," especially when this operating point is approached from a quiescent state below the firing range. The difference between the linear expectation and the findings in A and B is the intensive nonlinearity, and is plotted in C as the Wiener second-order intensive kernel, sometimes called the "diagonal" of the full second-order kernel, which is two-dimensional, as shown in Fig. 2C. This second-order intensive kernel is the correction one would have to add to the output of a linear representation of the neuron to better approximate the measured response. (For details, see Marmarelis & Marmarelis, 1978; the Appendix of Emerson *et al*, 1987b; and below.) Note that this measure is essentially zero except at positions 7 and 8, where the positive excursion indicates that we obtained a higher firing rate than would be expected from a linear system, another property of a threshold.

SIMPLE CELL: 1-BAR RESPONSE

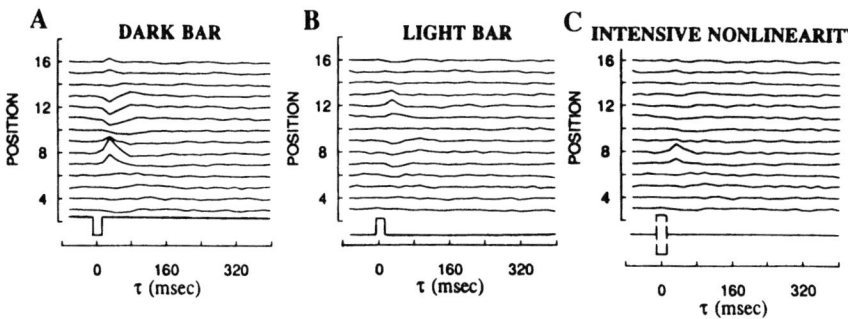

FIG. 1. Examination of the intensity domain for a simple cortical neuron (Cell 6-5). A: Time-course of responses to a 16-msec presentation of an optimally oriented dark-bar (see stimulus marker, below) at positions 3 through 16 of the receptive field (RF). Maximum response excursion was 4.5 spikes/sec at position 8. B: Responses across the same positions for a light bar. Both plots are Wiener second-order estimates of the responses to these stimuli (see text). C: The intensive nonlinearity of this cell, representing any response component that remains after averaging the dark- and light-bar responses of A and B (calculated as the "diagonal" of the Wiener second-order kernel). Note that position 8 appears to exhibit the only major nonlinearity, and provides the data for the remaining analysis.

This finding of an intensive nonlinearity at only one, or at most two positions in space, is common to all simple cells we have examined. The pattern is different from that in complex cells, in which intensive nonlinearities are distributed uniformly across space (Citron & Emerson, 1983; Emerson et al, 1987b). A way of interpreting the pattern of simple-cell intensive nonlinearities is that the response of the neuron to the dark bars presented at (auxiliary) position 8 was sufficiently strong to raise the average firing rate of the neuron, and "condition," or linearize (Spekreijse, 1969; Spekreijse & Oosting, 1970) the neuron's response at the other RF positions. Being able to classify the RF solely on the basis of the pattern of intensive nonlinearities is an advantage of our 16-bar random-grating stimulus. We can then analyze responses for single-position effects, because the sequence of luminances at each position is carefully randomized to ensure that it is independent of the luminance sequences at all other positions (see METHODS). From now on, we concentrate on the nature of nonlinearities at position 8, whose presence we have just demonstrated.

A Black-Box Model / External Model

A general and unbiased approach to describing a nonlinear system is to calculate the Wiener kernels of the system from responses to a white-noise input (Marmarelis & Marmarelis, 1978). Figure 2 shows the results of such analysis for position 8 of the simple cell examined above. Wiener-kernel analysis produces a black-box model, or external model of the system, in the sense that it places a box around the system, as in Fig. 2A, and describes the system in as much detail as desired in terms of the relationship between the input and output. The structure of the description bears no relationship to the structure of the system within the box; hence, it produces an "external" description or model. Still, it is usually a most useful first step, because it gives an overall picture of the system under study.

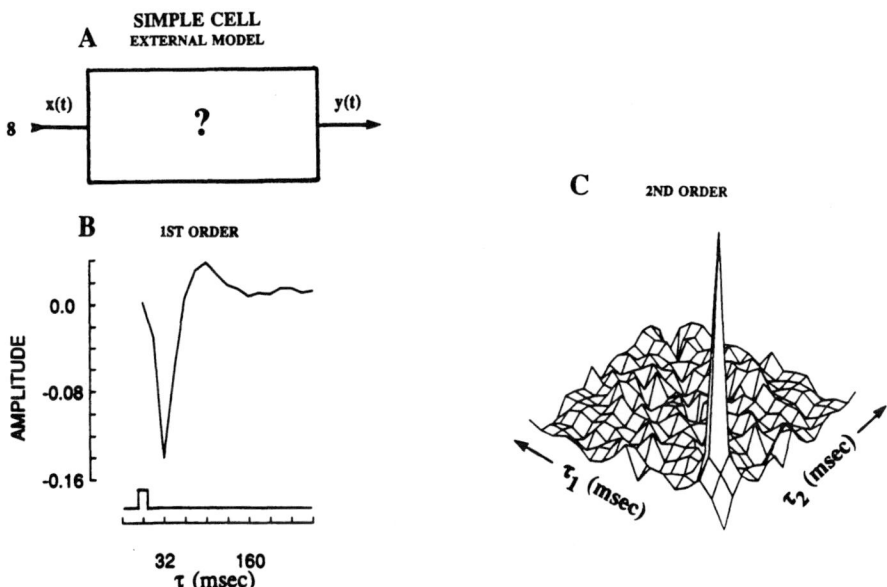

FIG. 2. External (Wiener-kernel) model of the simple cell. A: Cell is considered only in terms of its luminance input at position 8, and its firing-rate output. B: When the input is white noise, crosscorrelation of the output with occurrences of 16-msec bar presentations in the sequence produces the (Wiener) first-order kernel, a linear estimate of the light-bar impulse response. Units are (spikes/s)(222 cd/m^2)$^{-1}$(0.016 s)$^{-1}$. C: A crosscorrelation between the output and all temporal pairs of bars produces the 2D Wiener second-order kernel (see text). Range of data was -0.0336 to 0.0988 in units of (spikes/s)(222 cd/m^2)$^{-2}$(0.016 s)$^{-2}$. In reality, the kernels shown here were calculated with the "fast orthogonal algorithm," which produces more accurate kernel estimates than direct crosscorrelation.

In Fig. 2B is shown the Wiener first-order kernel of the system, as calculated by the fast orthogonal algorithm (Korenberg, 1988), which implicitly orthogonalizes over the actual experimental input stimulus. This procedure generates Wiener kernels of higher accuracy than crosscorrelation of the output directly with the raw stimulus (Lee & Schetzen, 1965), especially for measurement periods of intermediate length. The sharply downward response is to be interpreted as the response of the "best" linear approximant of the system to a brief light flash at position 8 of the RF, negative because position 8 is in a so-called OFF area (Emerson, In Press).

The Wiener second-order kernel is shown in the perspective view of Fig. 2C as a two-dimensional function of the two time variables, τ_1 and τ_2, beginning with the onset of each of the two bars that cause the nonlinear interaction. It provides the correction needed to the first-order estimate for a more accurate prediction of a temporal pair of bar presentations. The positive peak near the origin shows mainly that shortly after two bars occur closely in time, the cell fires more intensely than expected on the basis of a linear model, another indication of a possible threshold nonlinearity. A slice along the upward and leftward "diagonal" of this plot is equivalent to the intensive nonlinearity plotted in Fig. 1C, and represents the case where the two bars are presented simultaneously.

Peeking Inside the Black Box: a Cascade Model

Although the set of Wiener kernels provides a complete description of a system, it is not in a very useful form for studying structure, because the model implied by the kernels is not isomorphic with the system. However, information about the visual system that we can gain independently of the kernel measurements shown here may be useful in constraining that structure. We know that the stream of information from the retina through at least the lateral geniculate nucleus (LGN) is processed in a sequence of serial steps. Adding the early stages of the cortex is not likely to change this overall structure. Therefore, a model embodying a series, or "cascade" of information-processing steps is a good first approximation to the structure of the retino-cortical pathway.

Furthermore, the psychophysical equivalence among the appearance of light flashes whose amplitude-duration product is constant for durations below about 50 msec, called Bloch's Law (Bloch, 1885), indicates that the first operation in the human visual system after a linear transduction of light is a temporal integration. In fact, Tyler and Hamer (1988) have reported strict reciprocity for intensity vs durations that are less than 10 msec. This psychophysical reciprocity over an intensity range of 10^8 guarantees that the first nonlinearity occurs after temporal integration. Therefore, a minimal nonlinear cascade model to represent early events in the visual system consists of a dynamic linear filter, followed by a static nonlinearity (see INTRODUCTION). Figure 3 incorporates this structure, plus an added dynamic linear stage that, with methods used here, is almost as easy to identify as the LN model, and provides a more accurate representation of the neuron.

CASCADE MODEL

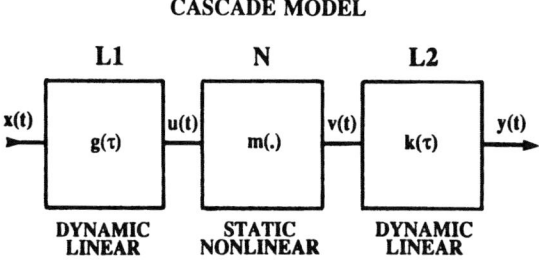

FIG. 3. Proposed internal structure of the external model of Fig. 2A. A linear transduction with memory (defined by temporal integration), L1, is followed by the major point-, or static-nonlinearity, N, which is followed by an optional second dynamic linear filter, L2.

It is known (Bussgang, 1952; Korenberg & Hunter, 1986; Spekreijse & Oosting, 1970) that in a LNL cascade, the convolution of the two linear filters, $g(\tau)$ and $k(\tau)$, is proportional to the first-order Wiener kernel, shown here in Fig. 2B. Fortunately, the Wiener second-order kernel of Fig. 2C provides the constraint that allows us to deconvolve the two linear kernels. Korenberg (1982, 1985) has shown that the first nonzero slice along one of the τ axes provides the shape of L1. As the two filters are not otherwise constrained by the model, their individual scales are arbitrary. However, we know that the inversion in sign in the visual system occurs in bipolar cells of the retina (Werblin & Dowling, 1969, their Fig. 7), and certainly before the major intensive nonlinearity we see in the cortex. Therefore, the excursion in L1 must be negative, which fully constrains the shapes of the two linear filters, as shown in Fig. 4. Note that both filters are essentially lowpass, with L2 showing slight ringing (a small bandpass component).

The Shape of the Intensive Nonlinearity

Once the linear filters of an LNL model are identified, full identification of the neural model requires only establishing the shape of the static nonlinearity in the cascade that best fits the responses to the rich white-noise experiment that was performed on this neuron. A fairly general form for representing the static nonlinearity is a polynomial, and it can be estimated by minimizing the mean square error for the entire model (Korenberg et al, 1988; Korenberg & Hunter, 1986). Figure 5 shows both third- and fourth-degree polynomial fits to the experimental data. The chief difference between these fits is not the sharpness of the transition between the positively sloped region on the right and the level region on the left, but rather, how flat the curve on the left remains for increasingly negative values of the input signal, $u(.)$. Therefore, higher-degree least-square-error polynomial fits will not sharpen the shape of the transition region, and are unnecessary to describe the function. The shape supports our hypothesis that N represents the floor- or threshold-nonlinearity in the cortical neuron being measured. Cortical neurons are notorious for their low background firing rate, and it is only because of the white-noise-driven high mean firing rate that not all of the positions in Fig. 1C show such nonlinearities. Note that the linear filters before and after N require that for each filter, zero input generate zero output. Therefore, any changes in maintained level of firing between input and output, i.e., any offsets, must be represented in N. The vertical elevation above zero for the flat region in Fig. 5 indicates that *under the conditions of the white-noise experiment*, at mean luminance there was a significant maintained firing rate.

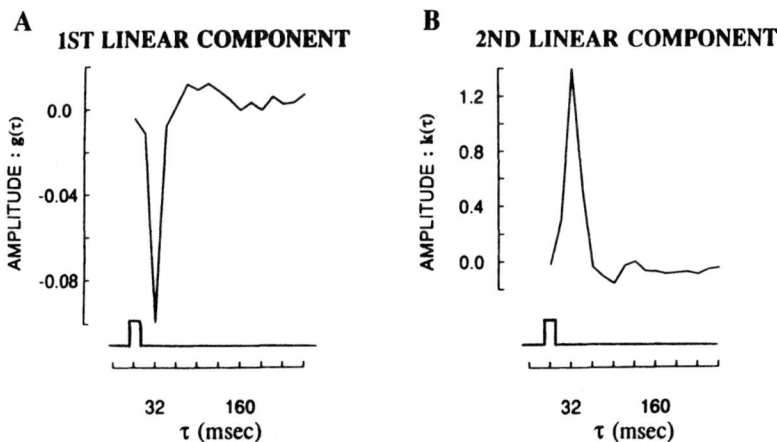

FIG. 4. First-order kernels (impulse responses) for the two linear filters, L1 and L2, in Fig. 3. The Wiener second-order kernel of Fig. 2C provides the information necessary to divide the first-order kernel of Fig. 2B into its two components. The scales of these plots are arbitrary, because the model does not constrain the absolute amplitude of two of the three LNL transformations.

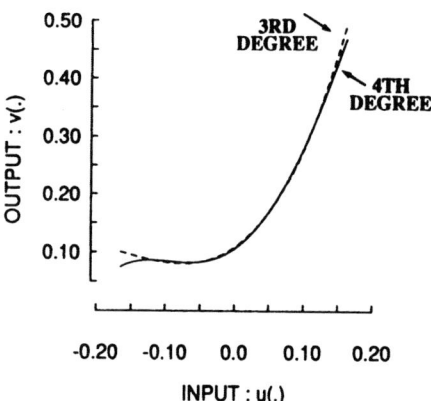

STATIC NONLINEAR COMPONENT

FIG. 5. Third- and fourth-degree polynomial estimates of the static nonlinearity, N, in Fig. 3. Note that the two polynomials for this input/output function differ slightly in their attempts to minimize the output for increasingly negative inputs to the static nonlinearity, a property that agrees well with a neural threshold.

To the extent that our LNL model mimics the operation of this cortical simple cell that has been stimulated at one position (see DISCUSSION for goodness-of-fit), the evidence seems to support a late-occurring intensive nonlinearity, probably the firing threshold of the neuron being measured. A surprising implication of this interpretation is that all the processing that leads to the intracellular voltage of a simple cortical neuron seems to be linear, although there is an indication of slightly larger dark-bar than light-bar responses, at upper positions of Fig. 1 A vs B where threshold nonlinearities seem to have been linearized out. In addition to this mild compressive nonlinearity, one might expect to see other intensive nonlinearities that have been reported in the retina. For example, Y ganglion cells in the cat's retina are known to show full-wave-rectifying nonlinearities, especially at high spatial frequencies (Shapley & Victor, 1980; Victor & Shapley, 1979); and in the catfish retina, type-C and type-N amacrine cells generate nonlinearities that appear in the ganglion-cell responses (Sakai & Naka, 1987a).

Apparently, however, these relatively subtle nonlinearities are overwhelmed by the threshold nonlinearities of neurons that transmit their output information in the form of action potentials. These include neurons in the cortex; in the lateral geniculate nucleus (LGN), which precedes the cortex; and possibly even in the retina itself. These ascending levels of processing are listed in reverse order because neurons with the highest threshold (lowest mean-luminance firing rate) will show the strongest threshold nonlinearities, and mean-luminance firing rate decreases from retina to cortex. As the static nonlinearity, N, in Figs. 3 and 5 accounts for the cortical nonlinearity, the threshold in the LGN is the most likely threshold to violate the model implication of precortical linearity. The following hypothesis about precortical connectivity would alleviate this expected violation.

A Closer Examination of the First Linear Filter, L1

We must consider the possibility that fundamentally nonlinear elements can be combined in a way that cancels their nonlinearities. Pollen and Ronner (1982) suggested that pairs of simple cells 180° out of phase would carry complete information for a linear representation of an image. But suppose the simple cell itself were to integrate on- and off-center inputs from the LGN. Figure 6 shows the LNL model for this simple cell, but with an expanded representation for L1. The retinal inversion that occurs only in bipolar cells of the OFF pathway is shown after an early-retinal temporal filter, L1.1. N1.1 shows the threshold nonlinearities that could occur in ganglion cells if their input signal were modulated sufficiently. In fact, we know that retinal ganglion cells in the cat exhibit strong threshold nonlinearities when they are tested with high-contrast white-noise-

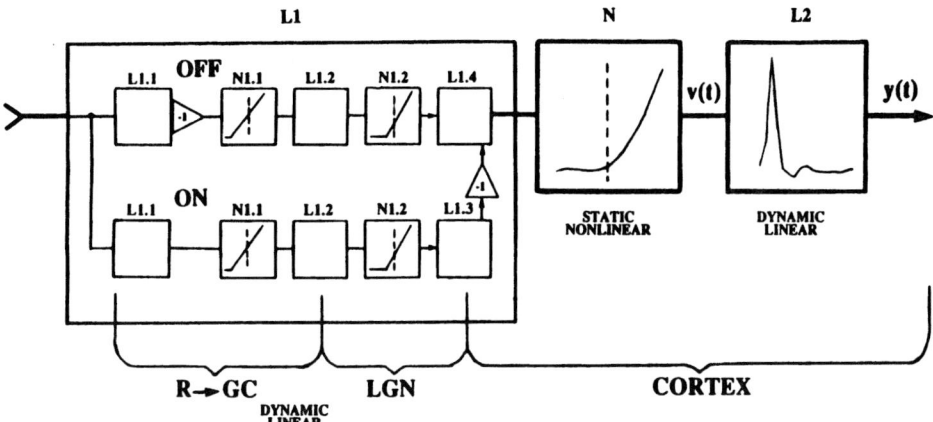

FIG. 6. An expanded version of the LNL model of Fig. 3 to show the components within L1 that could account for its overall linear representation of stimulus luminance in an OFF area of the cortical RF. The first three sub-boxes in L1 account for linear transduction of light to an internally transmitted signal, linear filtering of the signal, and (linear) conversion into a spike train, all occurring between receptors and the output of ganglion cells in the retina. The knee of the ganglion-cell threshold (N1.1) is shown far to the left to signify that the threshold is so low that a nonlinearity would appear only under the most extreme negative modulations of the steady-state signal at mean luminance (indicated by the dashed vertical line). The last three sub-boxes account for transmission of the spike train to the LGN, conversion to a voltage through synaptic input to a higher-threshold LGN neuron, nonlinear recoding into another spike train through principle cells of the LGN, and transmission to the cortex. The last sub-boxes, operations L1.3 and L1.4, include inversion of the ON-pathway signal and summing of that signal with the OFF-pathway signal to re-create a linear intracellular voltage in the cortical neuron. This "push-pull" approach has the effect of cancelling the even-ordered nonlinearities distributed along each of the complementary pathways.

modulated rings (unpublished data). But elongated stimuli such as those used here, and 2D checkerboard-like stimuli that we have used on ganglion cells (Citron et al, 1988) are not as likely to drive these cells below threshold, because the shapes of the image-elements are not well matched to the annular organization of the RF. L1.2 accounts for signal transmission to cells of the LGN, where the higher threshold (N1.2) will probably generate floor-like nonlinearities in both ON and OFF pathways. However, the cortical inversion of the ON-pathway signal and summing with the OFF-signal (L1.3 and L1.4) can create a linear representation by restoring active cortical inhibition in such a dark-on-excitatory area of a simple cell. The retinal inversion required within L1 is known to exist, and many classes of inhibitory interneurons are available to effect the cortical inversion (see Emerson et al, 1985 for a brief discussion of anatomical evidence). It seems likely that the visual cortex would use this standard push-pull technique to obtain transmission of a linear signal through inherently nonlinear components. We note that Segal (1973) has considered parallel arrangements of LNL models with threshold-plus-saturation static nonlinearities, where the input and output means of the overall system were linearly related.

The Gradual Threshold as an Element of the Complex-Cell Intensive Nonlinearity

Keeping in mind that one of the aims of this modeling was to determine whether the intensive nonlinearities in simple cells were of a form that could explain motion processing, we now consider how the shape of the N might be used to calculate the squaring function needed by the energy model, and apparently also by complex cells to effect their contrast sign independence (i.e.,

GENERATING EVEN-ORDER FUNCTION
(COMPLEX CELL)

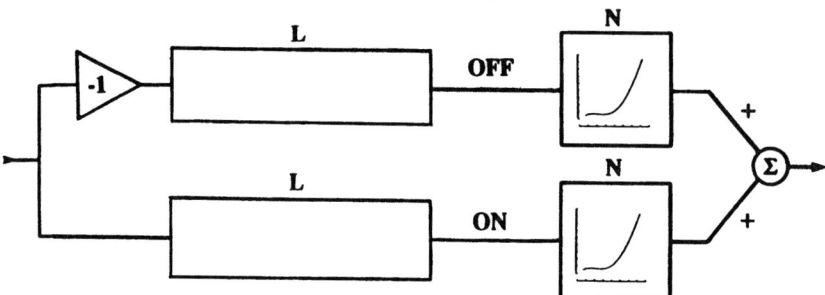

FIG. 7. Model for generating the squaring function needed by complex cells. A separate static nonlinear operation is provided for the linear ON and OFF pathways. Summation of the static nonlinear signals produces a close approximation to the output of a full squarer (see Fig. 8).

showing excitatory responses to dark or light bars). Consider a model of the form shown in Fig. 7. The front end is related to that of Fig. 6, in that the L's could be of the form L1, except that the linearized ON and OFF pathways remain separated, and then are transformed separately by the same static nonlinearity, N, as shown in Fig. 6. Because of the early inversion in the OFF pathway, the overall input/output functions at the outputs of the N transformations would appear as in Fig. 8A. Their juxtaposition in this figure illustrates that each is a rough approximation to a half squarer. Summation of these two signals, which in this case cancels the odd-ordered nonlinearities, generates an even-ordered transformation that is very nearly equal to a pure squarer. Their similarity, illustrated in Fig. 8B, follows from the small contribution of the fourth-degree polynomial term in the resulting "even" plot.

Apparently a nonlinear transformation that is as natural for a neuron as a threshold with a gradual onset is capable of explaining the full-wave rectifying properties of a complex cell, which in turn would explain its similar responses to light and dark stimuli. If the connections of Fig. 7

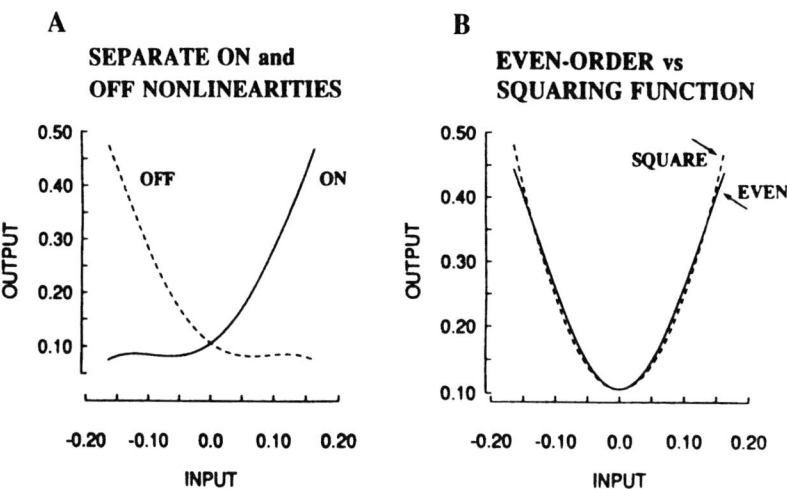

FIG. 8. Shape of nonlinear transformations in complex-cell model of Fig. 7. A: Overall input/output functions at the output of the N transformations in Fig. 7. These functions approximate half-squarers for positive signals (solid curve) and negative signals (dashed curve) at the input to the appropriate N transformation. B: After summation, the resultant "even" transformation (solid) is well approximated by a full-squarer, as shown by comparison with the dashed curve. (The squaring function was fit by eye.)

were arranged for the appropriate spatial and temporal inputs, they would be capable of calculating motion energy for a moving image (Adelson & Bergen, 1985).

DISCUSSION

The approach taken here has demonstrated threshold-like intensive nonlinearities in a cortical simple cell, characterized the transformation between its light input and its spike output in terms of its Wiener kernels, and illustrated an attempt to represent the system as an LNL cascade of operations.

Our conclusions about the order of the first two model transformations, the relatively low degree of the polynomial needed for N, and the likelihood that N represents a neural threshold agree with those of Mancini (1983/1984). He tested two-stage NL and LN combinations, eliminated the NL model, and then tested the adequacy of an LN model to predict responses to a pulse of duration four times that of the measuring stimulus. The L was identified as the Wiener first-order kernel of the system, and N was assumed to be a polynomial of degree up to two. A high correlation between measured and modeled responses indicated that a second-order LN model provides a good first approximation of single inputs to a cortical simple cell. Here, we have concentrated on identifying the transformations in a three-stage LNL model. We have used methods of Korenberg (1973b, 1982, 1985) to deconvolve the Wiener first-order kernel into its initial and final linear components, and then to generate a polynomial description of N directly from the physiological experiment by using a least-squares fit to measured white-noise responses. As we have seen, polynomial components of third but probably not of fourth degree seem to be important to account for simple-cell nonlinearities.

LNL Model Goodness-of-fit

Conclusions that can be drawn about the nature of transformations in the LNL model depend on how accurately this model captures the operation of the neuron. First we want to assure ourselves that the measured responses of the cell can be adequately represented by an LNL model. A necessary (but not sufficient) mathematical condition proposed by Korenberg and Hunter (1986, their Eq. (24)) correlates the shape of the Wiener first-order kernel (Fig. 2B) with the shape of the second-order kernel (Fig. 2C) integrated over either of the identical τ variables. Our data meet this criterion except for a small deviation at 200 msec (not shown). One possible candidate for this small discrepancy is the mild compressive nonlinearity in upper positions of Fig. 1, shown by the slightly larger responses of dark-bar responses in A vs the light-bar responses in B. Unfortunately, identification of a second static nonlinearity would require the more difficult task of evaluating a third-order kernel. A second possible reason for our slight discrepancy could be that our kernels were evaluated from a stimulus with a ternary, rather than a theoretically-more-correct Gaussian amplitude distribution. A very small, but consistent, estimation error in the second-order kernel will be magnified in the test when it is integrated over τ.

Given that the LNL model is approximately the correct form for the known connectivity between retina and cortex, the next question is whether we have identified a correct LNL model. The easiest way to assess how well the LNL model captures the operations of the simple cell is to test both under a variety of conditions. The white-noise stimulus includes a rich array of stimulus conditions, and therefore provides a thorough test of the measured neuron. In particular, the Wiener second-order kernel, although theoretically limited to the lowest level of nonlinearity, is an orthogonal measure of the system, and carries information about higher-order nonlinearities (Marmarelis & Marmarelis, 1978). Therefore, the Wiener second-order kernel of the synthesized model in Figs. 3 and 6 would provide a valuable comparison with the measured data. Figure 9 shows contour plots for the cortical measurement in A, and for the synthesized cascade model in B. Slicing such a function at equally spaced contour levels shows the shapes of subtle positive regions (solid contours) and negative regions (dashed contours) better than the perspective view of Fig. 2C, which shows the same data as in Fig. 9A.

The major features in A are captured in the model of B, except for the low-amplitude negative oscillations along the diagonal of A. Examination of Fig. 2C shows that the diagonal values were not much larger than the surrounding noise. Even a non-orthogonal measurement of the kernel-diagonal, which however used all 25,000 data points (Fig. 1C), was somewhat less noisy

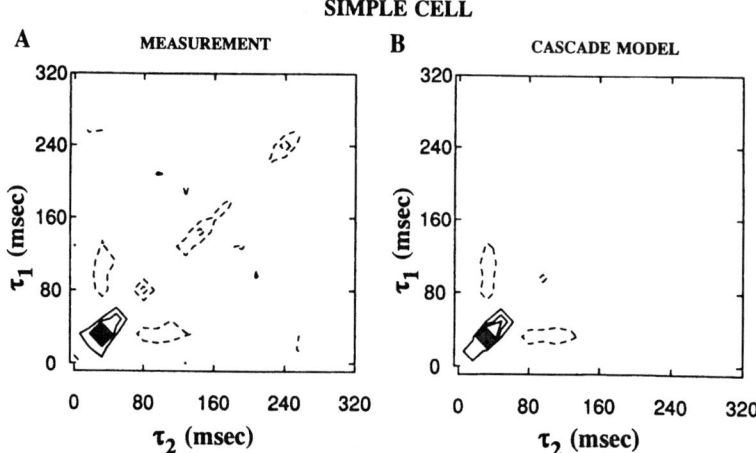

FIG. 9. Comparison between Wiener second-order kernels of measured simple cell and of the identified LNL model of Fig. 3. A: Contour plot of measured simple cell shows major intensive facilitation (solid contours) near the origin of the diagonal; and pattern of suppression (dashed contours) along the τ axes, and further along the diagonal. Data same as in Fig. 2C. The positive range in each of these plots is drawn with seven equally spaced contours, where the lowest contours occur at ± half the interval. This method avoids placing a contour at zero, because it would be dominated by noise. Lowest contours here were ±0.00706 in same units as in Fig. 2C. B: Contour plot of the synthesized system agrees with measurements of A except for negative oscillations along diagonal. Positive range covered by seven contours, as in A.

than the orthogonal estimate used here in Fig. 9A (limited by our calculation to 15,000 points) to compare with the model. Perhaps a more reliable estimate of the Wiener second-order kernel would have agreed better with the synthesized model. In addition, a deconvolution of L1 and L2 that employed information from more of the surface of the Wiener second-order kernel (Fig. 2C) might have provided a more reliable estimate of the model (Korenberg, 1973a, b, c). In spite of these shortcomings, we regard this initial attempt at identifying a cascade model of a cortical cell as a qualified success.

The Meaning of the Second Linear Component, L2

We were surprised to see any temporal dependence at all for L2. If N represents a neural threshold function, then there would not seem to be any processing necessary between generating a spike and our measuring it with a nearby extracellular electrode. However, as Korenberg and colleagues (1988) have found in the roach, the second linear filter integrates over a longer period than the first one, and shows mild oscillations. French and Korenberg (In Press) hypothesize that L2 reflects certain temporal processes of spike-generation and recovery, in particular, slow inactivation of sodium channels.

Future Directions

We have made a number of assumptions here to simplify the analysis. One is that the carefully counterbalanced sequence of luminances at each position allows us to treat each input independently of the others. In a linear system, this would be true, but in a nonlinear system one should calculate the "derived" kernels (Schetzen, 1980) to correct for crosstalk between the inputs. In the present case, the magnitude and spatial extent of the nonlinearities (Fig. 1C) is small enough that crosstalk should not be a significant problem. In the future, we intend to include measurements at a single position. Secondly, we have used a frame-time of 16 msec between adjacent stimuli, because an interval approaching the total integration time of the system delivers the most effective stimulus (highest effective power) to exercise the nonlinearities. This long frame-time limits the resolution for identification of the linear filters. Confining the experiment to a single position would provide higher signal-to-noise ratio, which would permit using a higher frame rate for better temporal resolution in our single-input identification.

The advantage of multiple-input experiments is that they permit identification of spatial and spatiotemporal mechanisms responsible for spatial processing and motion, respectively. The present single-input analysis has shown the probable existence of cortical squarers. Therefore, we now wish to know whether these elements are used in processing motion, as suggested by the close fit of the energy model to the visual cortex (Emerson et al, 1987a). A straightforward generalization to multiple inputs for the analysis shown here has been proposed by Korenberg (1985, Figs. 5 and 6), and subsequently reviewed by Hunter and korenberg (1986, Fig. 7). Identification of the added spatial cross-terms in the multidimensional polynomial would tell us directly the magnitude of the polynomial terms suggested by the energy model.

The current approach, though preliminary, has demonstrated the feasibility of applying cascade modeling in a structure as complex as the mammalian visual cortex. It has also suggested a need to account for surprisingly linear responses in membrane potentials of cortical simple cells. Finally, it has prompted us to consider how gradual thresholds could account for the low-order cortical nonlinearities needed both for generating responses that are independent of stimulus contrast sign, and for motion processing.

SUMMARY AND CONCLUSIONS

1. Several lines of evidence suggest that simple neurons in the cat's visual cortex show threshold-like intensive nonlinearities. To begin studying the nature of these nonlinearities, we can represent the chain of neurons from retina to cortex as a black-box (external) model through its set of Wiener kernels.

2. Information about the order of events in the visual system, obtained through other approaches, helps us cast the external model into a more useful form that is defined by a sequence, or cascade of transformations, here, a linear/nonlinear/linear (LNL) cascade. We have used systems identification techniques to define the nature of these transformations directly from responses of the neuron to a white-noise stimulus sequence.

3. The identified two dynamic linear filters are mainly low-pass, and the static nonlinearity a spike-firing characteristic with a gradual onset for increasingly positive input values. This transformation probably represents the firing threshold characteristic of the cortical neuron being measured. We suggest that linearization of events preceding this threshold occurs through a push-pull arrangement that depends on separate ON and OFF pathways leading to the cortex, and then a cortical inversion of one pathway.

4. The smooth firing transformation approximates a half-squarer for positive values, and makes the threshold a candidate mechanism for the squaring operation needed by the "motion energy" model, which seems to operate in the visual cortex.

5. The squaring operation, for intensity, can be effected by half-squaring each of the linear ON- and OFF-channel signals and then summing them. Such a process would account for the apparent full-wave rectification that purports to explain "ON-OFF" responses in cortical complex cells.

ACKNOWLEDGMENTS

We are greatly indebted to S. A. Klein for his development of kernel formulations, to W. J. Vaughn for programing of Wiener-kernel calculations in Fig. 1, and to C. M. Cappiello for preparing figures. Supported by EY06679 to RCE, EY01319 (Core Grant) to Center for Visual Science, NSERC grant (Canada) to MJK, and EY04711 and EY00250 (RCDA) to MCC.

REFERENCES

Adelson, E. H. and Bergen, J. R., 1985, Spatiotemporal energy models for the perception of motion. *J. Opt. Soc. Am. A* **2**:284-299.

Bloch, A. M., 1885, Experiences sur la vision. *Soc. Biol. Mem.* **37**:493-495.

Bussgang, J. J., 1952, Crosscorrelation functions of amplitude-distorted Gaussian signals. *MIT Res Lab Elect Tech Rep* **216**:1-14.

Citron, M. C. and Emerson, R. C., 1983, White noise analysis of cortical directional selectivity in cat. *Brain Res.* **279**:271-277.

Citron, M. C., Emerson, R. C. and Ide, L. S., 1981, Spatial and temporal receptive-field analysis of the cat's geniculocortical pathway. *Vision Res.* **21**:385-397.

Citron, M. C., Emerson, R. C. and Levick, W. R., 1988, Nonlinear measurement and classification of receptive fields in cat retinal ganglion cells. *Ann. Biomed. Engrg.* **16**:65-77.

Citron, M. C., Kroeker, J. P. and McCann, G. D., 1981, Nonlinear interactions in ganglion cell receptive fields. *J. Neurophysiol.* **46**:1161-1176.

Emerson, R. C., 1988, A linear model for symmetric receptive fields: Implications for classification tests with flashed and moving images. *Spatial Vision* **3**:159-177.

Emerson, R. C., Bergen, J. R. and Adelson, E. H., 1987a, Movement models and directionally selective neurons in the cat's visual cortex. *Soc. Neurosci. Abstr.* **13**:1623.

Emerson, R. C. and Citron, M. C., 1988, How linear and nonlinear mechanisms contribute to directional selectivity in simple cells of cat striate cortex. *Invest. Ophthalmol. & Vis. Sci., Suppl.* **29**:23.

Emerson, R. C., Citron, M. C., Felleman, D. J. and Kaas, J. H., 1985, A proposed mechanism and site for cortical directional selectivity. In *Models of the Visual Cortex*, (edited by Rose, D. and Dobson, V.),John Wiley & Sons, Sussex, England.

Emerson, R. C., Citron, M. C., Vaughn, W. J. and Klein, S. A., 1987b, Nonlinear directionally selective subunits in complex cells of cat striate cortex. *J. Neurophysiol.* **58**:33-65.

Emerson, R. C. and Coleman, L., 1981, Does image movement have a special nature for neurons in the cat's striate cortex? *Invest. Ophthalmol. & Vis. Sci., Suppl.* **20**:766-783.

Emerson, R. C. and Gerstein, G. L., 1977a, Simple striate neurons in the cat. I. Comparison of responses to moving and stationary stimuli. *J. Neurophysiol.* **40**:119-135.

Emerson, R. C. and Gerstein, G. L., 1977b, Simple striate neurons in the cat. II. Mechanisms underlying directional asymmetry and directional selectivity. *J. Neurophysiol.* **40**:136-155.

Emerson, R. C., Korenberg, M. J. and Citron, M. C., 1988, Measurement of a simple-cell threshold function in cat's striate cortex. *Soc. Neurosci. Abstr.* **14**:899.

French, A. S. and Korenberg, M. J., In Press, A nonlinear cascade model for action potential encoding in an insect sensory neuron. *Biophys. J.*

Henry, G. H., Lund, J. S. and Harvey, A. R., 1978, Cells of the striate cortex projecting to the Clare-Bishop area of the cat. *Brain Res.* **151**:154-158.

Hubel, D. H. and Wiesel, T. N., 1962, Receptive fields, binocular interaction and functional architecture in the cat's visual cortex. *J. Physiol., London* **160**:106-154.

Knuth, D. E., 1969, *The Art of Computer Programming*, Vol. 2. Addison-Wesley, Reading, MA.

Korenberg, M., 1973a, Cross-correlation analysis of neural cascades. *Proc. 10th Ann. Rocky Mountain Bioeng. Symp.*47-52.

Korenberg, M., 1973b, Identification of biological cascades of linear and static nonlinear systems. *Proc. 16th Midwest Symp. Circuit Theory* **XVIII.2**:1-9.

Korenberg, M. J., 1973c, Obtaining differential equation, functional expansion or cascade representation for nonlinear biological systems. *Proc. New England Bioeng. Conf.* **1**:237-245.

Korenberg, M. J., 1982, Statistical identification of parallel cascades of linear and nonlinear systems. *IFAC Symp Ident Sys Param Est* **1**:580-585.

Korenberg, M. J., 1985, Identifying noisy cascades of linear and static nonlinear systems. *IFAC Symp Ident Sys Param Est* **1**:421-426.

Korenberg, M. J., 1988, Identifying nonlinear difference equation and functional expansion representations: the fast orthogonal algorithm. *Ann. Biomed. Eng.* **16**:123-142.

Korenberg, M. J., French, A. S. and Voo, S. K. L., 1988, White noise analysis of nonlinear behavior in an insect sensory neuron: kernel and cascade approaches. *Biol. Cybern.* **58**:313-320.

Korenberg, M. J. and Hunter, I. W., 1986, The identification of nonlinear biological systems: LNL cascade models. *Biol. Cybern.* **55**:125-134.

Lee, Y. W. and Schetzen, M., 1965, Measurement of the Wiener kernels of a nonlinear system by cross-correlation. *Int. J. Control* **2**:237-254.

Mancini, M., 1984, Temporal properties of single cells in cat visual cortex. *(Doctoral dissertation, University of Rochester, 1983). Diss. Abst. Intern.* **44**:3234-B.

Marmarelis, V. Z., Citron, M. C. and Vivo, C., 1986, Minimum-order Wiener modeling of spike-output systems. *Biol. Cybernetics* **54**:115-123.

Marmarelis, P. Z. and Marmarelis, V. Z., 1978, *Analysis of Physiological Systems: The White-Noise Approach*, Plenum Press, New York.

Movshon, J. A., Thompson, I. D. and Tolhurst, D. J., 1978a, Spatial summation in receptive fields of simple cells in the cat's striate cortex. *J. Physiol., London* **283**:53-77.

Movshon, J. A., Thompson, I. D. and Tolhurst, D. J., 1978b, Receptive field organization of complex cells in the cat's striate cortex. *J. Physiol., London* **283**:79-99.

Pollen, D. A. and Ronner, S. F., 1982, Spatial computation performed by simple and complex cells in the visual cortex of the cat. *Vision Res.* **22**:101-118.

Sakai, H. M. and Naka, K., 1987a, Signal transmission in the catfish retina. IV. Transmission to ganglion cells. *J. Neurophysiol.* **58**:1307-1328.

Sakai, H. M. and Naka, K., 1987b, Signal transmission in the catfish retina. V. Sensitivity and circuit. *J. Neurophysiol.* **58**:1329-1350.

Schetzen, M., 1980, *The Volterra and Wiener theories of nonlinear systems*, John Wiley & Sons, New York.

Segal, B. N., 1973, *An analysis of multipath neural systems using random parameter models*, Unpublished Master's. Eng. Thesis, McGill University, Montreal.

Shapley, R. M. and Victor, J. D., 1978, The effect of contrast on the transfer properties of cat retinal ganglion cells. *J. Physiol., London* **285**:275-298.

Shapley, R. M. and Victor, J. D., 1980, The effect of contrast on the non-linear response of the Y cell. *J. Physiol., London* **302**:535-547.

Shapley, R. M. and Victor, J. D., 1981, How the contrast gain control modifies the frequency responses of cat retinal ganglion cells. *J. Physiol., London* **318**:161-179.

Sherman, S. M., Watkins, D. W. and Wilson, J. R., 1976, Further differences in receptive field properties of simple and complex cells in cat striate cortex. *Vision Res.* **16**:919-927.

Spekreijse, H., 1969, Rectification in the goldfish retina: analysis by sinusoidal and auxiliary stimulation. *Vision Res.* **9**:1461-1472.

Spekreijse, H. and Oosting, J., 1970, Linearizing: a method for analysing and synthesizing nonlinear systems. *Kybernetik* **7**:23-31.

Tyler, C. W. and Hamer, R. D., 1988, Linearity of human photoreception. *Invest. Ophthalmol. & Vis. Sci., Suppl.* **29**:409.

Victor, J. D. and Shapley, R. M., 1979, The nonlinear pathway of Y ganglion cells in the cat retina. *J. Gen. Physiol.* **74**:671-689.

Werblin, F. S. and Dowling, J. E., 1969, Organization of the retina of the mudpuppy, Necturus maculosus. II. Intracellular recording. *J. Neurophysiol.* **32**:339-355.

MODELING OF NEURONAL NETWORKS THROUGH EXPERIMENTAL DECOMPOSITION

Theodore W. Berger, T. Patrick Harty, German Barrionuevo
and Robert J. Sclabassi

Departments of Behavioral Neuroscience, Psychiatry,
Neurological Surgery and Electrical Engineering
University of Pittsburgh, Pittsburgh, PA 15260

INTRODUCTION

We recently have initiated a combined theoretical and experimental
analysis of nonlinear system properties of the hippocampal formation (Berger
et al., 1987, 1988a,b; Sclabassi et al., 1988a,b), a region of the mammalian
brain which plays a major role in learning and memory functions (Squire,
1982; Thompson et al., 1983; Berger et al., 1986; Berger and Bassett, 1989).
The hippocampal formation is composed of five subsystems: the entorhinal,
dentate, hippocampal (CA3 and CA1 regions), and subicular cortices (Fig. 1).
A single population of projection neurons provides the only output from each
subsystem (Lorente de No, 1934), and is the source of monosynaptic, excitatory input to the projection neurons of other subsystems (Andersen et al.,
1971b). The interconnectivity of the five cortical regions forms a closed
feedback loop (Steward, 1976; Swanson and Cowan, 1977; Swanson et al., 1978;
Sorensen and Shipley, 1979), such that activity within any one subsystem
modulates activity of the other subsystems. Interneurons within each subsystem also modulate output of the projection neurons through local feed-
forward and feedback pathways (Lorente de No, 1934; Amaral, 1978; Alger and
Nicoll, 1982; Lacaille et al., 1987).

We have been quantitatively characterizing the nonlinear dynamics
resulting from the interaction within this network structure of the many
elements of the hippocampal formation. Our strategy has been the following.
The nonlinear response properties of each subsystem are assessed experimentally by applying a random interval train of electrical impulses to the
fiber pathway which connects that subsystem to the remaining network. For
example, the majority of our studies to date have utilized stimulation of
the perforant path, which arises from the entorhinal cortex and terminates
within the dentate gyrus. Throughout random train delivery, evoked activity
is recorded from the projection neurons of the stimulated subsystem; in the
case of perforant path stimulation, evoked activity is recorded from granule
cells of the dentate gyrus. Nonlinear input/output properties of the subsystem are defined as the kernels of a functional power series expansion,
and computed as the relationship between inter-impulse intervals of the
input signal and magnitude of projection cell output.

Such an analysis is completed for the intact hippocampal formation, when
each subsystem is embedded in the larger network, and for experimentally

reduced preparations in which feedback from other subsystems and/or from intrinsic interneurons is eliminated. Through systematic "decomposition" of the network, the contribution of each feedback loop can be assessed quantitatively, and an open-loop characterization of the projection neuron population ultimately can be obtained. We will review here the initial results of applying this experimental strategy to the dentate gyrus of the in vivo hippocampal formation, and to the dentate gyrus of the in vitro hippocampal slice. In the case of the in vitro preparation, perforant path input to dentate granule cells is preserved (Skrede and Westgaard, 1971), but connectivity between other subsystems (e.g., CA1 and the entorhinal cortex) and connectivity with the contralateral hippocampus through commissural projections has been transected. In addition, associational connections among different subregions of the dentate have been eliminated, and feedback and feedforward inhibition of granule cells resulting from connections with local interneurons has been substantially reduced.

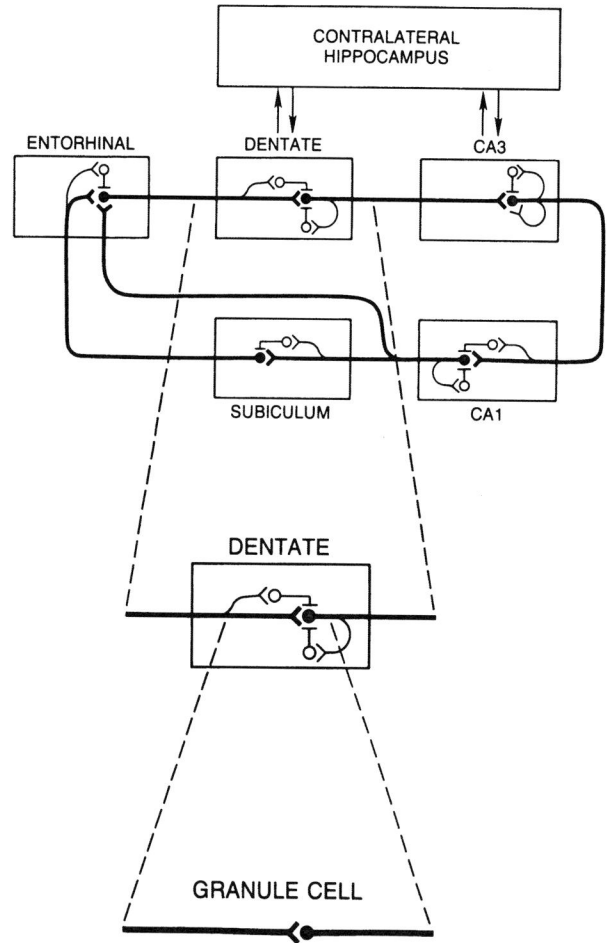

Fig. 1. Schematic representation of the five subsystems of the hippocampal formation (top diagram). Using the dentate gyrus as an example, experimental "decomposition" procedures are used to isolate each subsystem from the modulatory influence of the larger network (middle diagram), and to isolate the projection neurons (granule cells in the case of the dentate gyrus) from the modulatory influence of interneurons within the subsystem (bottom diagram).

METHODS

In Vivo Preparations

Studies of the in vivo hippocampus were conducted using male, New Zealand white rabbits chronically implanted with stimulating and recording electrodes. Using halothane anesthesia and standard surgical procedures (see Berger et al., 1988a), a bipolar stimulating electrode was positioned in the perforant path and a recording electrode was placed in the cell body layer of the ipsilateral dentate gyrus. Stimulating electrodes were constructed of stainless steel and insulated except for 500 μm at the tips. Recording electrodes also were constructed of stainless steel, and insulated except for 50 μm at the tips; tip impedances measured approximately 1 Mohm when tested in vitro using a 135 Hz input signal. After placement of the electrodes, leads from a reference skull screw positioned over the frontal lobe and from the stimulating and recording electrodes were connected to pins of a plastic headstage. The headstage and electrodes then were cemented to the skull using dental acrylic, and the skin surrounding the headstage was sutured closed. Animals were given 1-2 weeks to recover from surgery. Data for all experiments described here were collected while animals were unanesthetized, awake, and mildly restrained.

In Vitro Preparations

Slices of the hippocampal formation were prepared from male, New Zealand white rabbits. Animals first were anesthetized with halothane, and the skull overlying the parietal cortex was removed bilaterally. The hippocampal formation and overlying neocortex were extracted, gently separated, and the hippocampal formation of both hemispheres then blocked under cold, oxygenated medium. Using a vibratome, slices of tissue (600-800 μm in thickness) were cut with an orientation that was perpendicular to the longitudinal axis of the hippocampus. When sectioned in this manner, slices contain the dentate gyrus, the CA3 and CA1 regions of the hippocampus, and the intrinsic circuitry that connects these subsystems. This "trisynaptic pathway" is maintained because it is oriented perpendicular to the longitudinal axis of the hippocampus, and thus, parallel to the plane of section (Andersen et al., 1971b; Skrede and Westgaard, 1971). In contrast, all connections with the contralateral hippocampal formation are transected, and the entorhinal cortex is removed. The distal ends of perforant path axons and their synaptic contacts onto dentate granule cells are intact, however, so that entorhinal input to the dentate remains functionally viable and experimentally accessible. Thus, slicing transverse to the longitudinal axis creates a preparation in which open-loop properties of the dentate, CA3, and CA1 subsystems each can be examined.

Also transected are two sources of interconnectivity intrinsic to the dentate: the axons of interneurons that interconnect different subregions of the dentate ipsilaterally (the "associational" system; Zimmer, 1971; Berger et al., 1981), and the axons of interneurons that provide feedforward and feedback inhibition of granule cells (Lorente de No, 1934; Amaral, 1978). This intrinsic connectivity is affected because the axons of both systems project parallel to the longitudinal axis of the hippocampus, and thus, perpendicular to the plane of section (Struble et al., 1978).

Immediately following sectioning and throughout testing, the slices were maintained in an incubation medium consisting of the following: 125 mM NaCl; 5 mM KCl; 1.25 mM NaH_2PO_4; 2 mM $MgCl_2$; 2 mM $CaCl_2$; 26 mM $NaHCO_3$; and 10 mM glucose. The medium was warmed to 34° and oxygenated with 95% O_2/5% CO_2 (pH = 7.3-7.4). Slices were allowed to equilibrate for 90 minutes before experiments began.

Random Impulse Train Stimulation

The perforant path of both _in vivo_ and _in vitro_ preparations was stimulated with the same series of 4064 impulses with inter-impulse intervals drawn from a Poisson distribution having a mean inter-event interval of 500 ms and a range of inter-event intervals of 1-5000 ms. Field potentials evoked to each stimulus in the random train were amplified and band-pass filtered using low and high frequency limits of 10 Hz and 10K Hz, respectively. Data then were digitized continuously and stored for later analysis. Amplitudes of the "population spike" components of each field potential, which reflect the number of granule cells activated by each impulse (see Fig. 2), were determined by measuring the difference between the midpoint of a line connecting the peaks of the initial and secondary positivities and the peak of the negativity. Latencies of population spikes were computed from stimulus onset to peak negativity of the population spike.

Analytical Procedures

The transformational properties of the perforant path-dentate projection were determined using computational procedures described in previous reports (Berger et al., 1987, 1988a; Sclabassi et al., 1988b) and in a companion chapter (see Sclabassi et al., this volume). Briefly, the population output

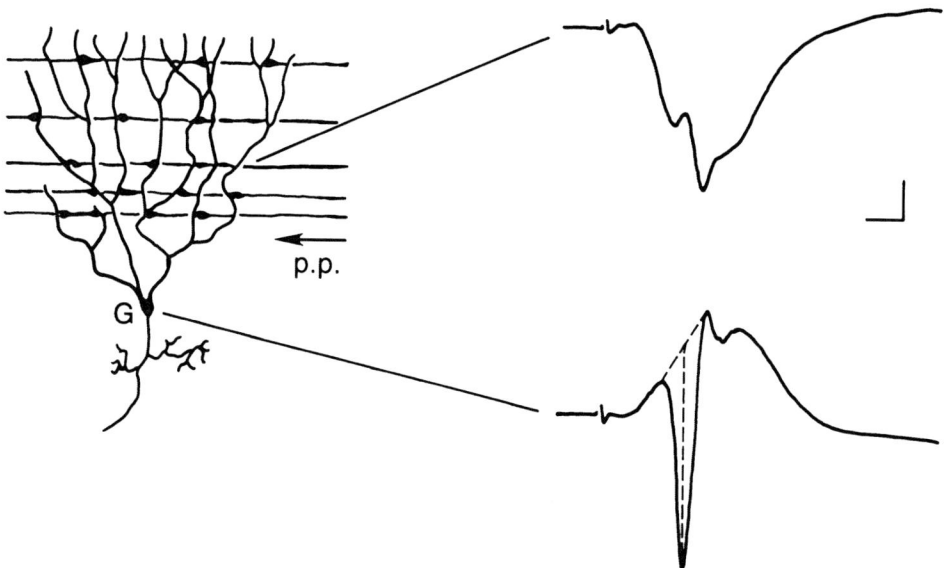

Fig. 2. Examples of extracellular field potentials evoked in the dendritic (right upper trace) and granule cell body (right lower trace) layers of the dentate gyrus in response to the delivery of a single electrical impulse (0.1 ms duration) to fibers of the perforant path. The negative-going wave recorded in the cell body region, termed the "population spike", reflects the population action potential discharge of granule cells; amplitude of the population spike correlates positively with the number of granule cells discharged by the impulse (Andersen et al., 1971a; Lomo, 1971). Procedures for measuring spike amplitude are shown schematically.

of dentate granule cells, y(t), was approximated by a functional power series:

$$y(t) = G_0 + G_1[h_1, x(t)] + G_2[h_2, x(t)] + G_3[h_3, x(t)] + E \qquad (1)$$

where (G_i) is a set of mutually orthogonal functionals, (h_i) is a set of symmetric kernels which characterize the relationship between the input and output, and E is an error term due to truncation. The train of discrete input events defined by x(t) is a set of delta functions representing the stimulus train. The first four kernels of the series were obtained by the process of orthogonalization (Wiener, 1958; Ogura, 1972) using the cross-correlation technique of Lee and Schetzen (1965) applied to point process events (Krausz, 1975; Sclabassi et al., 1977).

Because the kernels were computed only for the population spike components of the extracellular field potentials, all kernels are interpreted with respect to granule cell output. The zero order kernel, h_0, is a constant that represents the average output of granule cells, both evoked and non-evoked, over the total observation time; for the experimental conditions used, the majority of activity is evoked, and consequently $h_0 \to 0$. The first order kernel, $h_1(\tau)$, is the average of all evoked dentate population spike responses (with a latency of τ) occurring during train stimulation. The second order kernel, $h_2(\tau, \Delta)$, represents the modulatory effect of a preceding stimulus occurring Δ ms earlier on the number of granule cells activated (with a latency of τ) by the most current stimulation impulse, irrespective of any other impulses which may occur during that interstimulus interval. The third order kernel, $h_3(\tau, \Delta_1, \Delta_2)$, represents the modulatory effect of any two preceding stimuli occurring Δ_1 ms and Δ_2 ms earlier on the number of granule cells activated by the most current stimulation impulse, irrespective of any other impulses which may occur during either interval.

First, second and third order kernels were computed with a temporal resolution of 1 ms, which was equivalent to the smallest interval in the random stimulus train. These initially computed 1-ms kernels then were combined to obtain a 10 ms temporal resolution for first and second order kernels. A temporal resolution of 30 ms was used for third order kernels to obtain a sufficient number of cases for each pair of intervals, Δ_1 and Δ_2. Second and third order kernels were multiplied by two and six, respectively, to express the facilitation and suppression they predicted for paired and triplet impulse inputs (Sclabassi et al., 1988a; Balzer et al., 1989). Second and third order kernels also were normalized relative to the first order kernel, to express modulatory influences relative to the average population spike amplitude evoked during random train stimulation.

RESULTS

First Order Kernels

Latencies of granule cell population spikes evoked during random interval train stimulation rarely exceeded 10 ms for either in vivo or in vitro preparations, though spike latencies recorded from slices typically were longer than those recorded in vivo (Fig. 3, upper panels). Because of this distribution of latencies, the first bin of the first order kernel computed with a 10 ms resolution is equivalent to the average population spike amplitude to all impulses in the random train. The first bin values of first order kernels for in vitro preparations (0.3 $\pm$ 0.03 mV; mean $\pm$ S.E.M.; N = 12) consistently were smaller than those for in vivo preparations (2.2 $\pm$ 0.3 mV; N = 8). This result indicates that fewer numbers of granule cells were responding in slices compared to the intact dentate gyrus, which is consistent with the reduced nature of the _in vitro_ preparation (Fig. 3, lower right panel).

Second Order Kernels

Second order kernels represent the dependence of granule cell output on prior impulses within the train, as a function of the inter-impulse interval (Δ). Data from the intact dentate gyrus consistently revealed prominent nonlinearities in response to several ranges of interstimulus intervals (Fig. 4, upper left panel). For example, population spike amplitude was almost completely suppressed when preceding impulses occurred within 10-30 ms. In contrast, there was a marked facilitation of spike amplitude when preceding impulses occurred within 50-400 ms. Maximum facilitation, equivalent to 74 ± 4% of the first order kernel, was exhibited when Δ = 90-100 ms. Facilitation of a lower magnitude (approximately 10%

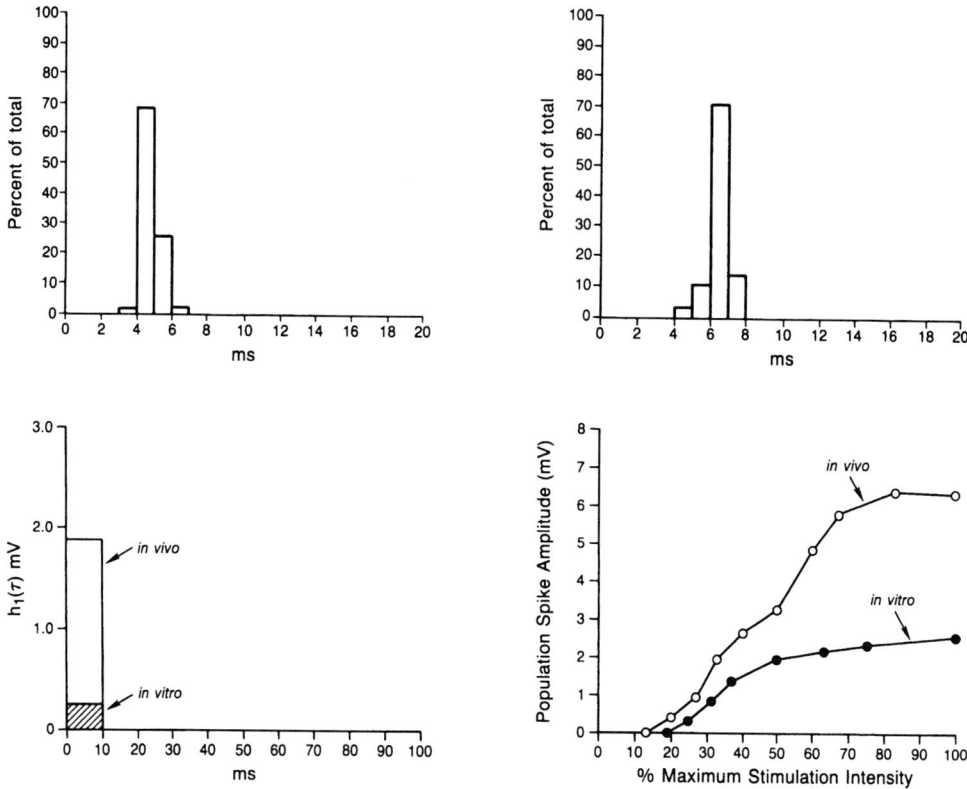

Fig. 3. Top panels: Latency distributions for granule cell population spikes elicited during random impulse train stimulation of the perforant path for representative _in vivo_ (left) and _in vitro_ (right) preparations. Bottom left panel: Representative first order kernels for population spike amplitude. Bottom right panel: Relationship between stimulation intensity of single impulses delivered to the perforant path and amplitude of the dentate granule cell population spike.

of the first order kernel) was evident for some _in vivo_ preparations when preceding impulses occurred within 400-1000 ms. Intervals greater than 1000 ms induced no consistent change in spike amplitude.

Second order nonlinearities of the _in vitro_ slice (Fig. 4, upper right panel) were equally prominent, but were qualitatively different than those exhibited by _in vivo_ preparations. Instead of suppression of granule cell output in response to short inter-impulse intervals, a robust facilitation

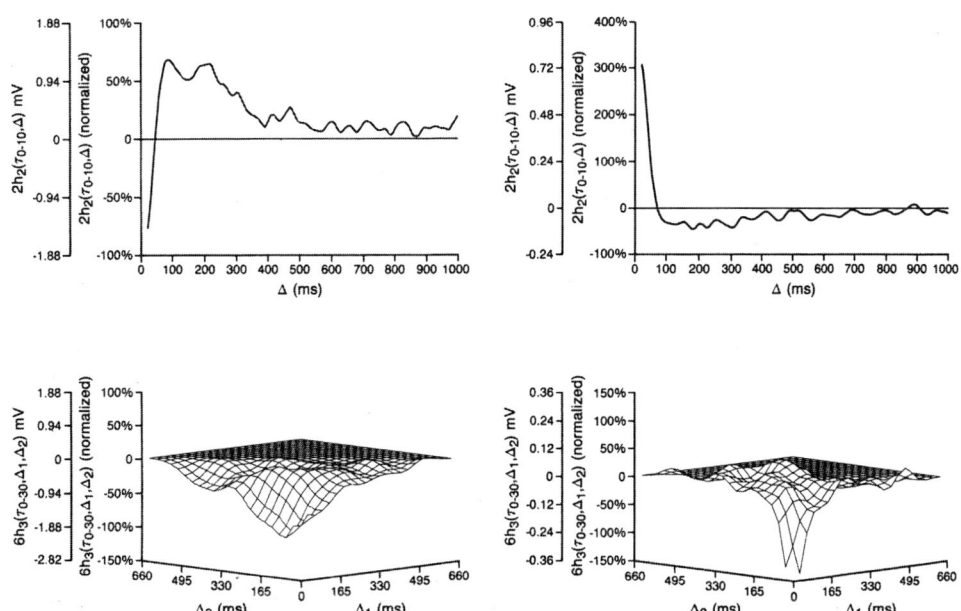

Fig. 4. Second (top panels) and third order kernels (bottom panels) for the dentate gyrus of representative _in vivo_ (left panels) and _in vitro_ (right panels) preparations. For both second and third order kernels, the dependent variable is expressed in two forms. The inside axis (normalized) expresses the change in population spike amplitude relative to the first order kernel value (see Fig. 3, bottom left panel). The outside axis (mV) expresses the change in population spike amplitude in absolute millivolts.

was observed. The facilitation was greater in magnitude (with respect to normalized kernel values, 241 $\pm$ 27%) and occurred maximally in response to shorter intervals (10-20 ms) compared to data from _in vivo_ preparations. In further contrast to results from intact preparations, facilitation in slices occurred only in response to intervals less than 100-150 ms, a much narrower range than observed _in vivo_. Finally, data from slices consistently exhibited suppression to intervals of 150-800 ms; interstimulus intervals within the same range produced facilitation or no effect for _in vivo_ preparations.

Third Order Kernels

Third order nonlinear response properties of the in vivo and the in vitro dentate also were computed, and represent granule cell output as a function of the inter-impulse intervals (Δ_1 and Δ_2) for all pairs of preceding impulses (Fig. 4, bottom panels). Third order kernels from in vivo preparations revealed that when both preceding intervals were _200-300 ms, granule cell output was reduced by an average maximum of approximately 60% of the first order kernel, with magnitude of the suppression inversely related to interval length. Third order kernels from slices revealed suppression of a larger magnitude (normalized values, average maximum of 90%), and in response to a much narrower range of intervals (90 ms) than was observed for the in vivo dentate gyrus. Third order nonlinearities for in vitro slices also included a robust facilitation in response to input patterns defined approximately by $\Delta_1 = 100-200$ ms and $\Delta_2 = 300-400$ ms, which was not observed in vivo.

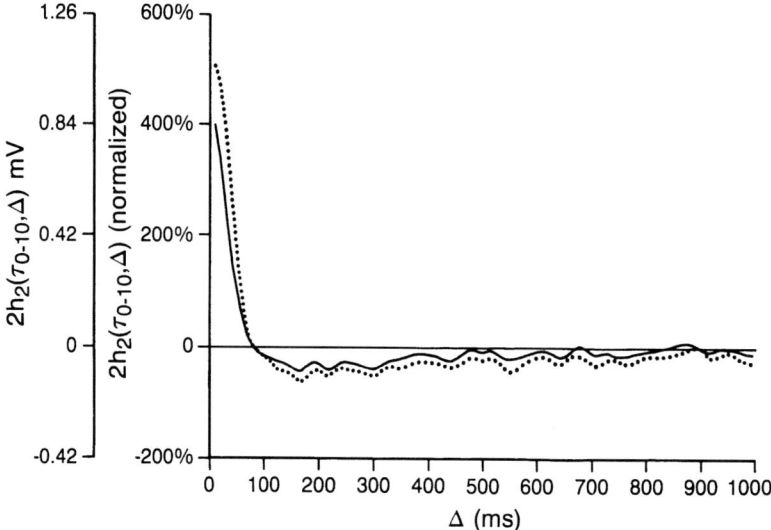

Fig. 5. Effects of the $GABA_A$ receptor antagonist bicuculline on second order kernel properties of the in vitro dentate gyrus. Kernels obtained in the absence and presence of bicuculline are represented by a solid line and a dotted line, respectively.

Contribution of Inhibitory Interneurons to Differences in Nonlinear Properties of the In Vivo and In Vitro Dentate Gyrus

Differences between the second and third order properties of the in vivo and in vitro dentate gyrus could not be accounted for on the basis of differences in the number of granule cells tested in each preparation (measured as the absolute magnitude of the evoked population spikes; see Fig. 3). The magnitudes of first order kernels for several of the in vitro preparations were within the range of in vivo preparations, and yet both second and third order kernels exhibited characteristics typical of other slices; the converse was true for several in vivo preparations.

In considering issues of circuitry, however, GABAergic interneurons of the dentate gyrus are known to be connected in a feedforward and feedback manner with granule cells, providing a powerful inhibitory influence through dense terminations onto granule cell somata (Lorente de No, 1934; Andersen et al., 1964; Ribak and Seress, 1983). GABAergic influence on granule cell output occurs relatively rapidly following an impulse to the perforant path, because of the short axonal length of the interneurons. As a result, the inhibitory postsynaptic potential induced through $GABA_A$ receptors has a short latency to onset and a rapid time to peak (within 10 ms); decay is exponential. Thus, the inhibitory effect of this interneuron population within second order kernels should be revealed by interstimulus intervals corresponding to such a time course. The suppression of the population spike when Δ = 10-30 ms for in vivo preparations is consistent with this hypothesis.

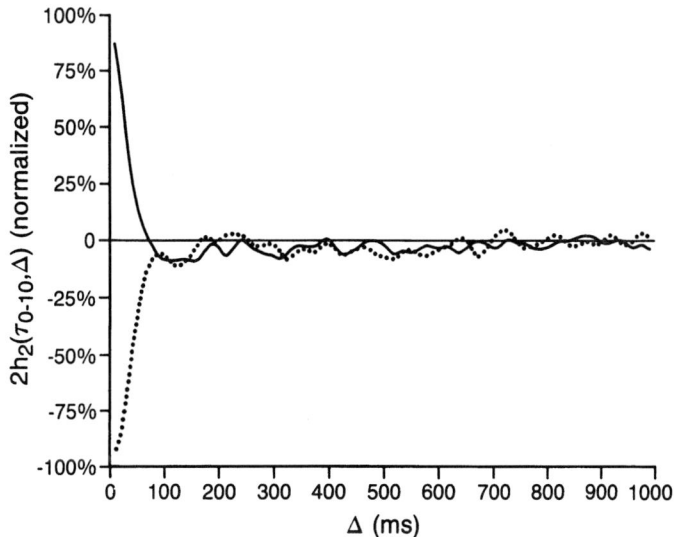

Fig. 6. Effects of the $GABA_A$ receptor allosteric agonist alphaxalone on second order kernel properties of the in vitro dentate gyrus. Kernels obtained in the absence and presence of alphaxalone are represented by a solid line and a dotted line, respectively.

The lack of similar suppression in second order kernels of in vitro slices may reflect the fact that axons of GABAergic interneurons course parallel to the longitudinal axis of the hippocampus (Struble et al., 1978), and thus, may be transected as a consequence of preparing the slices. This hypothesis was tested by applying the $GABA_A$ receptor antagonist bicuculline (10-20 μM) to the in vitro media. Very little change in second order kernel properties was observed (Fig. 5), indicating that GABAergic interneurons exert only a small influence on granule cell output in vitro. In an attempt to enhance the postsynaptic effect of GABAergic processes contained in the slice, the $GABA_A$ allosteric agonist alphaxalone (1-3 μM) was added to the in vitro media. In the presence of alphaxalone, facilitation induced by short interstimulus intervals was replaced with a suppression equivalent to that observed in vivo (Fig. 6).

Although suppression to short interstimulus intervals could be reinstated by the presence of a GABAergic agonist, the facilitation of spike amplitude to longer interstimulus intervals exhibited by in vivo preparations (150-400 ms) was not evident at any dose of alphaxalone used. The latter finding suggests that other, non-GABAergic longitudinal pathways (e.g., associational projections) or feedback connectivity involving other hippocampal subfields (e.g., pyramidal cell projections to the entorhinal cortex) are the source of in vivo second order kernel facilitation.

Third order nonlinearities of the in vitro dentate gyrus also were altered by pharmacological enhancement of GABAergic inhibition (data not shown). Doses of alphaxalone that re-instated suppression in second order kernels of slices also reduced the magnitude of suppression of third order kernels. Although third order kernels of the in vitro dentate gyrus more closely approximated those of the in vivo dentate in the presence of alphaxalone, significant differences between the two preparations remained, again implicating a modulatory role for other feedback pathways.

Contribution of the Commissural System to Differences in Nonlinear Properties of the In Vivo and In Vitro Dentate Gyrus

One of the major feedback pathways arising from connectivity between the dentate gyrus and other hippocampal subsystems involves commissural projections to the contralateral hippocampus (Fig. 1). The commissural system of the dentate gyrus arises from polymophic neurons lying subjacent to granule cells (Berger et al., 1981). Polymorphic cells are excited by perforant path input and project in an excitatory manner to the contralateral dentate (West et al., 1975; Hjorth-Simonsen, 1977). Thus, the commissural pathways are a possible source of feedback to the dentate gyrus.

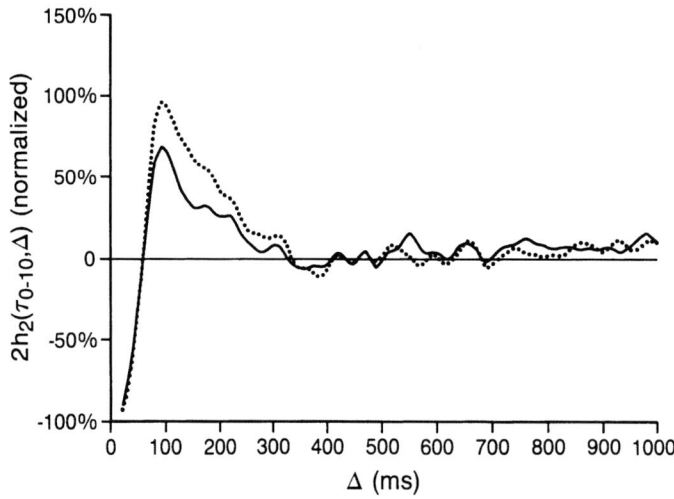

Fig. 7.   Effects of acute unilateral hippocampectomy on second order kernel properties of the in vivo dentate gyrus. Kernels obtained before and after the removal of the contralateral hippocampus are represented by a solid line and a dotted line, respectively.

We investigated the contribution of commissural projections to nonlinear properties of the dentate gyrus by conducting random impulse train stimulation of the in vivo perforant path before and after removal of the contralateral hippocampus. Animals first were implanted chronically with stimulating and recording electrodes and allowed to recover from the anesthesia and surgical procedures. Random impulse train characterization of the intact dentate gyrus then was completed when animals were awake and unanesthetized. The following day, animals were re-anesthetized, and the hippocampus (and overlying neocortex) contralateral to the implanted electrodes was removed by aspiration. After forty-eight hours of recovery from surgery and anesthesia, random train stimulation was conducted a second time. Second order kernels for both phases of a representative experiment are shown in Figure 7.

The predominant effect of acute removal of the commissural feedback loop was to enhance second order kernel facilitation of the population spike for $\Delta = 50-400$ ms. Facilitation is associated with the same range of interstimulus intervals prior to unilateral hippocampectomy. Thus, it is primarily the magnitude of facilitation which is influenced by loss of commissural feedback. Also note that suppression of the population spike to interstimulus intervals of 10-30 ms is slightly increased following hippocampectomy.

Both the increase in magnitude of facilitation and the much smaller increase in suppression are consistent with known physiological characteristics of the dentate commissural system. The majority of commissural fibers provide monosynaptic, excitatory input to granule cells (Laatsch and Cowan, 1966). However, some commissural elements also terminate onto inhibitory interneurons, contributing to feedforward inhibition of granule cells (Buszaki and Eidelberg, 1981; Douglas et al., 1983). The inhibitory influence predominates, and the increased facilitation after unilateral hippocampectomy may reflect loss of feedforward inhibition through this pathway. The much smaller change in suppression of spike amplitude in response to shorter interstimulus intervals may reflect loss of the weaker excitatory input. Further studies of the commissural system are investigating these interpretations, as well as the nature of interactions between perforant path and commissural input to granule cells using dual input random train stimuli (Sclabassi and Noreen, 1981).

Within the context of the present comparison of nonlinear response properties of the in vivo and the in vitro dentate gyrus, these results suggest that feedback provided by the commissural system cannot account for the lack of facilitation for $\Delta = 150-400$ ms in the second order kernels of slices. This possibility will need to be investigated further by examining in vivo the effects of a combination of unilateral hippocampectomy and pharmacological blockade of GABAergic receptors, which will better mimic the conditions of the in vitro slice. It appears likely, however, that the facilitation is mediated by transsynaptic feedback through the entorhinal cortex, or by feedforward influence involving perforant path activation of regions of the dentate distant from the recording site. Activity of these distant regions could propagate to the recording site through associational pathways. Experimental tests of both possibilities are possible. The contibution of feedback through the entorhinal can be investigated by transsecting efferents from the CA3 and CA1 pyramidal cell populations carried in the alveus. Characterization of the associational system can be studied in vitro using slices that are sectioned parallel to the longitudinal plane.

DISCUSSION

We have reviewed initial results of implementing an experimental strategy for developing an input/output model of the functional dynamics of a complex neuronal network, the hippocampal formation. The fundamental components of this strategy are to quantitatively characterize the dynamics of each subsystem of the larger network using nonlinear systems analytic procedures. Each subsystem is studied both when it is isolated from the network, so that dynamics intrinsic to the subsystem can be quantified, and when it is embedded in the network structure, so that the modulatory influence of other elements of the network can be quantified. The kernel functions describing the closed-loop and open-loop properties of each subsystem then can serve as the basis for a computer simulation through which global dynamics of the entire network are constructed and further investigated (see Sclabassi et al., this volume).

In the studies reviewed here we have focused on the contribution of connectivity to network dynamics. However, the same experimental and theoretical strategy applies equally well to studying the contribution of biophysical membrane properties. A single neuron also can be conceptualized as a network of several subsystems (Fig. 8): synaptic currents generated by ligand-dependent channels; passive membrane properties (resistance and capacitance) that determine the amplitude-time course of the resulting postsynaptic potentials; threshold for generation of action potentials, which introduces one of the most well-known nonlinearities in cell output; and voltage-dependent conductances, such as several classes of K^+ and Ca^{++}

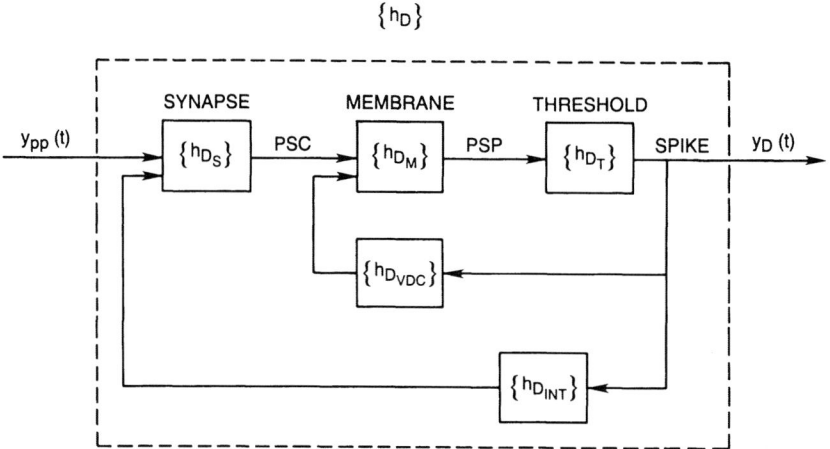

Fig. 8. Conceptualization of a network of subcellular variables that characterize a single cell h_D in the dentate gyrus. $y_{pp}(t)$ represents the input from the perforant path; PSC represents the postsynaptic currents resulting from synaptic events, h_{Ds}; PSP represents the postsynaptic potentials resulting from the translation of PSCs due to passive membrane properties, h_{Dm}; nonlinearities due to threshold events, h_{DT}, result in spike output; feedback modulation due to post-spike, voltage-dependent conductances and interneurons within the dentate are represented by h_{Dvdc} and h_{Dint}, respectively.

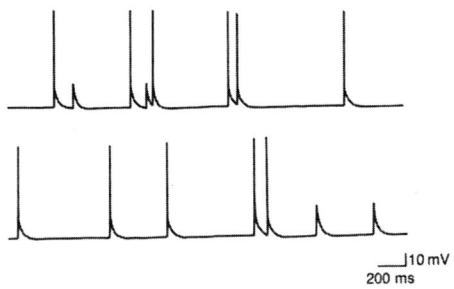

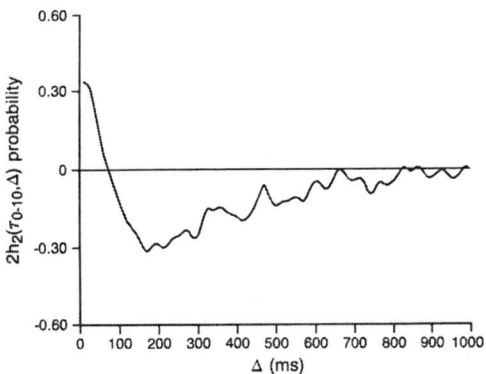

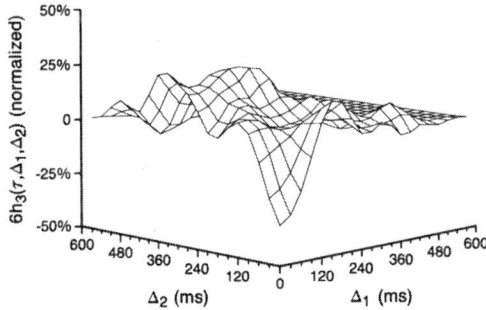

Fig. 9. Output characteristics of a single dentate granule cell in response
to random impulse train stimulation of the perforant path. The top
panel shows changes in intracellularly recorded membrane potential
evoked by stimuli as they occur in one segment of the train. The
cell responded with EPSPs, which produced action potentials for
eleven of the fifteen stimuli shown. Second and third order kernel
properties of a single granule cell are shown in the middle and
bottom panels, respectively. These kernels express the probability
of spike discharge as a function of interstimulus intervals.

currents, which are sources of feedback through which additional nonlinearities in cell output are introduced. To accomplish an analysis of biophysical membrane properties, utilization of the in vitro slice preparation can be expanded to include intracellular recording from single neurons during random impulse train stimulation, as illustrated in Fig. 9. Because blockade of many of the biophysical sources of feedback involve standard pharmacological and ion substitution procedures, the experimental "decomposition" of a network of subcellular variables also is feasible.

Our justification for analysis of only the population spike component of the evoked field potential has been detailed previously (Berger et al., 1987). In the context of the present discussion, the known underlying generator for the population spike provides a biologically interpretable link between observations made at the population level for studies of connectivity and observations made at the single cell level for studies of biophysical membrane properties. Through application of the same theoretical and experimental strategy to both levels of analysis, we are attempting to construct a biologically-based model of the hippocampal formation that integrates both cellular and network dynamics.

ACKNOWLEDGEMENTS

This research was supported by The Whitaker Foundation, the Office of Naval Research (N00014-87-K-0472), the Air Force Office of Scientific Research (89-0197), NIH (NS 52488), and NIMH (MH00343).

REFERENCES

Alger, B. E. and Nicoll, R. A., 1982, Feed-forward dendritic inhibition in rat hippocampal pyramidal cells studied in vitro, J. Physiol., 328:105-123.

Amaral, D. G., 1978, Golgi study of cell types in the hilar region of the hippocampus in the rat, J. Comp. Neurol., 182:851-914.

Andersen, P., Bliss, T. V. P. and Skrede, K. K., 1971a, Unit analysis of hippocampal population spikes, Exp. Brain Res., 13:208-211.

Andersen, P., Bliss, T. V. P. and Skrede, K. K., 1971b, Lamellar organization of hippocampal excitatory pathways, Exp. Brain Res., 13:222-238.

Andersen, P., Eccles, J. C. and Loyning, Y., 1964, Location of postsynaptic inhibitory synapses on hippocampal pyramids, J. Neurophysiol., 27:592-607.

Balzer, J. R., Sclabassi, R. J., and Berger, T. W., 1989, Effects of stimulation intensity on nonlinear response properties of hippocampal dentate granule cells to input from the medial perforant path, submitted.

Berger, T. W. and Bassett, J. L., 1989, System properties of the hippocampus, in: "Learning and Memory: The Biological Substrates," I. Gormezano, ed., Lawrence Erlbaum Associates, Hillsdale, NJ, in press.

Berger, T. W., Berry, S. D. & Thompson, R. F., 1986, Role of the hippocampus in classical conditioning of aversive and appetitive behaviors, in: "The Hippocampus," R. L. Isaacson and K. H. Pribram, eds., Plenum, New York.

Berger, T. W., Eriksson, J. L., Ciarolla, D. A. and Sclabassi, R. J., 1988a, Nonlinear systems analysis of synaptic transmission in the hippocampal perforant path-dentate projection. II. Effects of random train stimulation, J. Neurophysiol., 60:1077-1094.

Berger, T. W., Eriksson, J. L., Ciarolla, D. A. and Sclabassi, R. J., 1988b, Nonlinear systems analysis of synaptic transmission in the hippocampal perforant path-dentate projection. III. Comparison of random train and paired impulse stimulation, J. Neurophysiol., 60:1095-1109.

Berger, T. W., Robinson, G. B., Port, R. L. and Sclabassi, R. J., 1987, Nonlinear systems analysis of the functional properties of the hippo-campal formation, in: "Advanced Methods of Physiological System Modeling," V. Z. Marmarelis, ed., Biomedical Simulations Resource, Los Angeles.

Berger, T. W., Semple-Rowland, S. and Bassett, J. L., 1981, Hippocampal polymorph neurons are the cells of origin for ipsilateral association and commissural afferents to the dentate gyrus, Brain Res., 215:329-336.

Buzsaki, G. and Eidelberg, E., 1981, Commissural projection to the dentate gyrus of the rat: evidence for feed-forward inhibition, Brain Res., 230:346-350.

Douglas, R. M., McNaughton, B. L. and Goddard, G. V., 1983, Commissural inhibition and facilitation of granule cell discharge in fascia dentata, J. Comp. Neurol., 219:285-294.

Hjorth-Simonsen, A., 1977, Distribution of commissural afferents to the hippocampus of the rabbit, J. Comp. Neurol., 176:495-514.

Krausz, H., 1975, Identification of nonlinear systems using random impulse train inputs, Biol. Cyber., 19:217-230.

Laatsch, R. H., and Cowan, W. M., 1966, Electron microscopic studies of the dentate gyrus of the rat. I. Normal structure with special reference to synaptic organization, J. Comp. Neurol., 128:359-396.

Lacaille, J., Mueller, A., Kunkel, D. and Schwartzkroin, P., 1987, Local circuit interactions between oriens/alveus interneurons and CA1 pyram-idal cells in hippocampal slices: electrophysiology and morphology, J. Neurosci., 7:1979-1983.

Lee, Y. W. and Schetzen, M., 1965, Measurement of the kernels of a non-linear system by cross-correlation, Internat. J. Control., 2:237-254.

Lomo, T., 1971, Patterns of activation in a monosynaptic cortical pathway: the perforant path input to the dentate area of the hippocampal forma-tion, Exp. Brain Res., 12:18-45.

Lorente de No, R., 1934, Studies on the structure of cerebral cortex. II. Continuation of the study of the ammonic system, J. Psychol. Neurol., 46:113-177.

Ogura, H., 1972, Orthogonal functionals of the poisson process, IEEE Trans. Inform. Theory, It-18:473-481.

Ribak, C. E., and Seress, L., 1983, Five types of basket cell in the hippocampal dentate gyrus: a combined Golgi and electron microscopic study, J. Neurocyt., 12:577-597.

Sclabassi, R. J., Eriksson, J. L., Port, R. L., Robinson, G. B., and Berger, T. W., 1988a, Nonlinear systems analysis of the hippocampal perforant path-dentate projection. I. Theoretical and interpretational considera-tions, J. Neurophysiol., 60:1066-1076.

Sclabassi, R. J., Hinman, C. L., Kroin, J. S. and Risch, H. A., 1977, The modulatory effect of prior input upon afferent signals in the somatosen-sory system, Proc. Joint Auto. Control Conf., IEEE 12:787-795.

Sclabassi, R. J., Krieger, D. N. and Berger, T. W., 1988b, A systems theoretic approach to the study of CNS function, Ann. Biomed. Eng., 16:17-34.

Sclabassi, R. J., Krieger, D. N., Solomon, J., Samosky, J., Levitan, S., and Berger, T., 1989, Modeling of neuronal networks through theoretical decomposition, in: "Advanced Methods of Physiological System Modeling," V. Z. Marmarelis, ed., Plenum, New York, in press.

Sclabassi, R. J. and Noreen, G. K., 1981, The characterization of dual-input evoked potentials as nonlinear systems using random impulse trains, Proc. Pitts. Mod. Simul. Conf., 12:1123-1130.

Skrede, K. K. and Westgaard, R. H., 1971, The transverse hippocampal slice: a well-defined cortical structure maintained in vitro, Brain Research, 35:589-593.

Sorensen, K. E. and Shipley, M. T., 1979, Projections from the subiculum to the deep layers of the ipsilateral presubicular and entorhinal cortices in the guinea pig, J. Comp. Neurol., 188:313-334.

Squire, L. R., 1982, The neuropsychology of human memory, Ann. Rev. Neurosci., 5:241-273.

Steward, O., 1976, Topographic organization of the projections from the entorhinal area to the hippocampal formation of the rat, J. Comp. Neurol., 167:285-314.

Struble, R. G., Desmond, N. L. and Levy, W. B., 1978, Anatomical evidence for interlamellar inhibition in the fascia dentata, Brain Res., 152: 580-585.

Swanson, L. W. and Cowan, W. M., 1977, An autoradiographic study of the organization of the efferent connections of the hippocampal formation in the rat, J. Comp. Neurol., 172:49-84.

Swanson, L. W., Wyss, J. M. and Cowan, W. M., 1978, An autoradiographic study of the organization of intrahippocampal association pathways in the rat, J. Comp. Neurol., 181:681-716.

Thompson, R. F., Berger, T. W. and Madden, J., IV., 1983, Cellular processes of learning and memory in the mammalian CNS, Ann. Rev. Neurosci., 6: 447-491.

West, J. R., Deadwyler, S., Cotman, C. W., and Lynch, G., 1975, Time-dependent changes in commissural field potentials in the dentate gyrus following lesions of the entorhinal cortex in adult rats, Brain Research, 97:215-233.

Wiener, N., 1958, "Nonlinear Problems in Random Theory," Wiley Press, New York.

Zimmer, J., 1971, Ipsilateral afferents to the commissural zone of the fascia dentata, demonstrated in decommissurated rats by silver impregnation, J. Comp. Neurol., 142:393-416.

THEORETICAL DECOMPOSITION OF NEURONAL NETWORKS

Robert J. Sclabassi, Donald N. Krieger, Jackie Solomon, Joseph Samosky,
Steven Levitan and Theodore W. Berger

Departments of Neurological Surgery, Electrical Engineering, Behavioral
Neuroscience, and Psychiatry
Center for Neuroscience, Univ. of Pittsburgh, Pittsburgh, PA 15260

INTRODUCTION

Progress has been slow in satisfactorily defining relationships between the anatomical, biophysical, and biochemical characteristics of single neurons, and those properties emerging from their collective functioning as a network. This lack of progress can be attributed, at least in part, to the lack of a general theoretical approach for modeling neuronal systems. Such an approach would allow utilization of descriptors of individual neuronal elements at multiple levels, from the biophysical to the system level (Sejnowski et al., 1988). We have been investigating a solution to this problem, using a systems theoretic approach based on input/output analysis (Rugh, 1981; Sclabassi et al., 1988a). This allows the quantitative characterization of complex systems independent of the nature and number of processes responsible for the input/output transformation, or of the number and organization of subsystems contained within the network. We have been applying this approach, in an integrated theoretical and experimental fashion, to characterize the functional properties of the hippocampal formation (Berger et al., 1987), a brain structure known to be critically involved in learning and memory.

Anatomically, the hippocampal formation is composed of diverse populations of neurons, only some of which are known or observable, but all of which are organized into well defined systems of feedforward and feedback elements or nodes. We have previously established that the nonlinear input/output properties of major nodes within the hippocampus can be characterized experimentally by recording electrophysiological responses to random impulse train stimulation and by expressing the relationship between the node's input and output activity as the kernels of an orthogonalized functional power series (Sclabassi et al., 1988b; Berger et al., 1988a,b). The kernels may then be interpreted physiologically as generalized recovery functions and analytically as n^{th} order impulse responses, providing an external or input/output model of the hippocampal network.

Using both $in-vivo$ (Berger and Sclabassi, 1985) and $in-vitro$ (Harty et al., 1987) preparations, we have demonstrated that such procedures can be applied both when the network is intact and after experimental isolation of a node from one or more of the nodes influencing it. However, in the hippocampal formation, there are functional elements which cannot be isolated and whose input/output properties therefore cannot be measured directly. Rather, the transformational properties of these elements must be inferred from their effect on other directly observable groups of neurons.

This paper examines an approach to the characterization of the functional properties of these unobservable groups of neurons using the concept of network decomposition. We take advantage of the topology of the hippocampal formation to break it into a number of hierarchically ordered elements, some of which may be experimentally characterized, others of which cannot be isolated. Thus, these unobservable elements must be indirectly characterized using multiple experimental results coupled with the algebraic techniques described in this paper. The resulting complete set of characterizing functions may then be recombined to produce a global model of the hippocampal formation.

In order to implement our approach, n^{th} order multi-dimensional Laplace transforms of the nth order impulse are computed and manipulated algebraically to produce characterizing functions for the unobservable elements of the system. Based on the totality of these results, models of the observable and unobservable subsystems, as well as for the entire system, may be computed in terms of nth order convolution operators using n^{th} order impulse response functions.

An example is presented, which demonstrates the steps in the process of characterizing unobservable functions for the contralateral hippocampus. In this example, the intact hippocampal formation is conceptualized as a three element system - a feedforward element, which is the granule cell layer of the dentate gyrus; a feedback element, which is all other portions of the hippocampal formation except the granule cell layer; and the contralateral hippocampus. Multi-dimensional Laplace transforms are utilized to manipulate these system representations to arrive at computable forms for two unobservable elements, the feedback elements, and the contralateral hippocampus. Simulations are performed for these three networks to verify the capability of the experimental kernels to characterize the essential transformational features of the hippocampal formation.

THE HIPPOCAMPAL FORMATION

The hippocampal formation, a brain area implicated in the processes of learning and memory formation, has been the focus of intensive experimental and clinical research (Douglas, 1983; O'Keefe and Nadel, 1978; Squire, 1982). The intrinsic circuitry of the hippocampal formation consists of the entorhinal, hippocampal, and subicular cortices, and can be represented as a set of five major cell populations arranged in a serially organized, closed feedback loop (Fig. 1) (Lorente de No, 1934; Ramon y Cajal, 1911; Berger et al., 1980; Beckstead, 1978; Hjorth-Simonsen and Juene, 1972; Sorensen and Shipley, 1979; Swanson and Cowan, 1977; Swanson et al., 1978).

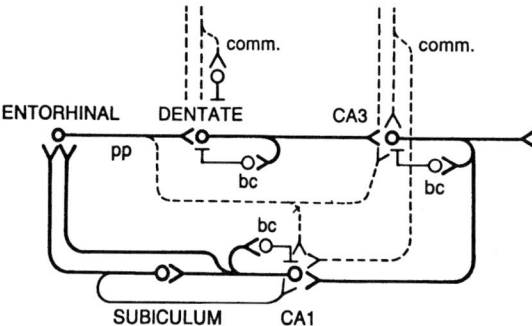

Fig. 1. Conceptualization of the hippocampus showing the intrinsic, trisynaptic pathway
involving excitatory input from the perforant path (PP) to granule cells
of the dentate gyrus, to pyramidal cells of the CA3 region inferior, and to
pyramidal cells of the CA1 region superior

Each of the major cell populations is excitatory to the next, so that the discharge of any one population in the loop results in sequential excitation of subsequent populations (Andersen et al., 1964; Andersen et al., 1971; Bartesaghi and Gessi, 1986; Van Groen and Lopes da Silva, 1986). Activity within each of these cell populations is modulated by feedforward and feedback connections with local interneurons (Alger and Nicoll, 1982; Andersen et al., 1964; Douglas et al., 1983; Buzsaki and Eidelberg, 1981; Miles and Wong, 1986), by commissural connections with the contralateral hippocampus (Berger et al., 1981; Hjorth-Simonsen, 1971), and by associational projections between lamella (Berger et al., 1981). Each of the major cell populations activates local, inhibitory feedback through axon collateral projections to basket cells (Alger and Nicoll, 1982; Andersen et al., 1964; Knowles and Schwartzkroin, 1981). A feedforward inhibitory pathway has been documented within the CA1 cell field (Alger and Nicoll, 1982), and may exist within the dentate gyrus as well (Buzsaki and Eidelberg, 1981; Donnegan et al., 1987; Ribak et al., 1978). Recurrent excitatory collaterals are prevalent among CA3 neurons of the hippocampus (Traub et al., 1987). A wide variety of other interneurons are known to reside within the different regions of the hippocampal formation, contributing to the output of each of the major projection neurons (Alger and Nicoll, 1982; Lorente de No, 1934). Thus, the hippocampal formation is composed of many different neuronal populations, which interact in a complex manner through feedforward and feedback circuitry.

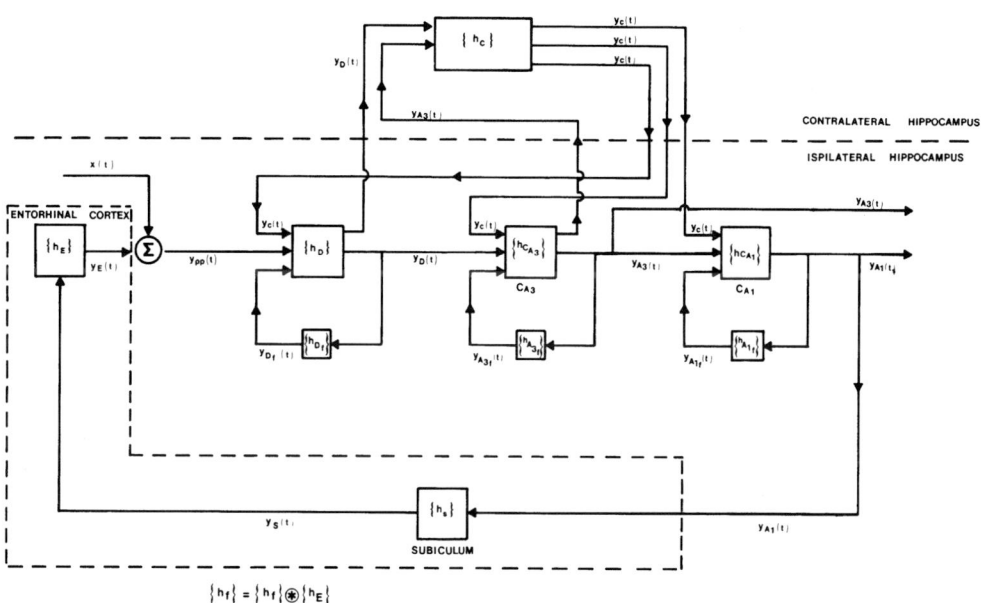

Fig. 2. Block diagram of the hippocampal formation

Figure 2 presents our current conceptualization of the hippocampal formation in block diagram form. In this formulation, the hippocampal formation is thought of as a multiloop feedback system. $X(t)$ is the experimental input to the system, and $\{h(.)\}$ are the nth order kernels, which are estimated either experimentally or through algebraic manipulation, and are used to quantitatively characterize the transformational properties of the system elements.

FUNCTIONAL POWER SERIES REPRESENTATION

An external model is a nonparametric representation of the input/output properties of a network element and treats each element as a black box (Casti, 1985). The dynamics of each network element may be quantitatively characterized by the kernels of the terms of a functional power series, which summarize the nonlinear transformational properties of the system's elements. Thus, each element's output, $y(t)$, may be given by a series of functionals, $G_i[h_i, x(t)]$, utilizing either experimentally or computationally estimated kernels $\{h_i(.)\}$,

$$y[x(t)] = \sum_{i=0}^{N} G_i[h_i; x(t)] \tag{1}$$

where $x(t)$ is the input to the element to be identified.

The first four terms of the orthogonalized power series are shown in Eqn. (2), (3), (4), and (5).

$$G_0 = h_0 \tag{2}$$

$$G_1[h_1; x(t)] = \int_{-\infty}^{\infty} h_1(\tau)x(t - \tau)d\tau - \lambda \int_{-\infty}^{\infty} h_1(\tau)d\tau \tag{3}$$

$$
\begin{aligned}
G_2[h_2; x(t)] = & \int_{-\infty}^{\infty} \int_{-\infty}^{\infty} h_2(\tau, \Delta)x(t - \tau)x(t - \Delta)d\tau d\Delta \\
& - 2\lambda \int_{-\infty}^{\infty} \int_{-\infty}^{\infty} h_2(\tau, \Delta)x(t - \tau)d\tau \Delta \\
& + \lambda^2 \int_{-\infty}^{\infty} \int_{-\infty}^{\infty} h_2(\tau, \Delta)d\tau d\Delta
\end{aligned}
\tag{4}
$$

$$
\begin{aligned}
G_3[h_3; x(t)] = & \int_{-\infty}^{\infty} \int_{-\infty}^{\infty} \int_{-\infty}^{\infty} h_3(\tau, \Delta_1, \Delta_2)x(t - \tau)x(t - \Delta_1)x(t - \Delta_2)d\tau d\Delta_1 d\Delta_2 \\
& - 3\lambda \int_{-\infty}^{\infty} \int_{-\infty}^{\infty} \int_{\infty}^{\infty} h_3(\tau, \Delta_1, \Delta_2)x(t - \tau)x(t - \Delta_1)d\tau d\Delta_1 d\Delta_2 \\
& + 3\lambda^2 \int_{-\infty}^{\infty} \int_{-\infty}^{\infty} \int_{-\infty}^{\infty} h_3(\tau, \Delta_1, \Delta_2)x(t - \tau)d\tau d\Delta_1 d\Delta_2 \\
& - \lambda^3 \int_{-\infty}^{\infty} \int_{-\infty}^{\infty} \int_{-\infty}^{\infty} h_3(\tau, \Delta_1, \Delta_2)d\tau d\Delta_1 d\Delta_2
\end{aligned}
\tag{5}
$$

The experimental kernels $\{h_i(.)\}$ are estimated by cross-correlating evoked activity with the system input $x(t)$, while the unobservable kernels are estimated from the algebraic manipulation of the system's n^{th} order Laplace transfer functions.

Experimentally, trains of electrical impulses are applied to the perforant path, and the amplitudes of the population spikes of the evoked field potentials are measured. These amplitudes reflect the number of granule cells discharging action potentials. Intervals between impulses vary randomly according to a Poisson distribution having a mean rate of λ. Cross-correlation techniques are used to compute the kernels of the power series expansion based upon these measures (Sclabassi et al.; 1977, Sclabassi et al., 1988a), using techniques similar to those employed by Lee and Schetzen (1965). These calculations require the assumptions of ergodicity and stationarity, allowing ensemble averages to be replaced by time averages, and utilize the principle of orthogonality. The results are:

$$h_n(\sigma_1, \ldots, \sigma_n) = \frac{1}{n!\lambda^n} \, \overline{y(t)z(t-\sigma_1)\ldots z(t-\sigma_n)} \qquad (6)$$

where $\overline{}$ indicates time average, and $z(t) = x(t) - \lambda$.

Variation in population spike amplitude provides a measure of the transformational properties of all the components of the hippocampal formation due to the system's anatomically closed loop structure. The $\{h(.)\}$ may be interpreted physiologically as generalized recovery functions and theoretically as n^{th} order impulse responses (Eykhoff, 1974), thus providing characterizing functions for system elements. The first order kernel reflects the average population spike amplitude to all impulses in the train, and is the best estimate of the network response if the system functions in a linear manner. The second order kernels represent the modulatory influence of the preceding impulse (occurring Δ ms in the past) on the system response to the present impulse. The third order kernels represent the modulatory influence of two preceding impulses (occurring Δ_1 and Δ_2 ms in the past) on the system response to the present impulse. These kernels provide the basis for the simulation of global network properties when the above equations are treated as convolution integrals.

EXAMPLE

Our approach is to model the functional properties of the hippocampal formation as the composite of input/output functions characterizing each physiologically meaningful elemental unit in the network. These relationships capture the dynamics of the elemental units in a common analytical framework, independent of the system level at which they are measured. This framework allows the models for the system to be manipulated and recombined, according to hypothesized topologies of the network under study. Computer simulations can then be utilized to study not only the dynamic properties of the individual elements and their relationship to the global properties of the network, but also the accuracy of the hypothesized network topology.

The example presented here requires the study of the network at three levels: (i) the intact hippocampal formation, with the dentate granule cell layer being considered a feedforward element, the contralateral hippocampus a parallel feedforward element, and the remainder of the ipsilateral hippocampus a feedback element (Fig. 3); (ii) a partially closed-loop system, where the contralateral hippocampus has been removed (Fig. 6); and (iii) an open-loop system, where the contralateral hippocampus has been removed, the ipsilateral trisynaptic pathway is open, and the local GABAergic pathways have been removed (Fig. 9). Four results are summarized: (1) the nth order Laplace transform of that network configuration; (2) the results of the algebraic manipulation of the nth order transforms used to calculate the unobservable element of interest (the contralateral hippocampus); (3) a set of experimental data obtained from the hippocampal formation for each network level; and (4) the results of simulations using the estimated kernels and the convolution integrals presented above. The solutions to these three cases allow computation of the external model of both the GABAergic pathways and the contralateral hippocampus, neither of which is available by direct measurement.

I. Intact Hippocampal Formation

A topographical model of the intact hippocampal formation is given in Fig. 3. In this model, C represents the effects of the non-observable contralateral hippocampus, A represents the feedforward dentate granule cell layer, B represents all other elements, while E represents the combined system properties which we have measured experimentally (Berger et al., 1988a).

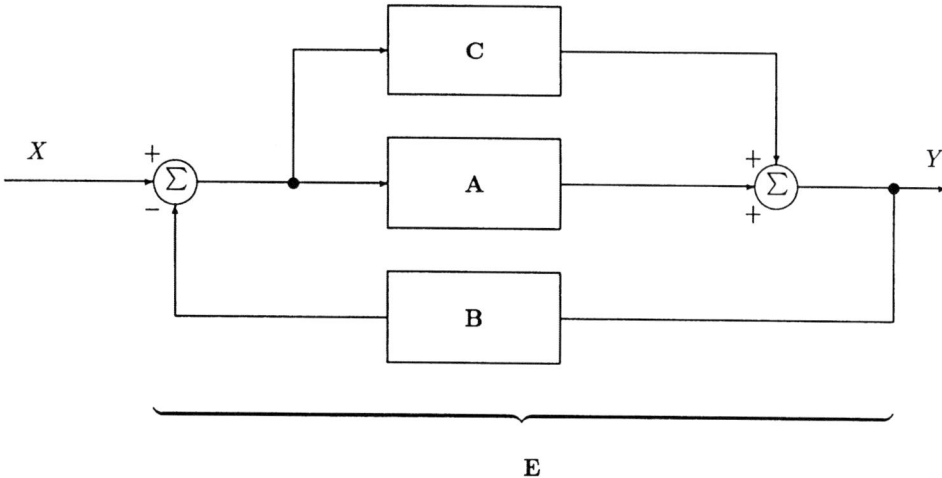

E

Fig. 3. Intact hippocampal formation where (A) represents dentate granule cell layer,
(B) represents all feedback elements, (C) represents contralateral hippocampus,
and (E) represents complete transfer characteristics of the intact formation

In order to test the adequacy of the experimental kernels characterizing E, a combined experimental and simulation study was performed. Data were obtained from the *in − vivo* intact hippocampal formation. The stimuli were applied to the medial bundle of the perforant path, and the evoked population spikes recorded from the granule cell layer of the dentate gyrus. The experimental kernels (Fig. 4) were computed as previously described.

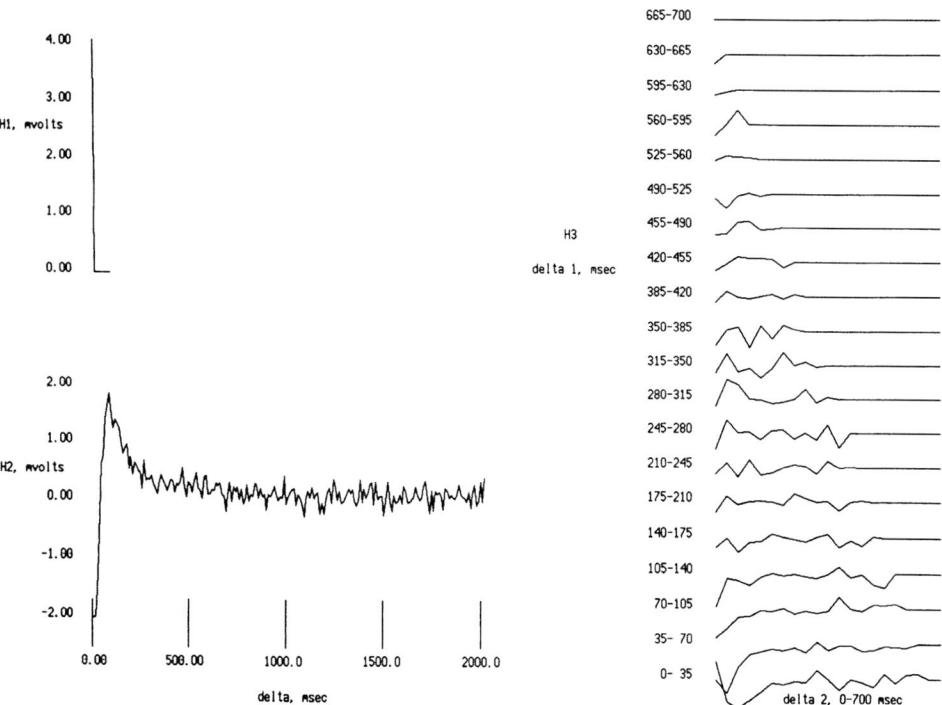

Fig. 4. First-, second- and third-order experimental kernels obtained from an intact
hippocampus

Paired impulse data were also obtained, where the separation between the impulses in a pair was systematically varied from 10 ms to 1200 ms, from the same animal from which the kernels were obtained. Utilizing the experimental kernels presented in Fig. 4, a third-order external model for the entire system (Fig. 3) was then evaluated computationally for an input signal, consisting of the same set of paired impulses which were used to obtain the experimental paired impulse data. For both the experimental and predicted paired pulse data, the average value of the response to the first impulse was subtracted from the amplitude of the population spike produced by the second impulse to give an estimate of the nonlinear residual. Figure 5 shows the match for this external model utilizing the first three terms of the power series. The experimental data is plotted as connected circles in these figures, while the predicted data is plotted as squares. In general, the predicted data is in excellent agreement with the experimental data; the regions where differences occur suggest that fourth-order terms may be required to adequately characterize the data.

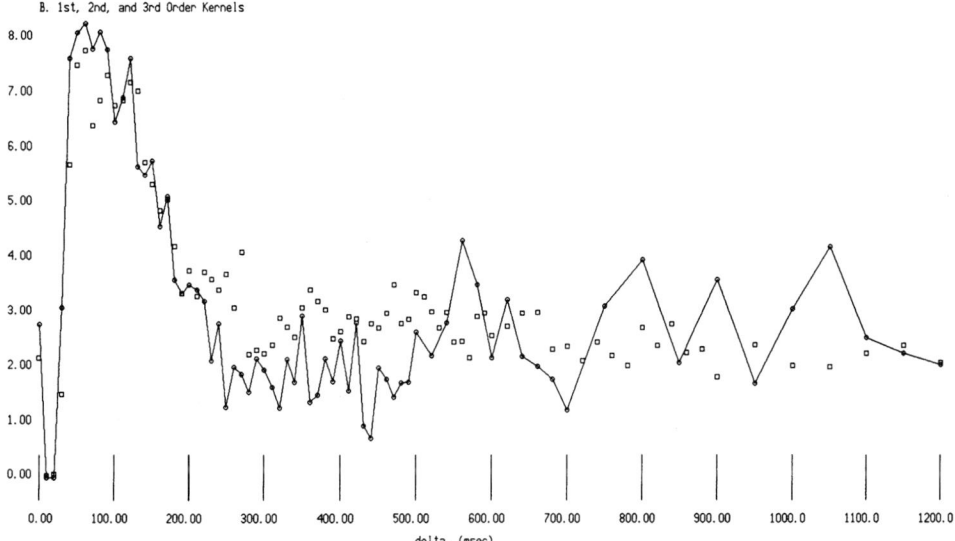

Fig. 5. Experimental and predicted twin-pulse data for third-order system

In order to evaluate the expressions for the contralateral hippocampus, the nth order impulse responses are transformed into n^{th} order Laplace transforms ($H_n(S_1, ...S_n)$) and manipulated algebraically (Brilliant, 1958; George, 1959). The nth order Laplace transforms for the intact system (E) as a function of the system elements (A, B, and C) are as follows:

$$E_1(S) = \frac{A_1(S) + C_1(S)}{1 - (A_1(S) + C_1(S))B_1(S)} \tag{7}$$

$$
\begin{aligned}
E_2(S_1, S_2) = \ & \left\{ [A_1(S_1) + C_1(S_1)] \cdot [A_1(S_2) + C_1(S_2)] \right. \\
& \cdot \ [A_1(S_1 + S_2) + C_1(S_1 + S_2)] \cdot B_2(S_1, S_2) \\
& + \ \left. A_2(S_1, S_2) + C_2(S_1, S_2) \right\} \Big/ [1 - (A_1(S_1) + C_1(S_1)B_1(S_1)]
\end{aligned}
$$

135

$$\cdot \quad [1 - (A_1(S_2) + C_1(S_2)B_1(S_2)]$$
$$\cdot \quad [1 - (A_1(S_1 + S_2) + C_1(S_1 + S_2))B_1(S_1 + S_2)] \tag{8}$$

and

$$
\begin{aligned}
E_3(S_1, S_2, S_3) \ = \ & \Big\{ [A_1(S_1 + S_2 + S_3) + C_1(S_1 + S_2 + S_3)] \\
\cdot \ & [\tfrac{2}{3}B_2(S_1, S_2 + S_3)E_1(S_1)E_2(S_2, S_3) \\
+ \ & \tfrac{2}{3}B_2(S_2, S_1 + S_3)E_1(S_2)E_2(S_1, S_3) \\
+ \ & \tfrac{2}{3}B_2(S_3, S_1 + S_2)E_1(S_3)E_2(S_1, S_2) + B_3(S_1, S_2, S_3)E_1(S_1)E_1(S_2)E_1(S_3)] \\
+ \ & \tfrac{2}{3}\{[A_2(S_1, S_2 + S_3) + C_2(S_1, S_2 + S_3)] \cdot [1 + B_1(S_1)E_1(S_1)] \\
\cdot \ & [B_1(S_2 + S_3)E_2(S_2, S_3) + B_2(S_2, S_3)E_1(S_2)E_1(S_3)] \\
+ \ & [A_2(S_2, S_1 + S_3) + C_2(S_2, S_1 + S_3)] \cdot [1 + B_1(S_2)E_1(S_2)] \\
\cdot \ & [B_1(S_1 + S_3)E_2(S_1, S_3) + B_2(S_1, S_3)E_1(S_1)E_1(S_3)] \\
+ \ & [A_2(S_3, S_1 + S_2) + C_2(S_3, S_1 + S_2)] \cdot [1 + B_1(S_3)E_1(S_3)] \\
\cdot \ & [B_1(S_1 + S_2)E_2(S_1, S_2) + B_2(S_1, S_2)E_1(S_1)E_1(S_2)]\} \\
+ \ & [A_3(S_1, S_2, S_3) + C_3(S_1, S_2, S_3)] \cdot [1 + B_1(S_1)E_1(S_1)] \cdot [1 + B_1(S_2)E_1(S_2)] \\
\cdot \ & [1 + B_1(S_3)E_1(S_3)] \Big\} \Big/ \{1 - [A_1(S_1 + S_2 + S_3) \\
+ \ & C_1(S_1 + S_2 + S_3)]B_1(S_1 + S_2 + S_3)\} \tag{9}
\end{aligned}
$$

These forward equations (7), (8), and (9) may be solved for the transfer characteristics of the contralateral hippocampus (C), given the transforms for A, B, and E, all of which are obtainable either experimentally or computationally, as shown in:

$$C_1(S) = \frac{E_1(S) - E_1(S)A_1(S)B_1(S) - A_1(S)}{[1 + E_1(S)A_1(S)]} \tag{10}$$

$$
\begin{aligned}
C_2(S_1, S_2) \ = \ & \{(E_2(S_1, S_2)[1 - (A_1(S_1) + C_1(S_1))B_1(S_1)] \\
\cdot \ & [1 - (A_1(S_2) + C_1(S_2))B_1(S_2)] \\
\cdot \ & [1 - (A_1(S_1 + S_2) + C_1(S_1 + S_2)]B_1(S_1 + S_2) \\
- \ & \{[A_1(S_1) + C_1(S_1)] \cdot [A_1(S_2) + C_1(S_2)] \cdot [A_1(S_1 + S_2) + C_1(S_1 + S_2)] \\
\cdot \ & B_2(S_1, S_2)\} - A_2(S_1, S_2)\} \tag{11}
\end{aligned}
$$

and finally,

$$
\begin{aligned}
C_3(S_1, S_2, S_3) \ = \ & \Big\{ E_3(S_1, S_2, S_3)\{1 - [A_1(S_1 + S_2 + S_3) + C_1(S_1 + S_2 + S_3)] \\
& B_3(S_1 + S_2 + S_3)]\} - [A_1(S_1 + S_2 + S_3) + C_1(S_1 + S_2 + S_3)] \\
\cdot \ & [\tfrac{2}{3}B_2(S_1, S_2 + S_3)E_1(S_1)E_2(S_2, S_3) \\
+ \ & \tfrac{2}{3}B_2(S_2, S_1 + S_3)E_1(S_2)E_2(S_1, S_3) \\
+ \ & \tfrac{2}{3}B_2(S_3, S_1 + S_2)E_1(S_3)E_2(S_1, S_2) \\
+ \ & B_3(S_1, S_2, S_3)E_1(S_1)E_1(S_2)E_1(S_3)] \\
- \ & \tfrac{2}{3}\{[A_2(S_1, S_2 + S_3) + C_2(S_1, S_2 + S_3)] \cdot [1 + B_1(S_1)E_1(S_1)]
\end{aligned}
$$

$$
\begin{aligned}
&\cdot \; [B_1(S_2+S_3)E_2(S_2,S_3)+B_2(S_2,S_3)E_1(S_2)E_1(S_3)] \\
+\; &[A_2(S_2,S_1+S_3)+C_2(S_2,S_1+S_3)]\cdot[1+B_1(S_2)E_1(S_2)] \\
\cdot \; &[B_1(S_1+S_3)E_2(S_1,S_3)+B_2(S_1,S_3)E_1(S_1)E_1(S_3)] \\
+\; &[A_2(S_3,S_1+S_2)+C_2(S_3,S_1+S_2)]\cdot[1+B_1(S_3)E_1(S_3)] \\
\cdot \; &[B_1(S_1+S_2)E_2(S_1,S_2)+B_2(S_1,S_2)E_1(S_1)E_1(S_2)]\} \\
-\; &A_3(S_1,S_2,S_3)[1+B_1(S_1)E_1(S_1)][1+B_1(S_2)E_1(S_2)] \\
\cdot \; &[1+B_1(S_3)E_1(S_3)]\Big\}\Big/[1+B_1(S_1)E_1(S_1)] \\
&[1+B_1(S_2)E_1(S_2)][1+B_1(S_3)E_1(S_3)]
\end{aligned}
\tag{12}
$$

Thus, the non-observable transformational properties (C) of the contralateral hippocampus may be calculated. However, to accomplish this we need information about each of the remaining elements (A and B), at least one of which (B) is itself unobservable, and thus requires additional steps to the solution.

II. Hippocampus After Contralateral Hippocampectomy

The second step in the determination of (C) is to obtain data from the ipsilateral hippocampus after a contralateral hippocampectomy to eliminate the contribution of both the pyramidal and granule cell commissural projections (Port et al., 1987).

Figure 6 shows the remaining system structure after the hippocampectomy. This consists of the granule cell layer (A) and all remaining hippocampal elements (B) treated as a combined feedback element.

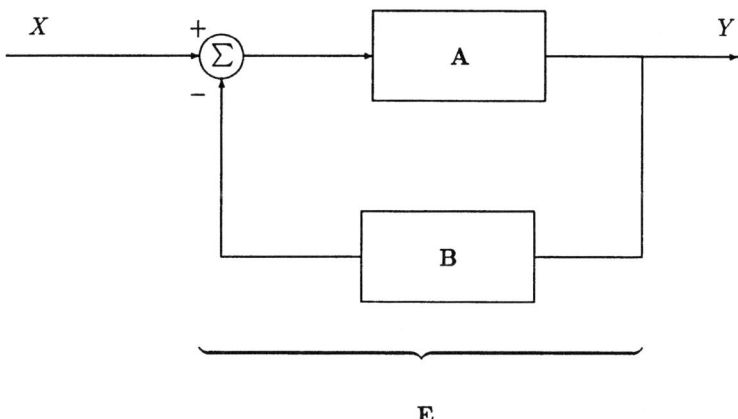

Fig. 6. Hippocampus after contralateral hippocampectomy

Experimentally, ablation of the contralateral hippocampus produced a modest, but significant, decrease in response amplitudes for short intervals in the range of 50-200 ms. The decreased responsivity to short intervals is consistent with a loss of direct excitatory input, and the increased responsivity to longer intervals is consistent with a loss of indirect input via inhibitory interneurons. Figure 7 shows the post-lesion kernels characterizing the functional properties of the system.

A simulation study was performed using these experimental data and a third-order model. Figure 8 illustrates the kernels computed from the simulated system. It can be seen that there is agreement between the experimental and theoretical kernels, demonstrating that the kernels do capture the essential formation concerning system function.

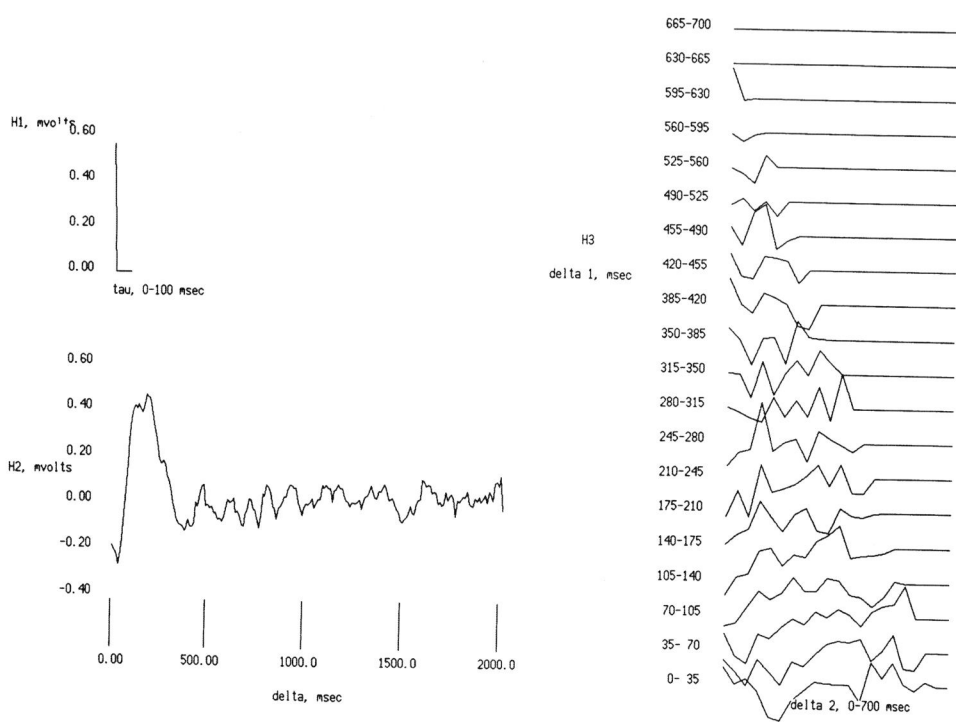

Fig. 7. Experimental kernels from *in − vivo* hippocampus after contralateral hippocampectomy

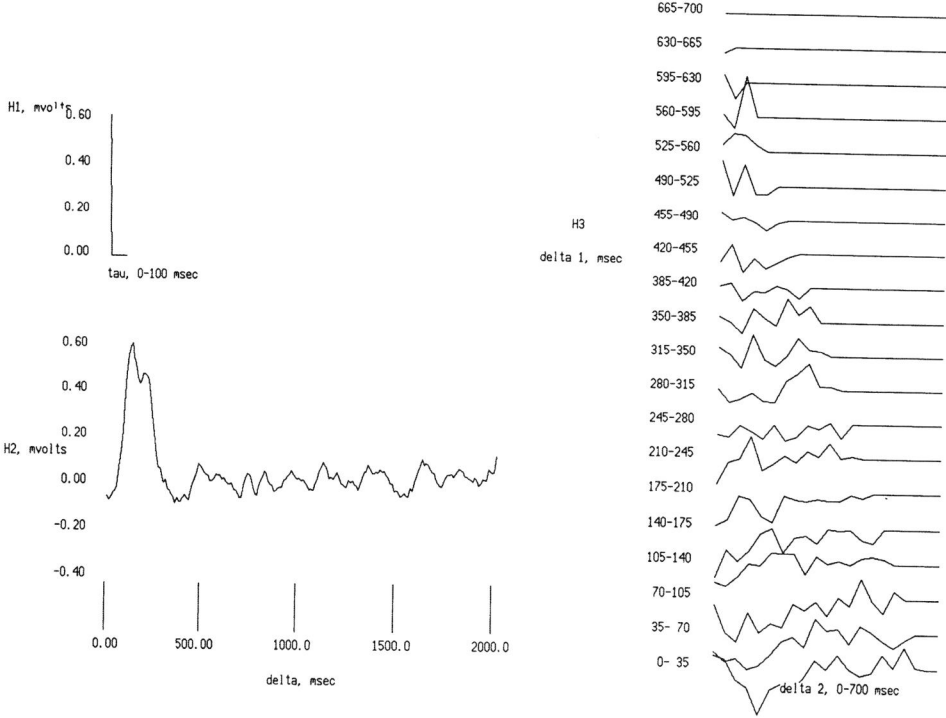

Fig. 8. Kernels obtained from simulation using third order model

The nth order transfer functions for the system (E) may be obtained experimentally and an algebraic expressions for E in terms of A and B may be directly written. A is also experimentally obtainable (next section); however, B is not directly observable. Thus, the next step is to solve the forward system equations for B.

The forward results for the equivalent system (E) are:

$$E_1(S) = \frac{A_1(S)}{1 - A_1(S)B_1(S)} \tag{13}$$

$$E_2(S_1, S_2) = \frac{A_1(S_1)A_1(S_2)A_1(S_1 + S_2)B_2(S_1, S_2) + A_2(S_1, S_2)}{[1 - A_1(S_1)B_1(S_1)][1 - A_1(S_2)B_1(S_2)][1 - A_1(S_1 + S_2)B_1(S_1 + S_2)]} \tag{14}$$

and

$$
\begin{aligned}
E_3(S_1, S_2, S_3) = & \left\{ A_1(S_1, S_2, S_3)\left[\frac{2}{3}B_2(S_1, S_2 + S_3)E_1(S_1) \cdot E_2(S_2, S_3) \right.\right. \\
& + \frac{2}{3}B_2(S_2, S_1 + S_3)E_1(S_2) \cdot E_2(S_1, S_3) \\
& + \frac{2}{3}B_2(S_3, S_1 + S_2)E_1(S_3) \cdot E_2(S_1, S_2) \\
& \left. + B_3(S_1, S_2, S_3)E_1(S_1)E_1(S_2)E_1(S_3)\right] \\
& + \frac{2}{3}\{A_2(S_1, S_2 + S_3)[1 + B_1(S_1)E_1(S_1)] \\
& \cdot [B_1(S_2 + S_3)E_2(S_2, S_3) + B_2(S_2, S_3)E_1(S_2)E_1(S_3)] \\
& + A_2(S_2, S_1 + S_3)[1 + B_1(S_2)E_1(S_2)] \\
& \cdot [B_1(S_1 + S_3)E_2(S_1, S_3) + B_2(S_1, S_3)E_1(S_1)E_1(S_3)] \\
& + A_2(S_3, S_1 + S_2)[1 + B_1(S_3)E_2(S_3)] \\
& \cdot [B_1(S_1 + S_2)E_2(S_1, S_2) + B_2(S_1, S_2)E_1(S_1)E_1(S_2)]\} \\
& + A_3(S_1, S_2, S_3)[1 + B_1(S_1)E_1(S_1)][1 + B_1(S_2)E_1(S_2)] \\
& \left. \cdot [1 + B_1(S_3)E_1(S_3)] \right\} \Big/ [1 - A_1(S_1 + S_2 + S_3)B_1(S_1 + S_2 + S_3)] \tag{15}
\end{aligned}
$$

Now, using equations (13), (14), and (15), B_1, B_2, and B_3 may be solved.

$$B_1(S) = \frac{E_1(S) - A_1(S)}{E_1(S)A_1(S)} \tag{16}$$

$$
\begin{aligned}
B_2(S_1, S_2) = & \left\{ E_2(S_1, S_2)[1 - A_1(S_1)B_1(S_1)] \cdot [1 - A_1(S_2)B_1(S_2)] \right. \\
& \left. \cdot [1 - A_1(S_1 + S_2)B_1(S_1 + S_2)] - A_2(S_1, S_2)\right\} \\
& \Big/ A_1(S_1)A_1(S_2)A_1(S_1 + S_2) \tag{17}
\end{aligned}
$$

$$
\begin{aligned}
B_3(S_1, S_2, S_3) = & \left\{ E_3(S_1, S_2, S_3)[1 - A_1(S_1 + S_2 + S_3)B_1(S_1 + S_2 + S_3)] \right. \\
& - A_3(S_1, S_2, S_3)[1 + B_1(S_1)E_1(S_1)][1 + B_1(S_2)E_1(S_2)][1 + B_1(S_3)E_1(S_3)] \\
& - \frac{2}{3}\{A_2(S_1, S_2 + S_3)[1 + B_1(S_1)E_1(S_1)]
\end{aligned}
$$

$$\cdot \quad [B_1(S_2 + S_3)E_2(S_2, S_3) + B_2(S_2, S_3)E_1(S_2)E_1(S_3)]$$
$$+ \quad A_2(S_2, S_1 + S_3)[1 + B_1(S_2)E_1(S_2)]$$
$$\cdot \quad [B_1(S_1 + S_3)E_2(S_1, S_3) + B_2(S_1, S_3)E_1(S_1)E_1(S_3)]$$
$$+ \quad A_2(S_3, S_1 + S_2)[1 + B_1(S_3)E_1(S_3)]$$
$$\cdot \quad [B_1(S_1 + S_2)E_2(S_1, S_2) + B_2(S_1, S_2)E_1(S_1)E_1(S_2)]\}$$
$$- \quad \frac{2}{3}A_1(S_1 + S_2 + S_3)[B_2(S_1, S_2 + S_3)E_1(S_1)E_2(S_2, S_3)$$
$$+ \quad B_2(S_2, S_1 + S_3)E_1(S_2)E_2(S_1, S_3)$$
$$+ \quad B_2(S_3, S_1 + S_2)E_1(S_3)E_2(S_1, S_2)\Big\}$$
$$\Big/ \quad [E_1(S_1)E_1(S_2)E_1(S_3)A_1(S_1 + S_2 + S_3)] \tag{18}$$

The solution of these equations requires the experimental determination of A (the granule cell layer transformations) and E (the ipsilateral hippocampal transformations), which are possible.

III. Granule Cell Layer

The final step in this example is the situation where the granule cell layer (A) is characterized with all the feedback components removed. In Fig. 9, A is represented as the cascade of two system elements G and H. For the example being investigated above, the determination of A is the stopping point experimentally; however, for theoretical completeness, the results for this simple cascade will also be presented.

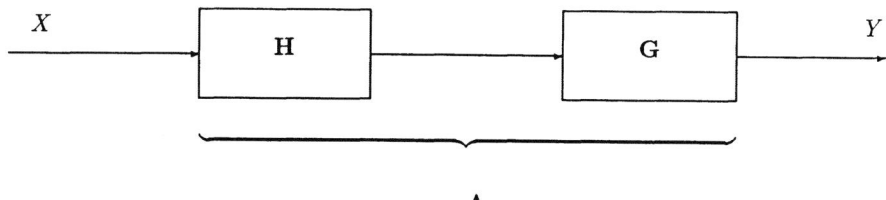

A

Fig. 9. Open loop hippocampal formation

Experimental data obtained from an $in-vitro$ hippocampal slice preparation (Harty et al., 1988), to which the GABA antagonist bicuculline has been applied, has been obtained to investigate this case. This preparation represents the open-loop system, where the contralateral hippocampus has been removed, the trisynaptic pathway is open, and the local inhibitory feedback (GABAergic pathways) has been removed, allowing the dynamics of the granule cell layer to be characterized. The second- and third-order nonlinearities of the $in-vitro$ hippocampal slice are markedly different than those for the intact hippocampal formation. The suppression of spike amplitudes seen in the $in-vivo$ preparation is replaced by a peak facilitation that is 80-90 ms earlier and 4-5 times greater. In addition, the facilitatory peak decays to baseline 100-200 ms earlier. Bicuculine blocks the remaining GABAergic effects of the local feedback neurons, and increases the early facilitation to an even greater degree.

Figure 10 shows the post-lesion, experimental, kernels characterizing the functional properties of the system.

Figure 11 shows the kernels computed from the simulation using the third-order model.

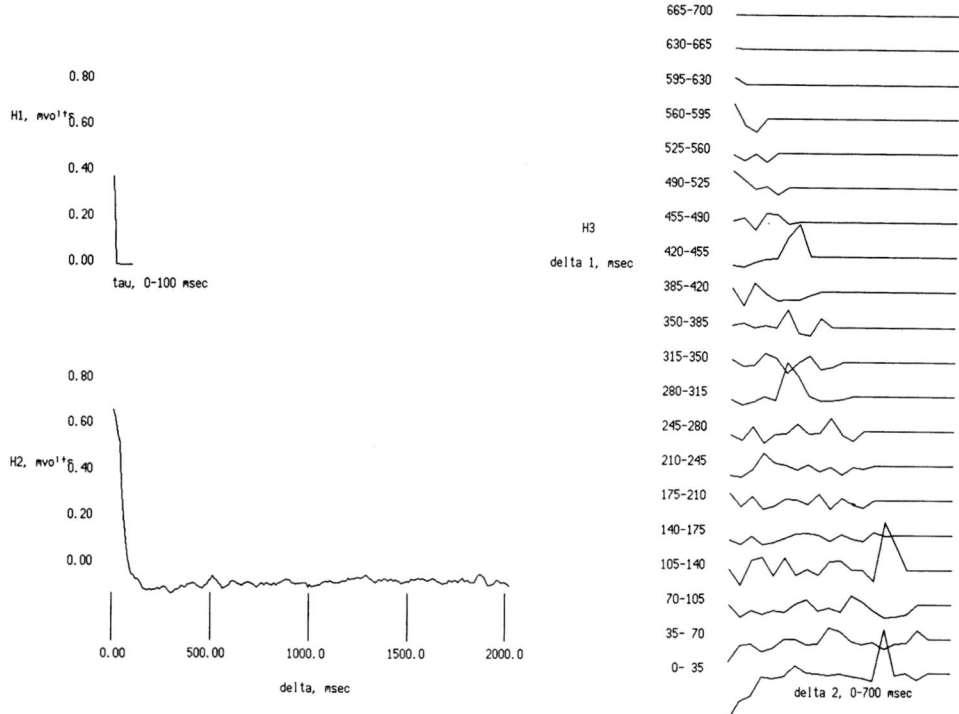

Fig. 10. Open loop experimental first-, second- and third-order kernels

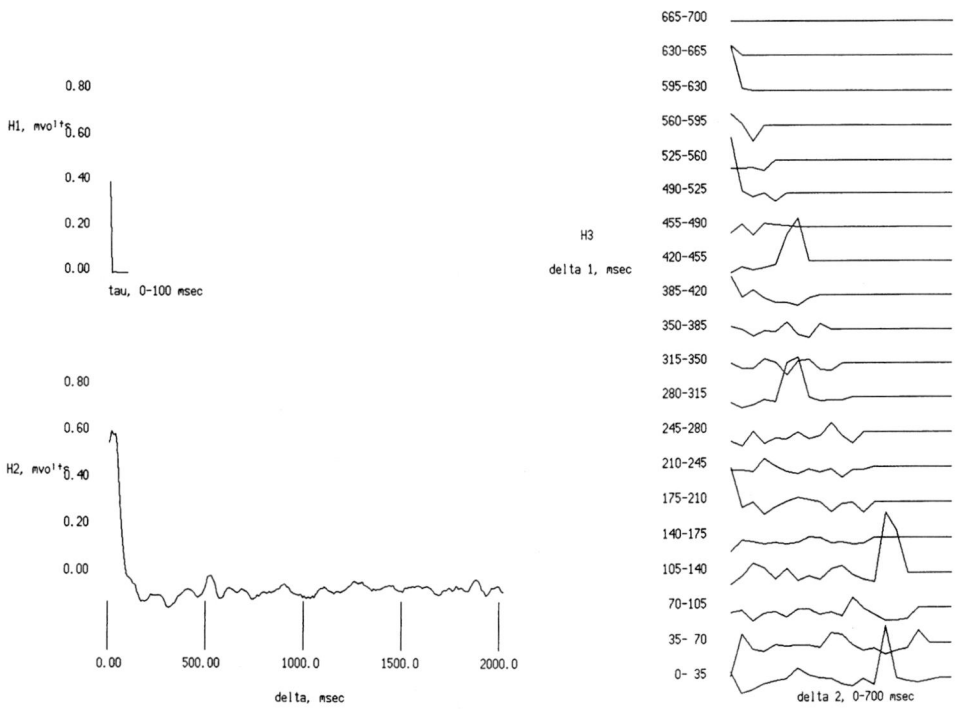

Fig. 11. Open loop simulated first-, second- and third-order kernels

The nth order kernel transformations for this cascade are given by:

$$E_n(S_1,\ldots,S_n) = \sum_{i=0}^{\infty} \sum H_i(\sigma_1,\ldots,\sigma_i)\Pi_{j=1}^{i} G_{m_j}(\ldots,S_i,\ldots) \quad (19)$$

where σ_j = the sum of the m_j S arguments. For the three kernels which we are currently evaluating experimentally, assuming $E_0 = G_0 = 0$, Eq. 19 gives:

$$E_1(S) = H_1(S)G_1(S) \quad (20)$$

$$E_2(S_1,S_2) = H_1(S_1 + S_2)G_2(S_1,G_2) + H_2(S_1,S_2)G_1(S_1)G_2(S_2) \quad (21)$$

and

$$
\begin{aligned}
E_3(S_1,S_2,S_3) &= H_1(S_1 + S_2 + S_3)G_3(S_1,S_2,S_3) \\
&+ \frac{2}{3}[H_2(S_1,S_2 + S_3)G_1(S_1)G_2(S_2,S_3) \\
&+ H_2(S_2,S_1 + S_3)G_1(S_2)G_2(S_1,S_3) \\
&+ H_2(S_3,S_1 + S_2)G_1(S_3)G_2(S_1,S_2)] \\
&+ H_3(S_1,S_2,S_3)G_1(S_1)G_1(S_2)G_1(S_3) \quad (22)
\end{aligned}
$$

Either G or H may be estimated utilizing the kernel transforms of the other subsystem and that of their cascade. Thus, by rearranging equations (23), (24), and (25), we may compute G_n given E_n (i.e., A_n) and H_n.

$$G_1(S) = \frac{E_1(S)}{H_1(S)} \quad (23)$$

$$G_2(S_1,S_2) = \frac{E_2(S_1,S_2) - H_2(S_1,S_2)G_1(S_1)G_1(S_2)}{H_1(S_1 + S_2)} \quad (24)$$

$$
\begin{aligned}
G_3(S_1,S_2,S_3) &= \left\{ E_3(S_1,S_2,S_3) - \frac{2}{3}[H_2(S_1,S_2 + S_3)\cdot G_1(S_1)G_2(S_2,S_3) \right. \\
&+ H_2(S_2,S_1 + S_3)G_1(S_2)G_2(S_1,S_3) \\
&+ H_2(S_3,S_1 + S_2)G_1(S_3)G_2(S_1,S_2)] \\
&+ \left. H_3(S_1,S_2,S_3)\cdot G_1(S_1)G_1(S_2)G_1(S_3) \right\} \Big/ H_1(S_1 + S_2 + S_3) \quad (25)
\end{aligned}
$$

DISCUSSION

We have been investigating the functional properties of the hippocampal formation using a systems theoretic approach. The transformational properties of the elemental components of the network are characterized by n^{th} order kernels of an orthogonalized functional power series. These kernels are interpreted as n^{th} order impulse responses for that network element, providing a nonparametric model of the underlying neural network.

These neural network models have several distinct advantages. Firstly, they are highly constrained by neurobiological data. Each node in the network is represented by a set of transformational properties obtained through experimental observation. As we decrease the number of feedback loops for a given node, the resulting characterization approaches isomorphism with the functional properties of the individual cell population. Secondly, the model can be expanded and modified as new findings emerge. This flexibility of the model originates from two sources: the kernels derived from closed-loop investigations of a given node contain the contributions of all constituents of the node, both known and unknown; and as more experimental detail becomes available, the modeled network may be reconfigured. Finally, the computational simulations and predictions are useful for formal hypothesis testing.

The problem being addressed in this chapter is the characterization of unobservable network elements. In particular, the question we asked was how to partition or decompose a neuronal network into the appropriate elemental units so that necessary and sufficient observable sets of input/output data may be experimentally measured to allow functional characterization of not only the observable but also the unobservable elements. The experimental characterizations provided by the kernels $\{h(.)\}$, obtained from "closed", "partially open", and "open" loop experiments, enable the calculation of kernels for "hidden" nodes; i.e., those nodes whose properties cannot be isolated and directly measured. We have formulated this problem as a decomposition problem.

In the engineering literature, the decomposition problem is focused on breaking a large system of equations into subsystems by searching for topological separations existing within the system of equations (Pichai et al., 1983). The subsystems are then solved separately and recombined to produce a global solution. The actual tearing of the system may be performed in any number of ways in order to achieve conceptual and computational simplifications.

In our problem, however, we have a hypothesized topology for the network, but do not know the system of equations. Thus, we would like to identify a system of equations describing the functioning hippocampal formation. This problem differs from the classic problem in two ways: first, we wish to determine a global system of equations rather than starting with these equations; and second, to accomplish this goal, we need to experimentally decompose the network, utilizing the known topology, in order to identify the functional characteristics of the elemental units. Thus, from our perspective, system decomposition requires either the input/output characterization of elemental units in their open-loop condition, or for elemental units that cannot be isolated directly, the calculation of input/output properties from combinations of characterizations of closed-loop, partial open-loop, and open-loop conditions. The use of the concept of decomposition is an attempt to develop a systematic strategy for breaking a large system apart into subsystems which may be separately experimentally characterized and then recombined to produce a global model of the system.

We conceive of the hippocampal formation as being composed of a number of interconnected elemental units, which are experimentally definable at multiple levels. However, not all of these elemental units are directly accessible to experimental manipulation; e.g. there are groups of neuronal elements whose input/output properties cannot be measured directly, but only inferred from their effect on other observable groups of neurons. Examples of these elements are basket cells providing an inhibitory loop around the granule cell layer, or the entire contralateral hippocampus modulating the activity of the ipsilateral hippocampus.

This raises the fundamental issue in the problem of network decomposition; i.e., what is the strategy for partitioning a network into the appropriate elemental units so that the necessary and sufficient observable sets of input/output data may be experimentally measured to allow characterization of all the directly nonobservable elements in the network? The corollary of this problem is: what is the strategy for recombining these elemental units once they are either measured experimentally or computed theoretically?

In order to develop a detailed understanding of these issues, we have examined a significant problem, at several hierarchical levels, with increasing degrees of complexity. The approach uses multi-dimensional Laplace transforms of the kernels rather than the time-domain representation of the kernels. This allows the equations to be algebraically manipulated, rather than requiring the solutions to integral equations (Marmarelis and Marmarelis, 1978). In the three cases examined here, we have assumed that the kernels for the intact network E_n are measurable, and that at least one of the elemental units was directly measurable.

We have implemented simulations to study the global network properties of the hippocampus using parallel processing technology, based on the system model and the experimentally derived constraints which the kernels represent (Sclabassi et al., 1988c). The simulations use a parallel structured digital computer, the transputer (INMOS, Inc.), with transputer element including a 10 MIPS processor with 4 KBytes of on-chip high-speed cache, private random access memory, and four 2-way 10 Mbit/sec links for establishing communication with other transputer elements.

The same three cases were simulated for which the transfer equations were investigated: the closed-loop system, with the dentate granule cell layer being the feedforward element, providing a summary of the entire system; the partially closed-loop system, where the contralateral hippocampus has been removed; and the open-loop system, using the hippocampal slice preparation, where the contralateral hippocampus was not present, the tri-synaptic pathway was open, and the GABAergic feedback has been removed using bicuculine.

The exciting point is that not only have we predicted the system response for alternative inputs to the systems, but also that we are now able to solve this system for the characterization properties of two components of the system which are not directly observable. These two components include the nth order kernels characterizing the contralateral hippocampus and the nth order kernels characterizing the inhibitory feedback properties at the granule cell layer.

ACKNOWLEDGMENTS

This research was supported by grants from the Whitaker Foundation, The National Institute of Mental Health (MH00343), the Office of Naval Research (N00014-87-K-0472), the Air Force Office of Scientific Research (AFOSR-89-0197), Apollo Computer Corporation, and Inmos Inc. The authors would like to acknowledge the assistance of H. B. Blue in the preparation of this manuscript.

REFERENCES

Andersen, P., Bliss, T.V.P., and Skrede, K.K., 1971, Unit analysis of hippocampal population spikes, Exp. Brain Res., 13:208-211.

Andersen, P., Eccles, J.C., and Loyning, Y., 1964, Location of post-synaptic inhibitory synapses on hippocampal pyramids, J. Neurophysiol., 27:592-607.

Alger, B.E., and Nicoll, R.A., 1982, Feed-forward dendritic inhibition in rat hippocampal pyramidal cells studies in vitro, J. Physiol., 328:105-123.

Bartesaghi, R., and Gessi, T., 1986, Hippocampal output to the subicular cortex: An electrophysiological study, Exp. Neurol., 92:114-133.

Beckstead, R.M., 1978, Afferent connections of the entorhinal area in the rat as demonstrated by retrograde cell- labeling with horseradish peroxidase, Brain Res., 152:249-264.

Berger, T.W., Robinson, G.B., Port, R.L., and Sclabassi, R.J., 1987, Nonlinear systems analysis of the functional properties of hippocampal formation, in: "Advanced Methods of Physiological System Modeling," V.Z. Marmarelis, ed., Los Angeles.

Berger, T.W., and Sclabassi, R.J., 1985, Nonlinear systems analysis and its application to study of the functional properties of the neural systems, in: "Memory Systems of the Brain: Animal and Human Cognitive Process," N.M. Weinberger, J.L. McGaugh, and G. Lynch, eds., Guilford, New York.

Berger, T.W., Semple-Rowland, S. and Bassett, J.L., 1981, Hippocampal polymorph neurons are the cells of origin for ipsilateral association and commissural afferents to the dentate gyrus, Brain Res., 215:3129-336.

Berger, T.W., Swanson, G.W., Milner, T.A., Lynch, G.S., and Thompson, R.F., 1980, Reciprocal anatomical connections between hippocampus and subiculum: evidence for subicular innervation of region superior, Brain Res., 183:265-276.

Berger, T.W., Eriksson, J.L., Ciarolla, D.A., and Sclabassi, R.J., 1988a, Nonlinear system analysis of the hippocampal perforant path-dentate projection: II. Effects of random pulse train stimulation, J. of Neurophysiol., 60:1077-1094.

Berger, T.W., Eriksson, J.L., Ciarolla, D.A., and Sclabassi R.J.,1988b, Nonlinear systems analysis of the hippocampal perforant path-dentate projection: III. Comparison of random train and paired impulse stimulation, J. of Neurophysiol., 60:1095-1109.

Brilliant, M.B., 1958, Theory of the Analysis of Nonlinear Systems, Technical Report #345, Research Laboratory of Electronics, Massachusettes Institute of Technology.

Buzsaki, G., and Eidelberg, E., 1981, Commissural projection to the dentate gyrus of the rat: Evidence of feed-forward inhibition, Brain Res., 230:346-350.

Casti, J.L., 1985," Nonlinear System Theory," Academic Press, New York.

Donnegan, N., Gluck, M., and Thompson, R. F., 1987, Model of changes in changes in cerebellar function during classical conditioning of rabbit eyelid reflex, Presented at the Conference on the Neurobiology of Learning and Memory, Park City, Utah.

Douglas, R.M., McNaughton, B.L., and Goddard, G.V., 1983, Commissural inhibition and facilitation of granule cell discharge in fascia dentata, J. Comp. Neurol., 219:285-294.

Eykhoff, P., 1974, "System Identification. Parameter and State Estimation," Wiley, New York.

George, D.A., 1959, Continuous nonlinear systems. Technical Report #335, Research Laboratory of Electronics, Massachusettes Institute of Technology.

Harty, T.P., Berger, T.W., Sclabassi, R.J., and Barrionuevo, G., 1987, Nonlinear response characteristics of the perforant path-dentate gyrus in the in vitro rabbit hippocampus, Soc. Neurosci. Abstr., 13:1330.

Hjorth-Simonsen, A. and Jeune, B., 1972, Origin and termination of the hippocampal perforant path in the rat studied by silver impregnation, J. Comp. Neurol., 144:215-232.

Knowles, W.D. and Schwartzkroin, P.A., 1982, A local circuit synaptic interactions in hippocampal brain slices, J. Neurosci., 1:318-322.

Lee, Y.W. and Schetzen, M., 1965, Measurement of the kernels of a non-linear system by cross-correlation, Internat. J. Control., 2:237-254.

Lorente de No, R., 1934, Studies on the structure of cerebral cortex. II. Continuation of the study of the ammonic system, J. Psychol. Neurol., 46:113-177.

Marmarelis, P.Z. and Marmarelis, V.Z., 1978, "Analysis of Physiological Systems: The White Noise Approach," Plenum, New York.

Miles, R., and Wong, R.K.S., 1986, Excitatory synaptic interactions between CA3 neurones in the guinea-pig hippocampus, J. Physiol. Lond., 373:397-418.

O'Keefe, J., and Nadel, L., 1978, " The Hippocampus as a Cognitive Map," Ciarendon Press, Oxford.

Port, R.L., Sclabassi, R.J., and Berger, T.W., 1987, Modulation of perforant path-dentate functional properties by commissural system: Acute effects of contralateral hippocampal ablation, Soc. Neurosci. Abstr., 13:1330.

Pichai, V., Sezer, M.E., and Siljak, D.D., 1983, A graph-theoretic algorithm for hierarchical decomposition of dynamic systems with applications to estimations and control, IEEE Transaction on Systems, Man, and Cybernetics SMC, 13:3 197-207.

Ramon y Cajal, S., 1955, " Histologie du Systeme Nerveux de l'Homme et des Vertebres," Instituto Ramon y Cajal, Madrid.

Ribak, C.E., Vaughn, J.E., and Saito, K., 1978, Immunocytochemical localization of glutamic acid decarboxylase in neuronal somata following colchicine inhibition of axonal transport, Brain Res., 140:315-332.

Rugh, W.J., 1981, "Nonlinear Systems Theory: The Volterra/Wiener Approach," John Hopkins University Press, Baltimore.

Sclabassi, R.J., Hinman, C.L., Kroin, J.S., and Risch, H., 1977a, The modulatory effect of prior input upon afferent signals in the somatosensory system, Proceedings of the 1977 Joint Automatic Control Conference, 2:787-795.

Sclabassi, R.J., Krieger, and Berger, T.W., 1988a, A systems theoretic approach to the study of CNS function, Annals of Biomed. Eng., 16:1 17-34.

Sclabassi, R.J., Eriksson, J.L., Port, R.L., Robinson, G.B., and Berger, T.W., 1988b, Nonlinear systems analysis of the hippocampal perforant path-dentate projection: I. Theoretical and interpretational considerations, J. of Neurophysiol., 60:3 1066-1076.

Sclabassi, R.J., Krieger, D., Solomon, J., Barrionuevo, G., and Berger, T.W., 1988c, An external network model of the hippocampal formation, First Annual Meeting, Internat. Neural Network Society, Boston, MA.

Sejnowski, T., Koch, C., and Churchland, P.S., 1988, Computational neuroscience, Science, 241:1299-1306.

Sorensen, K.E. and Shipley, M.T., 1979, Projections from the subiculum to the deep layers of the ipsilateral presubicular and entorhinal cortices in the guinea pig, J. Comp. Neurol., 188:313-334.

Squire, L.R., 1982, The neuropsychology of human memory, Ann. Rev. Neurosci., 5:241-273.

Swanson, L.W. and Cowan, W.M., 1977, An autoradiographic study of the organization of the efferent connections of the hippocampal formation in the rat, J. Comp. Neurol., 172:49-84.

Swanson, L.W., Wyss, J.M. and Cowan, W.M., 1978 An autoradiographic study of the organization of intrahippocampal association pathways in the rat, J. Comp. Neurol., 181:681-716.

Traub, R.D., Miles, R., and Wong, R.K.S., 1987, Models of synchronized hippocampal bursts in the presence of inhibition. I. Single population events, J. Neurophysiol., 58:739.

Van Groen, T. and Lopes da Silva, F.H., 1986, Organization of the reciprocal connections between the subiculum and the entorhinal cortex in the cat. II. An electrophysiological study, J. Comp. Neurol., 251:111-120.

THE GEOMETRY OF SYSTEM IDENTIFICATION:

FRACTAL DIMENSION AND INTEGRATION FORMULAE

Jonathan D. Victor

Department of Neurology, Cornell University Medical College
1300 York Avenue
New York City, New York 10021 USA

ABSTRACT

We consider a geometric interpretation of the orthogonal functional approach to the analysis of nonlinear systems. A deterministic nonlinear transducer is viewed as a functional on a phase space of input histories. To calculate the represention of the nonlinear transducer as an orthogonal expansion of standard functionals, one must perform certain integrations over the entire phase space. These integrations amount to cross-correlations of the response of the transducer under study with the responses of standard functionals. In the laboratory, this idealized pro- cedure cannot be carried out exactly. Rather, one chooses a particular input signal, which corresponds to choosing a path through the phase space of all possible inputs. Thus, laboratory implementations of the orthogonal functional approach correspond to approximations of particular integrals over the entire phase space by integrals over the path actually sampled by the test signal.

Two consequences of this viewpoint are discussed. (i) Ideas drawn from the technique of Gaussian quadrature may be used to improve the accuracy of experimental determinations of Wiener kernels. (ii) The "richness" of a test signal, as quantified by the fractal dimension of the path it traces in phase space, limits the number of terms in an orthogonal expansion that may legitimately be extracted (Victor 1987).

INTRODUCTION

One approach to the study of a nonlinear transduction is to represent this transduction as an orthogonal series of standard transductions. In this framework, the first few members of the orthogonal series are measured experimentally. These terms are used as a means to describe the transducer, or, inspected for clues to its internal structure. This approach, with origins in the functional Taylor series of Volterra (1932) and Wiener's theory of random processes (1958), is discussed in depth by Marmarelis and Marmarelis (1978).

There are several attractive aspects of this approach. (i) The first-order term in the orthogonal expansion is a natural generalization of the impulse-response of a linear transducer. (ii) Wiener (1958) showed that, in principle, a deterministic nonlinear transducer is completely described by its expansion in terms of functionals orthogonal with respect to a Gaussian white noise. (iii) An orthogonal series, when truncated at a finite order r, provides the transduction of order r which most closely approximates the transducer under study. (iv) For some simple model nonlinear transducers, the terms in the expansion have simple analytic forms (Korenberg 1973). (v) There are simple relationships between orthogonal series based on Gaussian test signals of different means and powers (Marmarelis and Marmarelis 1978). (vi) Orthogonal functional expansions based on a Gaussian input satisfy a maximum-entropy property (Victor and Johannesma 1986).

The main difficulty with this approach is that it is a theoretical analytic procedure, and several kinds of approximations must be made when it is reduced to practice. The most straightforward path to reducing the Wiener approach to laboratory practice requires choice of a (realizable) input signal which is meant to approximate an infinite ensemble of Gaussian white noise. Real laboratory apparatus has only a finite bandwidth and a finite amplitude range, and a real experiment has only finite duration. Thus, laboratory procedures must introduce departures from an idealized Wiener procedure.

Here, we develop a geometrical view of the Wiener procedure. This view helps to understand some of the errors which must be introduced in any implementation of the Wiener procedure. This viewpoint also suggests particular choices of test signals, along with computational procedures, which minimize some of these errors.

THE CASE OF THE STATIC NONLINEARITY

To begin, we recognize that in any laboratory implementation of the Wiener procedure, functions of time (such as test signals and responses) are represented by a discrete time-series. Once time is discretized, then many of conceptual difficulties with the Wiener procedure evaporate. In continuous time, a transducer is regarded as a functional which operates on a test signal. With time discretized, a transducer is regarded as an ordinary function of several variables; each variable represents the value of the test signal at a discrete number of time-steps in the past.

While the distinction between discrete time and continuous time does raise several technical issues, these are not relevant to the ideas we will discuss here. Indeed, the distinction between orthogonal (Wiener) series and Taylor (Volterra) series, as well as the relationship between kernel measurement procedures and integration formulae, will be seen most plainly for functions of a single variable.

We now consider the single-variable case in detail. This case corresponds to an "instantaneous", or "static" transduction, in which the value y of the output at a given time depends only on the value x of the input at the corresponding time. That is, the transducer under study is represented by an ordinary function

$$y = f(x) \tag{1}$$

In this simple context, the Volterra functional series corresponds to a Taylor series for the function f:

$$f(x) = \sum_{i=0}^{\infty} a_i x^i \tag{2}$$

To determine the coefficients a_i in the Taylor series, it is necessary to measure the transducer's response to infinitesimal inputs. Furthermore, the representation (2) only is valid for analytic functions f, and converges very slowly to functions which are not well-approximated by polynomials.

These considerations prompt re-arrangement of the Taylor series (2) into an orthogonal series

$$f(x) = \sum_{i=0}^{\infty} c_i h_i(x) \tag{3}$$

In this expansion, the $h_i(x)$ are the Hermite polynomials, which are orthogonal polynomials with respect to the Gaussian weight

$$w(x) = \frac{1}{\sqrt{2\pi}} e^{-x^2/2} \tag{4}$$

Each coefficient c_i in the orthogonal series (3) may be determined by integration of the transducer's response $f(x)$ against the corresponding Hermite polynomial:

$$c_i = \frac{1}{N_i} \int f(x) \, h_i(x) w(x) \, dx \qquad (5)$$

The normalization coefficients N_i are given by

$$N_i = \int [h_i(x)]^2 w(x) \, dx \qquad (6)$$

(Our convention here for the Hermite polynomials is that the leading term of $h_i(x)$ is x^i; in this convention, $N_i = i!$).

The integral (5) which determines the coefficients c_i may be interpreted as the expected value of the product of the response of the transducer of interest, f, and a standard transducer, h_i, for inputs drawn from a Gaussian ensemble. The customary laboratory approach to the evaluation of Wiener kernels is to estimate the expected value of this product measured over a sample of an input signal designed to approximate Gaussian white noise (Lee and Schetzen 1965).

However, there is no reason in principle that a stochastic signal must be used; the goal is experimental determination of the integral (5). The form of the integral (5) suggests that Gaussian quadrature (Johnson and Riess 1982) may be used to advantage. Gaussian quadrature is a numerical integration procedure which consists of selection of sample points θ_j and weights b_j which provide highly accurate estimates of integrals with respect to particular weight functions $w(x)$. For our application, the Gaussian quadrature formula is

$$\int g(x) \, w(x) \, dx = \sum_{j=1}^{m} b_j g(\theta_j) + R_m(g) \qquad (7)$$

where the remainder term $R_m(g)$ is given by

$$R_m(g) = \sqrt{\pi} \, \frac{m!}{(2m)!} \, \frac{d^{2m}}{dx^{2m}} g(u) \qquad (8)$$

for some value u (Abramowitz and Stegun 1964, formula 25.4.46). The remainder term (8) is zero for polynomial functions of degree $2m-1$ or less. Thus, the approximation of the integral of (7) by the weighted sum over the m sample points θ_j specified by (7) is exact for such polynomials.

This accuracy is achieved because there are 2m degrees of freedom in choosing the sample points θ_j and the weights b_j. To achieve this accuracy, the sample points θ_j are taken to be the zeroes of $h_m(x)$, and the weights b_j are given by

$$b_j = \frac{m!}{[\frac{d}{dx} h_m(\theta_j)]^2} \qquad (9)$$

The integral (7) is precisely the expression for the zeroth-order term c_0 in the orthogonal expansion (3). Integral expressions (7) for higher-order terms c_i are also of the form (7), with $g(x) = f(x)h_i(x)$. Thus, by applying Gaussian quadrature (7) to the general formula (5) for the coefficient c_i, we obtain

$$c_i = \frac{1}{i!} \left(\sum_{j=1}^{m} b_j f(\theta_j) h_i(\theta_j) + R_m(fh_i) \right) \tag{10}$$

The properties of the remainder term (8) imply that the formula (10) is exact for polynomial functions f of degree $2m-i-1$ or less.

Equation (10) thus provides a procedure to estimate the coefficients of the orthogonal expansion of an unknown function f. It does <u>not</u> call for observing the value of $f(x)$ for all values of x, with values of x weighted by $w(x)$. Rather, it calls for observing a finite set of values $f(\theta_j)$, and weighting these observations by the weights b_j specified by (9). Numerical values of the sample points θ_j and the weights b_j are given in Table 1. The sample points are almost, but not exactly, equally-spaced, and the weights are largest for sample points of low magnitude.

EXPANSIONS OF FUNCTIONS OF SEVERAL VARIABLES

We now extend the above analysis to expansions of functions of several variables. That is, we consider functions

$$y = f(\mathbf{x}) \tag{11}$$

where $\mathbf{x}$ denotes an L-tuple of variables $(x_1, x_2, \ldots, x_L)$. Considered as a description of a transducer's response to an input signal at the time t, the L-tuple of values $\mathbf{x}$ represents values of the input at L times t, $t-\Delta T$, ..., $t-(L-1)\Delta T$.

The several-variable analog of the orthogonal series (3) is

$$f(\mathbf{x}) = \sum_{i=0}^{\infty} c_i h_i(\mathbf{x}) \tag{12}$$

In this expansion, we have replaced quantities indexed by the scalar i by quantities indexed by the vector index $\mathbf{i} = (i_1, i_2, \ldots, i_L)$. The Hermite polynomials are replaced by their several-variable analogs

$$h_i(\mathbf{x}) = h_{i_1}(x_1) \cdot h_{i_2}(x_2) \cdot \ldots \cdot h_{i_L}(x_L) \tag{13}$$

In essence, (12) is the Wiener expansion of the functional f, regarded (via time-discretization) as a function of values of its inputs at multiple discrete times. The "Wiener kernel" of order n is essentially the collection of coefficients c_i with $i_1+i_2+\ldots+i_L = n$.

Table 1. Sample points and weights for m-point Gaussian quadrature
(m = 2 through 7) with respect to a Gaussian of unit variance

	j	b_j	θ_j
m = 2	$h_2(x) = x^2 - 1$		
	1	-1.000000	0.5000000
	2	1.000000	0.5000000
m = 3	$h_3(x) = x^3 - 3x$		
	1	-1.732051	0.1666666
	2	0.000000	0.6666667
	3	1.732051	0.1666666
m = 4	$h_4(x) = x^4 - 6x^2 + 3$		
	1	-2.334414	0.0458758
	2	-0.741964	0.4541241
	3	0.741964	0.4541241
	4	2.334414	0.0458758
m = 5	$h_5(x) = x^5 - 10x^3 + 15x$		
	1	-2.856970	0.0112574
	2	-1.355626	0.2220759
	3	0.000000	0.5333334
	4	1.355626	0.2220759
	5	2.856970	0.0112574
m = 6	$h_6(x) = x^6 - 15x^4 + 45x^2 - 15$		
	1	-3.324258	0.0025558
	2	-1.889176	0.0886158
	3	-0.616707	0.4088284
	4	0.616707	0.4088284
	5	1.889176	0.0886158
	6	3.324258	0.0025558
m = 7	$h_7(x) = x^7 - 21x^5 + 105x^3 - 105x$		
	1	-3.750440	0.0005483
	2	-2.366760	0.0307571
	3	-1.154405	0.2401231
	4	0.000000	0.4571429
	5	1.154405	0.2401231
	6	2.366760	0.0307571
	7	3.750440	0.0005483

Sample points (θ_j) and weights (b_j) for one-dimensional Gaussian quadrature
(eq. 7). For an m-point approximation, the sample points θ_j are the
zeroes of $h_m(x)$.

In a manner analogous to (5), the coefficients c_i in the orthogonal series (12) are again determined by integration of the transducer's response $f(x)$ against the corresponding multivariate Hermite polynomial:

$$c_i = \frac{1}{N_i} \int f(x) \, h_i(x) w(x) \, dx \tag{14}$$

where

$$N_i = N_{i_1} \cdot N_{i_2} \cdot \ldots \cdot N_{i_L} \tag{15}$$

and

$$w(x) = w(x_1) \cdot w(x_2) \cdot \ldots \cdot w(x_L) \tag{16}$$

A multivariate extension of the Gaussian quadrature formula (7) provides an efficient way to calculate the integrals required by (14). The multivariate analog of (7) is

$$\int g(x) \, w(x) \, dx = \sum_j b_j g(\theta_j) + R_m(g) \tag{17}$$

where the sum is over all vector indices $j = (j_1, j_2, \ldots, j_L)$ with integer entries in the range $[1,m]$. The sample points θ_j are L-tuples of one-dimensional sample points

$$\theta_j = (\theta_{j_1}, \theta_{j_2}, \ldots, \theta_{j_L}) \tag{18}$$

where each θ_{j_k} is a zero of $h_m(x)$. The weights b_j are products of the one-dimensional weights given by (9):

$$b_j = \left(\frac{m!}{[\frac{d}{dx} h_m(\theta_{j_1})]^2} \right) \cdot \left(\frac{m!}{[\frac{d}{dx} h_m(\theta_{j_2})]^2} \right) \cdot \ldots \cdot \left(\frac{m!}{[\frac{d}{dx} h_m(\theta_{j_L})]^2} \right) \tag{19}$$

The remainder term $R_m(g)$ is given by

$$R_m(g) = \left(\sqrt{\pi} \, \frac{m!}{(2m)!} \right)^L \frac{\partial^{2m}}{\partial x_1^{2m}} \cdots \frac{\partial^{2m}}{\partial x_L^{2m}} g(u) \tag{20}$$

for some value u. Formulae (17) – (20) are derived from their one-dimensional analogs (7) – (9) by considering the multidimensional integral in (17) to be an iterated integral.

We now use multidimensional Gaussian quadrature to estimate the coefficients of the multivariate orthogonal expansion (12). We apply (17) with $g(x) = f(x)h_i(x)$ to the formula (14) for the coefficients c_i:

$$c_i = \frac{1}{N_i} \left(\sum_j b_j f(\theta_j) \, h_i(\theta_j) + R_m(fh_i) \right) \tag{21}$$

The properties of the remainder term (20) imply that the formula (21) is exact for functions f whose dependence on <u>any</u> variable x_k is a polynomial of degree $2m-1-i_k$ or less.

Comparison of the one-variable and the several-variable case

The equation (21) estimates the terms in the orthogonal expansion (12) of a function f from values of the function on a discrete lattice of test inputs. As in the single-variable (L = 1) case, the estimation procedure prescribes a specific set of test inputs θ_j and particular weights b_j, which are not drawn from a Gaussian test ensemble. Rather, the m^L test inputs and m^L weights are chosen to make estimates of certain correlations related to (14) coincide with their exact values over an infinite ensemble of Gaussian inputs.

For both the single-variable and the multidimensional case, the estimation procedure is parametric in the integer m. As m increases, coefficients in the orthogonal series are determined more accurately. However, as m increases, more test inputs must be used.

An important difference between the multi-dimensional case and the single-variable case arises when one turns to laboratory application of the technique. In the single-variable (static nonlinearity) case, it is necessary to observe the output of the transducer function f to each of m input values θ_j. These input values may be applied in any order, with no effect on efficiency. In the multi-dimensional case, there are m^L input L-tuples $(\theta_{j_1}, \ldots, \theta_{j_L})$ to test, and dramatic improvements in efficiency may be realized if they are presented in an appropriate order. This is because each of these L-tuples corresponds to a sequence of input values at L consecutive time-intervals. With an appropriate ordering, such as that of a memory wheel (Stein 1969), their overlap may be exploited.

DISCUSSION

A geometrical viewpoint

It is helpful to view the above analysis in a geometrical manner. A transducer may be thought of as a function on the phase space of its inputs. In the single-variable (static) case, the phase space is the space of possible instantaneous input values: the real line. With time discretized, the phase space of a dynamic transducer is the set of all input histories. That is, the phase space is an L-dimensional space, where the L is the number of time-lags under consideration. In seeking an orthogonal decomposition of the transducer, we seek an orthogonal expansion of the function which defines the transducer on this phase space.

The formula (14) for a coefficient in this orthogonal series is an inner product of a standard orthogonal function and the function under study. This inner product is a crosscorrelation (Lee and Schetzen 1965), whose exact value is given by an integral over the entire phase space. In an experimental situation, only the points in phase space actually sampled by the test signal provide data for the calculation of the integral.

Fractal dimension

This viewpoint provides an immediate connection with the concept of fractal dimension (Mandelbrot 1977). The test signal is a trajectory which is parametrized by time, and thus is (in some sense) one-dimensional, although its path through phase space may be quite irregular. To determine the coefficients in an orthogonal series representation, we need to estimate integrals over the L-dimensional phase space by integrals over this one-parameter trajectory. Intuitively, such an estimation only makes sense if the test signal comes close to a set of higher dimension in the phase space.

The dimension of the space which is "filled" by the test signal is essentially its capacity dimension (Farmer et al. 1983), one of the prototypical fractal dimensions. The capacity dimension is calculated by determining how many spheres $N(\varepsilon)$ of a given radius ε are needed to completely cover the set of interest. For a line segment, the number of covering spheres $N(\varepsilon)$ is proportional to ε^{-1}, and for an ordinary surface, the number of covering spheres $N(\varepsilon)$ is proportional to ε^{-2}. Capacity dimension generalizes this relation between the dimension of a set and the number of covering spheres. For a set whose capacity dimension is d, the number of covering spheres $N(\varepsilon)$ is proportional to ε^{-d}.

Notions of fractal dimension derive from scaling behavior of sets, as some parameter of interest becomes small. For this reason, fractal dimensions of empirical objects cannot be determined exactly, but should be viewed merely as a descriptor of approximate scaling behavior of over a range of the small parameter (Mayer-Kress 1987). Precisely the same considerations apply to determination of the dimension of a test signal. With these understandings, it may be shown that the capacity dimension of the test signal is an upper bound for the order of the orthogonal kernels that may be unambiguously determined (Victor 1987).

Integration formulae

From a more practical point of view, we see that the choice of a test signal is tantamount to a choice of an algorithm for the numerical integration of the integrals (14). Because of the special form of these integrals, Gaussian quadrature is a natural choice for an integration formula. Above we have shown how Gaussian quadrature may be naturally extended to the multivariate setting. In this extension, its most attractive feature is retained: kernels of polynomial systems of high order are determined exactly.

Note that no attempt is made to match the cross-sectional distribution of the inputs with those of the white noise upon which the orthogonal expansion is based. Indeed, the variance of the test signal for the m-point approximation is m-1, while the variance of the white noise for the orthogonal expansion is 1. In the m-point approximation, the input signal consists of m discrete levels θ_j. These test inputs are approximately equally-spaced in magnitude, and are presented with equal frequency. Responses to these test inputs are then weighted by coefficients b_j, so that as many of the integrals (14) as possible are computed exactly.

Comparison with other implementations of the Wiener procedure

To compare the Gaussian quadrature procedure with other implementations of the Wiener procedure, we examined their relative efficiencies for estimation of terms in the orthogonal expansion (3) for several test functions. We present only the results for the first term c_0 (Table 2).

Four methods were compared. The first method (referred to as "Gaussian quadrature") used the proposed Gaussian quadrature algorithm (10). This formula is a sum over samples at unequally-spaced points which cover the range θ_1 to θ_m, weighted by the quantities b_j (eq. 9).

The second method (referred to as "trapezoid rule") used a set of m equally-spaced inputs θ^{tr}_j, which cover the range θ_1 to θ_m and are weighted by the "intuitive" Gaussian weights $w(x)$.

$$c_i \approx \frac{1}{i!} \sum_{j=1}^{m} w(\theta^{tr}_j) f(\theta^{tr}_j) h_i(\theta^{tr}_j) \qquad (22)$$

Table 2. Errors incurred by several methods for calculation of the coefficient c_0 of an orthogonal series

number of points (m)	Gaussian quadrature	trapezoid rule	equal spacing, unit variance	Gaussian white noise

$$f(x) = x^2; \quad c_0 = 1$$

number of points (m)	Gaussian quadrature	trapezoid rule	equal spacing, unit variance	Gaussian white noise
2	0	-0.03212	0	1.00000
3	0	-0.07492	0	0.81650
4	0	-0.00088	0	0.70711
8	0	0.00002	0	0.50000
16	0	*	0	0.35355
∞	0	0	0	0

$$f(x) = x^4; \quad c_0 = 3$$

number of points (m)	Gaussian quadrature	trapezoid rule	equal spacing, unit variance	Gaussian white noise
2	-2.00000	-2.03212	-2.00000	6.92820
3	0	-0.22475	-1.50000	5.65685
4	0	-0.24600	-1.36000	4.89898
8	0	-0.00149	-1.23810	3.46410
16	0	*	-1.20941	2.44949
∞	0	0	-1.20000	0

$$f(x) = e^x; \quad c_0 = \sqrt{e} \simeq 1.64872$$

number of points (m)	Gaussian quadrature	trapezoid rule	equal spacing, unit variance	Gaussian white noise
2	-0.10564	-0.15520	-0.10564	1.52820
3	-0.01053	-0.05899	-0.08301	1.24777
4	-0.00075	-0.01515	-0.07620	1.08060
8	*	-0.00007	-0.07008	0.76410
16	*	*	-0.06862	0.54030
∞	0	0	-0.06813	0

$$f(x) = |x|; \quad c_0 = \sqrt{2/\pi} \simeq 0.79789$$

number of points (m)	Gaussian quadrature	trapezoid rule	equal spacing, unit variance	Gaussian white noise
2	0.20212	0.17000	0.20212	0.42625
3	-0.22053	-0.26379	0.01861	0.34803
4	0.09019	0.10600	0.09654	0.30141
8	0.04295	0.05381	0.07498	0.21313
16	0.02098	0.02798	0.06984	0.15070
∞	0	0	0.06815	0

Errors quoted for the first three methods are deterministic; the error quoted for the "Gaussian white noise" method is a root-mean-squared average. The symbol * indicates a nonzero error of absolute value less than 0.00001. For details, see text.

The third method (referred to as "equal spacing, unit variance") used a set of m equally-spaced inputs θ^{esuv}_j, each weighted by $1/m$, positioned so that the mean of the test signal is zero and its variance is unity. That is,

$$c_i \simeq \frac{1}{i!} \sum_{j=1}^{m} \frac{1}{m} f(\theta^{esuv}_j) h_i(\theta^{esuv}_j) \tag{23}$$

where θ^{esuv}_j are equally-spaced and span the range from $-[3(n-1)/(n+1)]^{1/2}$ to $[3(n-1)/(n+1)]^{1/2}$.

The fourth method (referred to as "Gaussian white noise") used m samples drawn at random from a true Gaussian distribution $w(x)$. The first three methods are deterministic, and therefore Table 2 displays the exact error incurred by these methods. The fourth method is stochastic; Table 2 therefore displays the root-mean-squared error incurred by this approach. This quantity rms_m is given by

$$[rms_m]^2 = \frac{1}{m} \left(\frac{1}{N_i^2} \int f(x)^2 h_i(x)^2 w(x) \, dx - c_i^2 \right) \tag{24}$$

The reason these methods were selected is as follows: The "Gaussian white noise" method is essentially the stochastic technique most frequently used by laboratory investigators. The "equal spacing, unit variance" method includes the three-level method of Emerson and co-workers (1987), as well as multi-level extensions (e.g., Marmarelis 1975). Finally, comparison of the "trapezoid rule" method with Gaussian quadrature reveals the effect of non-uniform spacing of the sample points θ_j and non-intuitive choices of the weights b_j.

First we examine the simple nonlinear function $f(x) = x^2$. All methods provide the correct result for sufficiently high values of the number of points m. However, the "Gaussian quadrature" and the "equal spacing, unit variance" methods are exact at $m = 2$, while the other two methods only approach the correct value as the number of points increases. For the "trapezoid rule", the convergence is rapid, while for "Gaussian white noise", convergence is proportional to $m^{-1/2}$.

For the higher-order polynomial nonlinearity $f(x) = x^4$, "Gaussian quadrature" provides the exact result for $m \geq 3$, as expected from the formula (8) for the error term. The "trapezoid rule" does not begin to show rapid convergence until $m > 4$, and the "equal spacing, unit variance" method converges slowly to an incorrect result.

Because the error estimate for Gaussian quadrature ((8) or (20)) is related to a high-order derivative of the function to be integrated, one might worry that the advantage of Gaussian quadrature for the integration of polynomials turns into a disadvantage if the function being integrated is not a polynomial. As seen from the remaining two examples of Table 2, $f(x) = e^x$ and $f(x) = |x|$, this potential drawback does not appear to arise for functions likely to be encountered in the laboratory. In all but one case ($f(x) = |x|$, $m = 2$), "Gaussian quadrature" provides a more accurate result than the "trapezoid rule". As with $f(x) = x^4$, the "equal spacing, unit variance" method converges to the wrong value, and "Gaussian white noise" converges to the correct value, but very slowly.

We summarize these numerical results. For sufficiently high values of m, "Gaussian quadrature", "trapezoid rule", and "Gaussian white noise" all converge to the correct result. In nearly all cases, "Gaussian quadrature" provides the most accurate estimate of c_0. Convergence of "Gaussian quadrature" is most rapid, even for the transcendental and the non-analytic example. Convergence of the "Gaussian white noise" algorithm is the slowest, limited by $m^{-1/2}$, according to equation (24), and is particularly poor for accelerating nonlinearities. The "equal spacing, unit variance" method only converges to the correct result for the purely quadratic nonlinearity. This is because only its first two moments agree with that of the Gaussian weight $w(x)$; for more complex nonlinearities, this signal fails to separate the terms of the Wiener expansion.

The origin of this problem is that the "equal spacing, unit variance" method is in reality calculating orthogonal kernels with respect to a non-Gaussian weight (i.e., the discrete distribution actually used). While these certainly constitute legitimate orthogonal kernels in their own right, they do not share most of the useful analytic properties enjoyed by kernels orthogonal with respect to a Gaussian weight (see Introduction). For transductions whose Taylor expansion includes terms such as x^4 or higher even order, these two sets of kernels differ beginning at order 0.

Comparison with M-sequence methods

For $m = 2$, the procedure advanced here is closely allied with that of Sutter (1987). In Sutter's procedure, the input signal is an "M-sequence": a sequence of length $2^L - 1$ which contains all L-tuples except the L-tuple of all zeroes. M-sequences are nearly orthogonal to cyclic shifts of themselves, and this provides the basis for an efficient means for calculation of the requisite cross-correlations.

Any M-sequence may be converted into a memory wheel by insertion of a zero adjacent to the longest run (L-1) of zeroes in the M-sequence. In the limit of a large number of lags, the difference between the memory wheel signal and that of the M-sequence is negligible: the only difference between the stimulus sequences is the L-tuple of all zeroes, which occurs only once in 2^L inputs. Insertion of this extra zero removes a small but systematic error in the estimation of the Wiener kernels, at the expense of spoiling a highly efficient algorithm for calculation of the required cross-correlations. For large L, the small but systematic error becomes negligible and is probably outweighed by the computational advantages of Sutter's technique. But for small L (e.g., $L \leq 6$) the computational saving is small, and the systematic error in kernel estimation resulting from the omitted L-tuple may well dominate.

Emerson et al. (1987) used a three-level pseudorandom sequence to analyze spatial nonlinearities in complex cells of cat visual cortex. Other than the omission of the all-zero L-tuple, this stimulus sequence corresponds to an m=3 memory wheel. However, there is an important difference between the m=2 case and the m=3 case. The weights for Gaussian quadrature are not equal in the m=3-case, while they are equal in the m=2 case (see Table 1). Thus, the "Gaussian quadrature" and the "equal-weight, unit variance" methods are distinct for m=3, while they coincide for m=2. The implications of this difference were discussed above, both in terms of the accuracy of the expansion and the nature of the orthogonal weight.

Other test ensembles

Orthogonal functional expansions may be constructed with respect to any input ensemble, and not just with respect to white noise (Victor and Knight 1979). In this more general setting, Gaussian quadrature may be applied as well. In general, the test ensemble (and the integrals analogous to (14) that it induces) determines a set of sample points θ_j and weights b_j that provide for efficient estimation of the integrals (14). As in the case considered here, a stimulus which passes through the sample points θ_j need not look like a typical signal from the test ensemble, but will be more highly structured.

If, as in the present case, the test ensemble consists of "white" (uncorrelated) signals, then the sample points θ_j and weights b_j will have product forms analogous to (18) and (19), and memory wheels provide an efficient way to sample the m^L L-tuples θ_j. The coordinates of θ_j will always be the zeroes of the mth-order orthogonal polynomials corresponding to the cross-sectional distribution w(x) of the test ensemble. However, the weights b_j depend on $w(x)$ in a more complex manner (Johnson and Riess, 1982).

Thus, from the point of view of the computational efficiency of a quadrature algorithm, there is no reason to select a Gaussian white noise over other input ensembles. However, the standard Wiener orthogonal expansion, which assumes a Gaussian white noise as the weighting for the inner product (14), does have a number of advantages, as outlined in the Introduction.

Room for improvement

Although the improved separation of low-order and high-order nonlinearities provided by this technique represents an advantage over previous experimental approaches to the measurement of orthogonal kernels, we do not claim that this scheme is an "optimal" one.

One obvious area in which there is potential room for improvement is evident on examination of the table of weights (Table 1). For reasonably high values of m, values of the weights for the extreme sample values are quite small. From a point of view of optimizing signal-to-noise in the experimental values of the expression (21), it makes little sense to spend the same amount of time measuring responses to L-tuples θ_j with low values of the weights b_j as is spent measuring responses to L-tuples with high values of the weights. Rather, one might imagine a modified scheme in which begins with a memory wheel of length m^L to measure responses to all L-tuples. However, rather than repeat this sequence, the next test signal might consist of a memory wheel of length $(m-2)^L$ which contains all L-tuples that do not include the extreme values θ_1 and θ_m, etc.

Relation to the sum-of-sinusoids technique

Another potential shortcoming of the scheme presented above is that by considering time to be discrete, the periods between stimulus transitions is "wasted". Consider in contrast analysis with a sinusoidal input at a fixed frequency. The "orthogonal kernels" correspond to the Fourier components in the response at the various harmonics of the input frequency. Since the response may be measured continuously through this cycle, Fourier analysis of the response provides for measurement of the "orthogonal kernels" without any systematic error. This is because the test signal sweeps through all of phase space (which is literally a phase space, being parametrized by the phase of the sinusoid). However, these orthogonal kernels, which are based on a test signal which is very different from Gaussian white noise, are quite distinct from the Wiener kernels, and do not share many of their properties listed above.

The sum-of-sinusoids technique (Victor and Knight 1979) aims to bring some of the advantages of continuous time to an orthogonal functional series which is close to that of Wiener's. We consider a test signal which is composed of a superposition of q sinusoids. In the limit that q is large in comparison with the order of nonlinearity of interest, orthogonal kernels with respect to this input signal approach the orthogonal kernels measured with respect to a filtered Gaussian noise. The phase space of this signal is a q-dimensional torus which is parametrized by the phases of the component sinusoids. When the frequencies are incommensurate, the trajectory of a test signal is a line which spirals through this space, eventually becoming dense in it. In principle, integrals of all orders corresponding to (14) would be exactly evaluated by their values on this trajectory. However, this trajectory corresponds to an experiment of infinite duration.

Thus, it is more practical to use a sum of sinusoids which are not incommensurate, but which are all common multiples of a long repeat period (Victor and Shapley 1980). The trajectory of a test signal is no longer dense in phase space, but closely approaches every point of phase space. Because not all points in phase space are sampled, there is an inevitable confounding of high-order nonlinearites and low-order ones. This problem may be minimized by employing several test inputs, with carefully-chosen initial phases. Particularly efficient schemes of this variety (Victor and Shapley 1980) correspond to numerical integration on a close-packed lattice in eight dimensions with 30 neighbors for each node, supplemented by continuous integration along an oblique direction.

SUMMARY

In choosing an experimental approach to measure orthogonal kernels, the experimenter must consider the intrinsic noise of the system, the order of the nonlinearity of interest, and the goals of the study.

A test input chosen at random to approximate Gaussian white noise will not have any systematic error, but will in general produce a substantial non-systematic error because the sample of noise has moments which are unlikely to equal those of the ideal ensemble. This error due to "noise in the noise" is distinct from additional errors due to noise generated within the transducer under study.

An equally-spaced, equally-weighted signal derived from a shift register sequence or a memory wheel eliminates the "noise in the noise", but introduces systematic errors which correspond to confounding of high-order and low-order nonlinearities.

For a given number m^L of distinct measurements, a Gaussian quadrature scheme (21) provides better separation of low-order and high-order non-linearities, in comparison with the more intuitive equally-spaced, equally-weighted test inputs. A high value of m decreases systematic error (by sorting out high-order and low-order nonlinearities), but suffers from greater sensitivity to internal noise within the transducer. A low value of m (e.g., m = 2) is less sensitive to transducer noise (since there are fewer L-tuples to test), but may fail to resolve high-order intensive nonlinear-ites and low-order ones.

The sum-of-sinusoids approach takes advantage of the continuous nature of time. Since the power spectrum of the input may be tailored to the system under study, the sum-of-sinusoids approach may be advantageous if the system under study has timescales of interest that vary widely. The number of sinusoids, q, plays the same role as the number of sample points m: as q increases, the systematic deviation from true Wiener kernels decreases, but sensitivity to intrinsic transducer noise increases.

ACKNOWLEDGMENTS

J. Victor is supported in part by NIH grants EY6871 and NS877 and the McKnight Foundation.

REFERENCES

Abramowitz, M., and Stegun, I.A., 1964, "Handbook of Mathematical Functions," National Bureau of Standards. Reprinted by Dover, New York, 1970.

Emerson, R.C., Citron, M., Vaughn, W.J., and Klein, S., 1987, Nonlinear directionally selective subunits in complex cells of cat striate cortex, J. Neurophysiol., 58:33-65.

Farmer, J.D., Ott, E., and Yorke, J.A., 1983, The dimension of chaotic attractors, Physica, 7D:153-180.

Johnson, L.W., and Riess, R.D., 1982, "Numerical Analysis," Addison-Wesley, Reading.

Korenberg, M.J., 1973, Cross-correlation analysis of neural cascades, Proc. 10th Ann. Rocky Mountain Bioeng. Symp., 47-51.

Lee, Y.N., and Schetzen, M., 1965, Measurement of the kernels of a nonlinear system by cross-correlation, Int. J. Control, 2:237-254.

Mandelbrot, B., 1977, "Fractals: Form, Chance, and Dimension," Freeman, San Francisco.

Marmarelis, V.Z., 1975, Identification of nonlinear systems through multi-level random signals, in: "Proc. 1st Symp. on Testing and Identification of Nonlinear Systems," V.Z. Marmarelis, ed., California Institute of Technology, Pasadena, CA.

Marmarelis, P.Z. and Marmarelis, V.Z., 1978, "Analysis of Physiological Systems: The White-Noise Approach," Plenum, New York.

Mayer-Kress, G., 1987, Application of dimension algorithms to experimental chaos, Los Alamos preprint LA-UR-87-1030, Los Alamos National Laboratory, Los Alamos.

Stein, S.K., 1969, "Mathematics: The Man-Made Universe," W.H. Freeman, San Francisco.

Sutter, E., 1987, A practical nonstochastic approach to nonlinear time-domain analysis, in: "Advanced Methods of Physiological System Modelling," V. Z. Marmarelis, ed., Biomedical Simulations Resource, Los Angeles.

Victor, J.D., 1987, The fractal dimension of a test signal: implications for system identification procedures, Biological Cybernetics, 57:421-426.

Victor, J.D., and Johanessma, P.I.M., 1986, Maximum-entropy approximations of stochastic nonlinear transductions: an extension of the Wiener theory, Biological Cybernetics, 54:289-300.

Victor, J.D., and Knight, B.W., 1979, Nonlinear analysis with an arbitrary stimulus ensemble, Q. Appl. Math., 37:113-136.

Victor, J.D., and Shapley, R., 1980, A method of nonlinear analysis in the frequency domain, Biophys. J., 29:459-484.

Volterra, V., 1932, "Theory of Functionals and of Integral and Integro-Differential Equations," Blackie, London and Glasgow.

Wiener, N., 1958, "Nonlinear Problems in Random Theory," Wiley, New York.

FAST ORTHOGONAL ALGORITHMS FOR NONLINEAR SYSTEM

IDENTIFICATION AND TIME-SERIES ANALYSIS

Michael J. Korenberg

Department of Electrical Engineering
Queen's University
Kingston, Ontario, K7L 3N6, Canada

INTRODUCTION

In this paper we consider methods of obtaining parsimonious models of physiological systems and of biological time-series data. For the application to system identification, our model takes the form of a nonlinear difference equation, whose significant terms are to be determined and whose coefficients estimated. The system may also be modelled by a functional expansion whose kernels (a constant, and one- and multi-dimensional weighting functions) are to be measured. For the application to time-series analysis, our model generally takes the form of a sinusoidal series, whose component frequencies need not be commensurate. However, we also consider series approximations using other functions, such as exponentials. For both system identification and time-series applications, we use orthogonal approaches (Korenberg, 1987, 1988) which include rapid searches for significant terms to incorporate into the model.

Wiener (1958) pioneered the representation of nonlinear systems via a series of orthogonal functionals, which he created using the Gram-Schmidt process. These functionals were mutually orthogonal for the particular Gaussian process used as input. System identification reduced to measuring the Wiener kernels in his orthogonal functionals. These kernels were determinable directly by cross-correlation (Lee and Schetzen, 1965) or via the frequency domain (French and Butz, 1973). Wiener's orthogonal approach has been carefully analyzed and extended (Palm and Poggio, 1977, 1978). Important biological applications of the Wiener theory have been made, particularly in neurophysiology (Marmarelis and Naka, 1972; Marmarelis and Marmarelis, 1978; Sakuranaga et al., 1986).

The advantages afforded by Wiener's orthogonal approach prompted the development (Korenberg, 1985; McIlroy, 1986), of an orthogonal search method for selecting difference equation model terms. This method enabled accurate economical difference equation models to be systematically and rapidly constructed for a nonlinear system with unknown structure. The model term selection can be carried out still more efficiently by a fast orthogonal search (Korenberg, 1987, 1988) which is examined below.

The term selection procedures can also be adapted (Korenberg, 1987, 1988) to choose significant sinusoidal frequencies for accurately approximating biological time-series data. As examined below, these procedures can attain significantly finer resolution than achievable by a conventional Fourier series analysis. Certainly there exist (Kay and Marple, 1981; Mohanty, 1986) well-known parametric procedures for estimating power spectral density and resolving a sum of sinusoidal signals in noise, e.g., the maximum likelihood estimator, the maximum entropy method, the Prony and Pisarenko methods. However, as discussed below, spectral analysis by the fast orthogonal search has some important advantages. In addition to offering high resolution, our methods do not require solution of a polynomial equation, readily cope with noisy, missing and unequally-spaced data, and automatically determine model order. Indeed, a simple statistical test for deciding when to terminate model development is presented below.

We first consider difference equation modelling of nonlinear systems, followed by our application to time-series analysis.

SYSTEM IDENTIFICATION

Model

Consider approximating a nonlinear system by the difference equation model (Haber and Keviczky, 1976; Billings and Leontaritis, 1982)

$$y(n) = F[y(n-1),..., y(n-K), x(n),..., x(n-L)] + e(n) \tag{1}$$

where $x(n)$ and $y(n)$ are respectively the system input and output, F is a polynomial and $e(n)$ is the equation error. Let the data record be defined for $n = 0,..., N$. Equation 1 can be expressed more concisely as

$$y(n) = \sum_{m=0}^{M} a_m p_m(n) + e(n) \tag{2}$$

where

$$p_0(n) = 1, \tag{3}$$

and for $m \geqslant 1$

$$p_m(n) = y(n-k_1) ... y(n-k_i)x(n-\ell_1) ... x(n-\ell_j). \tag{4}$$

In Eq. 4,

$$i \geqslant 0, \quad 1 \leqslant k_1 \leqslant K,..., 1 \leqslant k_i \leqslant K$$

$$j \geqslant 0, \quad 0 \leqslant \ell_1 \leqslant L,..., 0 \leqslant \ell_j \leqslant L$$

and $i = 0$ denotes that the term $p_m(n)$ contains no factors y. Similarly $j = 0$ indicates the term contains no factors x. (A discrete-time finite-memory, finite-order functional expansion is a special case of Eq. 1 where no y terms appear on the right side of that equation.)

We next consider how the model terms may be selected.

Selecting Model Terms By Orthogonal Search Method

Suppose that the nonlinear system to be identified has unknown structure. Then a difference equation model can be systematically constructed for the system via the orthogonal search method (Korenberg, 1985; McIlroy, 1986). Using this method, we first build up a model

$$y(n) = \sum_{m=0}^{M} g_m \, w_m(n) + e(n). \tag{5}$$

Here the $w_m(n)$ are orthogonal functions constructed (by Gram-Schmidt orthogonalization) from the $p_m(n)$ in Eq. 4. (Once the model of Eq. 5 is constructed, it can of course be rearranged into the form of Eq. 2.) The orthogonal expansion coefficients g_m are chosen to achieve a least-squares fit, i.e., to minimize the mean-square error

$$\text{m.s.e.} = \overline{(y(n) - \sum_{m=0}^{M} a_m \, p_m(n))^2} \tag{6}$$

or equivalently

$$\text{m.s.e.} = \overline{(y(n) - \sum_{m=0}^{M} g_m \, w_m(n))^2}. \tag{7}$$

The overbar on the right sides of Eqs. 6 and 7 denotes the time-average from $n = 0$ to $n = N$.

Suppose that $a_M \, p_M(n)$ was the last term added to the model in Eq. 2. Then it can be shown that the addition of this term reduced the m.s.e. by the quantity

$$Q(M) = g_M^2 \, \overline{w_M^2(n)}. \tag{8}$$

Consider selecting the term $p_M(n)$ out of a set of candidate terms, via the orthogonal search method (Korenberg, 1985; McIlroy, 1986). For each candidate term, evaluate the quantity Q in Eq. 8. Choose the candidate for which Q is greatest since the selection of this term will result in the greatest reduction in mean-square error. Continuing to select model terms in this way, we can rapidly build up an accurate difference equation approximation for a real system, particularly if a threshold is used to reject unsuitable candidate terms (McIlroy, 1986).

However, the construction of the orthogonal functions $w_m(n)$ wastes computing time and memory, and is avoided in the fast orthogonal search.

Selecting Model Terms By Fast Orthogonal Search

To circumvent the creation of orthogonal functions, the fast orthogonal search uses a Cholesky factorization as follows. The orthogonal expansion coefficients in Eq. 5 can be obtained from the relation

$$g_m = \frac{C(m)}{D(m,m)} \ , \quad m = 0,\ldots, M. \tag{9}$$

where

$$D(0,0) = 1 \tag{10}$$

$$D(m,0) = \overline{p_m(n)} \ , \quad m = 1,\ldots, M \tag{11}$$

$$D(m,r) = \overline{p_m(n)\, p_r(n)} - \sum_{i=0}^{r-1} \alpha_{ri}\, D(m,i)$$

$$m = 1,\ldots, M; \quad r = 1,\ldots, m \tag{12}$$

$$\alpha_{mr} = \frac{D(m,r)}{D(r,r)} \ , \quad m = 1,\ldots, M; \quad r = 0,\ldots, m-1 \tag{13}$$

$$C(0) = \overline{y(n)} \tag{14}$$

$$C(m) = \overline{y(n)\, p_m(n)} - \sum_{r=0}^{m-1} \alpha_{mr}\, C(r), \quad m = 1,\ldots, M \tag{15}$$

The g_m in Eq. 5 are related to the a_m in Eq. 2 as follows (Korenberg et al., 1988):

$$a_m = \sum_{i=m}^{M} g_i\, v_i \tag{16a}$$

where

$$v_m = 1 \tag{16b}$$

$$v_i = - \sum_{r=m}^{i-1} \alpha_{ir}\, v_r, \quad i = m+1,\ldots, M \tag{16c}$$

Begin constructing the difference equation model by using Eq. 3 to introduce a constant term. Consider choosing the remaining terms by fast orthogonal search. Suppose that $a_M\, p_M(n)$ was the last term added to the model of Eq. 2. Then it can be shown (Korenberg, 1987) that adding this term reduced the mean-square error by the quantity

$$Q(M) = g_M^2\, D(M,M) \tag{17}$$

Equation 17 can in fact be utilized to choose $p_M(n)$, where M is successively set equal to 1, 2,... . Thus, use Eqs. 9-15 and 17 to evaluate Q for each candidate for $p_M(n)$. Choose the candidate with largest Q value, optionally subject to exceeding a specified positive threshold level. (Avoid choosing a candidate for which D(M,M) is very small: this prevents division by neglegibly small numbers in Eqs. 9 and 13.) Continuing to select model terms this way, we can rapidly build up a difference equation approximation without constructing orthogonal functions as an intermediate step.

In choosing $p_M(n)$, we can avoid repeating calculations performed in the earlier rounds of searching. Thus, for each candidate for $p_M(n)$, we can use the following pseudocode:

```
        D(0,0) = 1
Calculate D(M,0) from Eq. 11
FOR r = 0  TO  M-1
Calculate α_Mr from Eq. 13
Calculate D(M,r+1) using Eq. 12
NEXT r
```

This yields both the α_{Mr} and $D(M,M)$ corresponding to the candidate for $p_M(n)$. Then set m = M in Eq. 15 to obtain C(M), and hence g_M from Eq. 9. This enables us to evaluate Q from Eq. 17 for the candidate.

After selecting $p_M(n)$, carry out the above pseudocode again for the chosen candidate, and also re-evaluate g_M. This is done to properly set the values of the α_{Mr}, $D(M,M)$, $C(M)$ and g_M before continuing the process.

It can be shown (Korenberg, 1987) that the mean-square error remaining, after selecting model terms up to and including $p_M(n)$, is

$$\text{m.s.e.} = \overline{y^2(n)} - \sum_{m=0}^{M} g_m^2 \, D(m,m) \tag{18}$$

Hence, cease searching for model terms if the m.s.e. given by Eq. 18 is acceptably small, or if a pre-set limit on the number of model terms has been reached. Model development may also be terminated if no remaining candidate term can cause a reduction in mean-square error exceeding a specified threshold level. These criteria can be used to terminate automatically the searching, thereby determining the final value for M (i.e., the model order). Equations 16 can then be employed to calculate the a_m in the model of Eq. 2.

A simple statistical test for deciding when to terminate model development is developed in the next section.

Determining Model Order

Suppose that model terms up to $p_{M-1}(n)$ have been selected, and, in searching for $p_M(n)$, we have found the candidate for which Q(M) (Eq. 17) is largest. For this candidate, consider the measure

$$r = \left[\frac{Q(M)}{\overline{y^2(n)} - \sum_{m=0}^{M-1} g_m^2 \, D(m,m)} \right]^{1/2} \tag{19}$$

Since we have already fitted model terms up to $p_{M-1}(n)$, the residue is

$$R(n) = y(n) - \sum_{m=0}^{M-1} a_m \, p_m(n) \tag{20}$$

It can be shown that r in Eq. 19 is the correlation coefficient of $R(n)$ and $w_M(n)$, $n = 0,\ldots, N$. Here $w_M(n)$ is the orthogonal function corresponding to the candidate under consideration for $p_M(n)$.

Clearly, if the residue $R(n)$ is merely noise, then we do not wish to fit the $p_M(n)$ term to the model. Suppose that $R(n)$ is independent Gaussian noise. Then it can be shown that, for sufficiently large N, r is approximately normally distributed with a mean of zero and a standard deviation of $1/(N+1)^{1/2}$. Hence, for sufficiently large N,

$$|r| < 2/\sqrt{N}$$

with probability of about 0.95.

We can use the 95% confidence interval in conjunction with Eq. 19 as follows. Before selecting a candidate for $p_M(n)$, require that

$$\left[\frac{Q(M)}{\overline{y^2(n)} - \sum_{m=0}^{M-1} g_m^2 D(m,m)} \right]^{1/2} > \frac{2}{\sqrt{N}}$$

or equivalently,

$$g_M^2 D(M,M) > \frac{4}{N} \left(\overline{y^2(n)} - \sum_{m=0}^{M-1} g_m^2 D(m,m) \right) \tag{21}$$

For smaller values of N, a more precise criterion is readily obtained. Simply replace the factor 4/N (on the right side of the immediately above inequality) with the square of the critical correlation coefficient value for sample size N+1. If there are no remaining candidates which can meet the above criterion, then model development is terminated. This test provides a convenient alternative to use of the Akaike (1974) information criterion in determining model order.

TIME-SERIES ANALYSIS

Approximation by a Sum of Exponentials

The fast orthogonal search can be used to fit time-series data by a sum of exponentials. In this case, the time-series is treated as the system output, while the candidate functions for approximating the output are the exponential functions. In particular, Eq. 3 still holds, and for $m \geqslant 1$,

$$p_m(n) = \exp(-n/\lambda_m) \tag{22}$$

The λ_m are selected by searching a set of candidate "time-constants". This selection can be accelerated as follows. First use Eq. 17 to evaluate the function Q(1) for each candidate time-constant. Then select those candidates occurring at the maximum or the "relative maxima" of Q(1) which exceed a specified threshold level. The selected candidates can then serve as the time-constants in the exponential series under development, or alternatively form a reduced set of candidate time-constants to be searched.

However, a time-series which is to be fit by a sum of exponentials can readily be transformed into a time-series to be fit by a sinusoidal series. The frequencies in the sinusoidal series are related simply to the required time-constants in the desired exponential series. For this reason, we defer further examination of approximations by exponential series until we have considered development of sinusoidal series representations for time-series data. Note that fast orthogonal search can also be used to fit time-series data with exponentially-decaying sinusoids, logarithmic functions, splines, polynomials, and a wide variety of other functions. We will next concentrate on developing parsimonious sinusoidal series approximations for time-series data (Korenberg, 1987).

Approximation by a Sum of Sinusoids

In this section, Eq. 2 represents a sinusoidal series which is to be developed by fast orthogonal search to approximate a given time-series $y(n)$, $n = 0,\ldots, N$. Equation 3 still holds and for $i = 1, 2,\ldots$

$$p_{2i-1}(n) = \cos \omega_i n \qquad (23)$$

$$p_{2i}(n) = \sin \omega_i n \qquad (24)$$

The sinusoidal frequencies ω_i are to be selected by searching through a set of candidate frequencies ω_A, ω_B,... which need not be commensurate. The sinusoidal series is systematically built-up by adding term pairs of the form

$$T_i = a_{2i-1} p_{2i-1}(n) + a_{2i} p_{2i}(n) \qquad (25)$$

In Eq. 25, a_{2i-1} and a_{2i} are respectively the cosine and sine amplitudes used in defining the i-th sinusoidal component T_i. It can be shown (Korenberg, 1987) that adding the term pair T_i to the model reduces the mean-square error by the quantity

$$Q_1(i) = g_{2i-1}^2 D(2i-1,2i-1) + g_{2i}^2 D(2i,2i) \qquad (26)$$

Begin constructing the sinusoidal series model by using Eq. 3 to introduce a constant term, and calculate g_0 from Eqs. 9, 10 and 14. Consider using fast orthogonal search to find the value for ω_i, $i \geqslant 1$. Set $M = 2i$. Use Eqs. 9-15 and 26 to evaluate $Q_1(i)$ for each candidate frequency for ω_i. Choose the candidate frequency with largest Q_1 value, optionally subject to exceeding a threshold level.

For example, before selecting a candidate frequency for ω_i, one may require that

$$g_{M-1}^2 D(M-1,M-1) + g_M^2 D(M,M) > \frac{4}{N} \left(\overline{y^2(n)} - \sum_{m=0}^{M-2} g_m^2 D(m,m) \right)$$

where $M = 2i$.

Again, for smaller values of N, one should replace the factor 4/N on the right side of the above inequality with the square of the critical correlation coefficient value for sample size N+1. Avoid choosing a candidate frequency with neglegibly small $D(m,m)$, m = M-1 and m = M. Continuing to select model frequencies this way, we can rapidly construct a sinusoidal series approximation for the given time-series.

Recall that, at the stage of searching for ω_i, M = 2i. In choosing ω_i, we can avoid repeating calculations performed in the earlier rounds of searching. Thus, for each candidate for ω_i, we can use the following pseudocode:

```
D(0,0) = 1
FOR m = M-1 TO M
Calculate D(m,0) from Eq. 11
NEXT m
FOR m = M-1 TO M
FOR r = 0 TO m-1
Calculate α    from Eq. 13
         mr
Calculate D(m,r+1) using Eq. 12
NEXT r
NEXT m
```

This pseudocode yields both the α_{mr} and the $D(m,m)$, m = 2i-1, 2i, corresponding to the candidate frequency for ω_i. Then use Eq. 15 to obtain $C(m)$ and Eq. 9 to obtain g_m, merely for m = 2i-1, 2i. This enables us to evaluate $Q_1(i)$ from Eq. 26 for the candidate frequency.

After selecting ω_i, carry out the above pseudocode again for the chosen candidate frequency, and also recalculate g_m, m = 2i-1, 2i. This is done to properly set the values of the α_{mr}, $D(m,m)$, $C(m)$ and g_m, m = 2i-1, 2i, before continuing the process.

Terminate model development if the mean-square error given by Eq. 18 (with M = 2i) is acceptably small, or if a limit on the number of model terms has been reached. Also cease searching for model frequencies if no remaining candidate frequency can cause a reduction in mean-square error exceeding a specified threshold level.

Related time-series analysis techniques of the present author have been applied to real data in projects supervised by him in the Department of Electrical Engineering, Queen's University, e.g., Ho et al. (1987). In the latter study, to cope with unequally-spaced data, the variable n on the right sides of Eqs. 23, 24 was replaced (Ho et al., 1987) by the function t(n) which measured the actual timing of the n-th data sample. The same strategy can be used to cope with missing or unequally-spaced data in the fast orthogonal search examined above.

Finally, note that the search for model frequencies can be accelerated as follows. First use Eq. 26 to evaluate $Q_1(1)$ for each candidate frequency. Then select those candidate frequencies occurring at the maximum or the "relative maxima" of $Q_1(1)$ which exceed a specified threshold level. The selected candidate frequencies can then serve directly as the model frequencies, or alternatively, define a reduced set of candidate frequencies to be searched.

Performance of Fast Orthogonal Search

Extensive testing of time-series analysis by fast orthogonal search has consistently shown it to afford greater resolution than a conventional Fourier series analysis. For example, fast orthogonal search was able to correctly resolve all 10 component frequencies, and sine and cosine amplitudes, in a 128 point test signal. In particular, two of the test frequencies resolved were separated only by about 1/8 of the resolution width of the fast Fourier transform (FFT). While fast orthogonal search is not always 8 times more accurate than the FFT (as it was in this example), it usually offers significantly higher resolution than the FFT, especially with short data records, and noisy or missing data.

Moreover, fast orthogonal search was tested on a 64 point time-series used by Kay and Marple (1981) in a study of various spectral estimation techniques. Fast orthogonal search showed significantly better resolution of frequencies than was reported (Kay and Marple, 1981) for any of the studied techniques except for the Prony spectral line method. Compared to the latter method, fast orthogonal search showed equivalent frequency resolution and estimation of amplitudes, but significantly more accurate estimation of phases.

Obtaining Exponential Series Approximations Via Sinusoidal Series Models

One means of fitting a time-series with a sum of exponentials is, first, to transform the time-series into one which is best-fit by a sinusoidal series using the fast orthogonal search. The frequencies selected for the sinusoidal series model reveal the required time-constants for fitting the original time-series by a sum of exponentials.

For example, suppose that a signal $u(t)$, $t \geqslant 0$, is a sum of two decaying exponentials:

$$u(t) = K_1 \exp(-b_1 t) + K_2 \exp(-b_2 t), \qquad b_1, b_2 > 0 \qquad (27)$$

Then, using the z-transform, consider

$$(z - \exp(-b_1))(z - \exp(-b_2)) \, U(z) = 0 \qquad (28)$$

which leads to

$$u(n+2) - d_1 \, u(n+1) + d_2 \, u(n) = 0 \qquad (29)$$

where

$$d_1 = \exp(-b_1) + \exp(-b_2)$$
$$\qquad (30)$$
$$d_2 = \exp(-b_1 - b_2)$$

Suppose that $y(t)$, $t \geqslant 0$, is the sum of two sinusoids with frequencies β_1 and β_2. Again using the z-transform, consider

$$(z^2 - 2z \cos\beta_1 + 1) (z^2 - 2z \cos\beta_2 + 1) Y(z) = 0 \qquad (31)$$

If

$$\cos\beta_i = \exp(-b_i), \qquad i = 1, 2, \qquad (32)$$

then Eq. 31 leads to

$$y(n+4) - 2 d_1 y(n+3) + (4 d_2+2) y(n+2) - 2 d_1 y(n+1) + y(n) = 0 \quad (33)$$

Eqs. 29 and 33 can be used to transform a time-series which is the sum of two decaying exponentials into a time-series which is the sum of two sinusoids. One method for doing this is to use the original time-series to find d_1 and d_2 by linear regression on Eq. 29, then to utilize Eq. 33 (with non-zero initial conditions) to generate the sum of sinusoids. Fast orthogonal search can then be used to find the significant frequencies in the sinusoidal series. Finally the required time-constants to fit the original time-series can be found using Eq. 32. Note that the step of finding the constants d_1, d_2 in Eq. 29 is also used in the Prony method for fitting a sum of exponentials. However, in the present approach, no polynomial equation need be solved to obtain the required time-constants.

An alternate method for transforming the exponential series into a sinusoidal series is via a nonlinear difference equation. From Eqs. 29 and 33 it follows readily that

$$y(n+4) = 2(y(n+3) + y(n+1)) \left(\frac{u(n) u(n+3) - u(n+1) u(n+2)}{u(n) u(n+2) - u^2(n+1)} \right)$$

$$- y(n+2) \left(2 + \frac{4(u(n+1) u(n+3) - u^2(n+2))}{u(n) u(n+2) - u^2(n+1)} \right) - y(n) \qquad (34)$$

Equation 34, with non-zero initial conditions, will generate a sum of two sinusoids if the input u is a sum of two decaying exponentials.

Example

Figure 1 shows the signal u when, in Eq. 27, $K_1 = 1$, $K_2 = 2$, $b_1 = 0.1$, $b_2 = 0.05$, so that the time-constants are 10, 20. Figure 2 shows the signal y generated from Eq. 34, with initial conditions $y(0) = 0$, $y(1) = 1$, $y(2) = 2$, $y(3) = 3$. Fast orthogonal search applied to the signal y found only two significant frequencies: radian frequencies of 0.3142 and 0.4398. These are the estimates for β_1 and β_2. Equation 32 then estimates b_1 and b_2 to equal 0.0502 and 0.1000 respectively. Note that, since the right side of Eq. 32 is always positive, the candidate frequencies to be searched can be limited to radian frequencies between 0 and $\pi/2$. In the present example, 100 candidate frequencies equally spaced in this interval were searched.

174

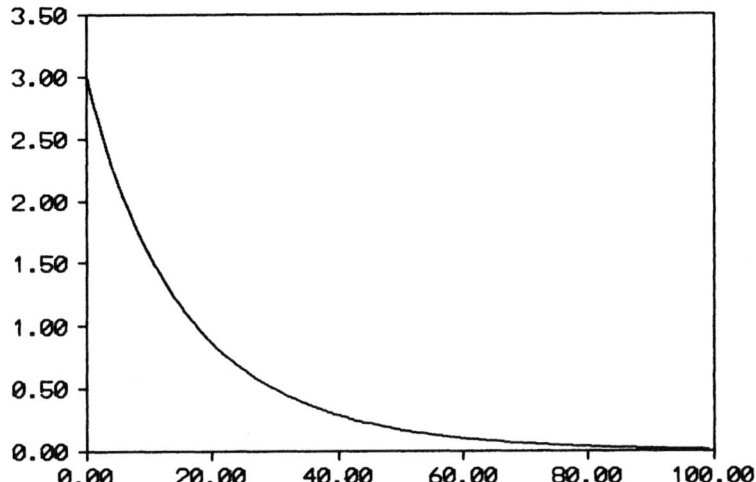

Fig. 1. Sum of two decaying exponentials, forming the test time-series for the Example.

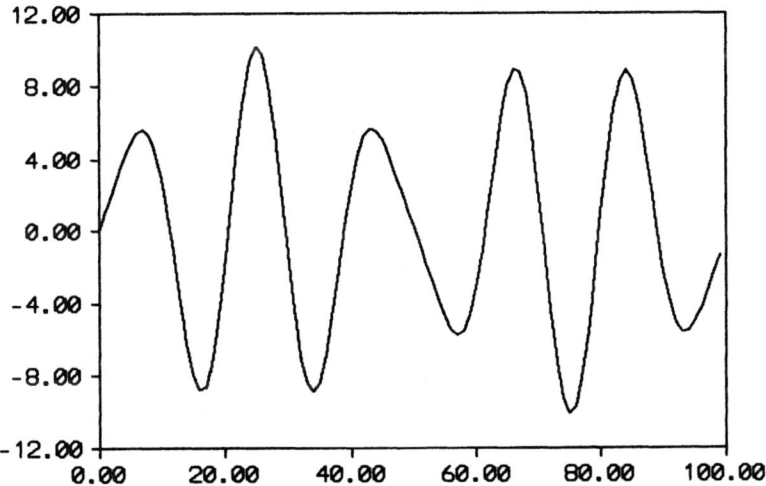

Fig. 2. Sum of two sinusoids, generated by Eq. 34 with the signal of Fig. 1 as input u. The frequencies of the above signal are related via Eq. 32 to the exponentials in the signal of Fig. 1.

The approach shown here can clearly also be utilized to fit a greater number of exponentials to time-series data.

CONCLUSION

Methods were described for obtaining concise sinusoidal series representations of biological time-series data. The methods were also applied to identification of systems with unknown structure. In both applications, the methods use rapid orthogonal searches for significant terms to include in the model for the time-series or the nonlinear system. The methods include a simple statistical test for deciding when to terminate model development. In collaboration with Dr. Ken-Ichi Naka and Dr. Hiroko M. Sakai of the National Institute for Basic Biology, Okazaki, Japan, we are presently using the time-series analysis to study the evolution of kernel shape observed in penetrating from outer to inner catfish retina.

ACKNOWLEDGEMENTS

Research supported by the Natural Sciences and Engineering Research Council of Canada, and the Advisory Research Committee of Queen's University.

REFERENCES

Akaike, H., 1974, A new look at the statistical model identification, IEEE Trans. AC-19: 716.

Billings, S.A., and Leontaritis, I.J., 1982, Parameter estimation techniques for nonlinear systems, IFAC Symp. Ident. Sys. Param. Est., 1: 427.

French, A.S., and Butz, E.G., 1973, Measuring the Wiener kernels of a nonlinear system using the fast Fourier transform algorithm, Int. J. Control, 17: 529.

Haber, R., and Keviczky, L., 1976, Identification of nonlinear dynamic systems, IFAC Symp. Ident. Sys. Param. Est., 1: 79.

Ho, T., Kwok, J., Law, J., and Leung, L., 1987, "Nonlinear System Identification," ELEC 490 Project Rept., Department of Electrical Engineering, Queen's University, Kingston, Ontario, Canada.

Kay, S.M., and Marple, S.L., 1981, Spectrum analysis - a modern perspective, Proc. IEEE, 69: 1380.

Korenberg, M.J., 1985, Orthogonal identification of nonlinear difference equation models, Proc. Midwest Symp. Circuit. Sys., 1: 90.

Korenberg, M.J., 1987, Fast orthogonal identification of nonlinear difference equation and functional expansion models, Proc. Midwest Symp. Circuit. Sys., 1: 270.

Korenberg, M.J., 1988, Identifying nonlinear difference equation and functional expansion representations: the fast orthogonal algorithm, Ann. Biomed. Eng., 16: 123.

Korenberg, M.J., Bruder, S.B., and McIlroy, P.J., 1988, Exact orthogonal kernel estimation from finite data records: extending Wiener's identification of nonlinear systems, Ann. Biomed. Eng., 16: 201.

Lee, Y.W., and Schetzen, M., 1965, Measurement of the Wiener kernels of a nonlinear system by cross-correlation, Int. J. Control, 2: 237.

Marmarelis, P.Z., and Marmarelis, V.Z., 1978, "Analysis of Physiological Systems. The White Noise Approach", Plenum Press, New York.

Marmarelis, P.Z., and Naka, K.-I., 1972, White noise analysis of a neuron chain: an application of the Wiener theory, Science, 175: 1276.

McIlroy, P.J.H., 1986, "Applications of nonlinear systems identification", M.Sc. Thesis, Queen's University, Kingston, Ontario, Canada.

Mohanty, N.C., 1986, "Random Signals Estimation and Identification. Analysis and Applications", Van Nostrand, New York.

Palm, G., and Poggio, T., 1977, The Volterra representation and the Wiener expansion: validity and pitfalls, SIAM J. Appl. Math., 33: 195.

Palm, G., and Poggio, T., 1978, Stochastic identification methods for nonlinear systems: an extension of the Wiener theory, SIAM J. Appl. Math., 34: 524.

Sakuranaga, M., Sato, S., Hida, E., and Naka, K.-I., 1986, Nonlinear analysis: mathematical theory and biological applications, CRC Crit. Rev. Biomed. Eng., 14: 127.

Wiener, N., 1958, "Nonlinear Problems in Random Theory", Wiley, New York.

NEW ALGORITHM FOR KORENBERG-BILLINGS MODEL OF NONLINEAR SYSTEM IDENTIFICATION

Hun H. Sun and J. H. Shi

Dept. of Electrical and Computer Engineering
Drexel University
Philadelphia, Pa. 19104

INTRODUCTION

A new algorithm is presented here for nonlinear system identification based on input and output relationship by a cascade structure of dynamic linear(L), static nonlinear(N), and dynamic linear(L) subsystems, or the Korenberg-Billings(K-B) Model. It uses a series of multilevel inputs to separate the Volterra components in the output signal and then uses Flethcer-Reeves method to minimize a cost function to separate the two linear subsystems from the nonliear subsystem. It does not employ any correlation functions as were used by both Korenberg and Billings. Therefore the input is not restricted to white Gaussian noise, the computation is simpler, and the result is much more accurate than the previous methods. Three computer programs, namely, IDENT, ODDIDENT, P_LIDENT, have been developed to identify systems with even, odd, and piece-wise types of nonlinearity. Numerical example is presented to show computation result. In addition, a nonlinear piecewise saturation model has been developed for the interfacial impedance of metallic bioelectrodes based on overpotential signals from several levels of current density input. The result shows remarkable similarity to the model developed previously by using fractal dimension study.

KORENBERG-BILLINGS(K-B) MODEL OF LNL CASCADE STURCTURE

Korenberg(1973a,b) was the first to introduce a general method to identify nonlinear system by cascaded LNL structure as shown in Fig.1, where $g_1(t)$ and $g_2(t)$ are the two linear subsystems and $F(\cdot)$ is the static nonlinear (non memory) subsystem which is represented by a polynomial to an arbitary degree of accuracy. Both the Wiener model and Hammerstein model can therefore be considered as special cases of this general model. Korenberg's algorithm uses Gaussian white noise input to compute the first and second order cross-correlatins and then applies Fourier Transform to obtain the two linear and the nonlinear subsystems. Billings and Fakhour(1978, 82) independently developed the similar equations of the correlatin functions and applied a parametrization procedure to obtain the linear and nonlinear subsystems. Both methods reply on white

Gaussian noise input, therefore the computation time is lengthy and accuracy is poor due to the case of high order cross-correlation functions. In addition, a prior knowledge of the linear subsystems are required, such as an assumption of the order of system or a 1st

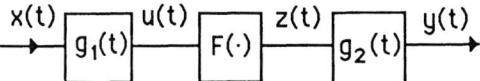

Fig.1. Block-structured LNL model.

approximation for $g_1(t)$ and $g_2(t)$. Korenberg(1985,86) has recently developed a fast orthogonal algorithm to improve the accuracy which enabled one also to compute the second order Wiener's kernel (Korenberg, 88), but the computation time can be very lengthy.

The method presented here does not require a prior knowledge of any of the subsystems and the input is not restricted and can be either step, sinusoidal or white noise and the nonlinear subsystem can be either continuous as represented by polynomial approximation or abrupt type as represented by piece-wise approximation.

SEPARATION OF THE VOLTERRA COMPONENTS IN THE OUTPUT SIGNAL BY MULTILEVEL INPUT METHOD

Gardiner(1973) employed multilevel inputs to the block structured system to separate the linear part from the output signal. He used step, sinusoidal and random signal as input and the number of the amplitude levels of the test signal is dependant on the order of the highest odd power of the static nonlinear element. However, the method does not completely identify the linear and nonlinear subsystems and is applicable to the cascaded system of a single linear block instead of the two linear blocks as described here. Tuis(1978) developed an algorithm to identify a Water-turbine control unit by a Hammerstein model using a multilevel pseudo-random input. The method employs high order correlation functions and for a fourth power nonlinearity, it requires up to 8th order autocorrelation.

We shall use the technique of multi level inputs to obtain the impulse response of the two linear subsystems and also to completely identify the nonlinear subsystem in terms of polynomial approximation and also the piecewise approximation.

According to Volterra Theory (Schetzen, 1980), the Nth order approximation of the output signal in discrete form can be represented by x(n), the input signal, and $h_k(t_1, t_2, \ldots, t_k)$, k=1,2,...N, the Volterra kernels, as:

$$y^*(n) = \sum_{\tau=-\infty}^{\infty} h_1(\tau)x(n-\tau) + \sum_{\tau_1=-\infty}^{\infty}\sum_{\tau_2=-\infty}^{\infty} h_2(\tau_1,\tau_2)x(n-\tau_1)x(n-\tau_2) + \ldots +$$

$$\sum_{\tau_1=-\infty}^{\infty} \ldots \sum_{\tau_N=-\infty}^{\infty} h_N(\tau_1,\ldots,\tau_N)x(n-\tau_1)\ldots x(n-\tau_N) \qquad (1)$$

180

or

$$y^*(n) = y_1(n) + y_2(n) + \ldots + y_N(n) \tag{2}$$

where $y_k(n)$, $k=1,2,\ldots$, N are the kth order Volterra components of the response of the system.

Let the input be a series of multilevel signal or x(t) be scaled as $c_i x(n)$, $i=1,2,\ldots$, N-1, where $c_i \neq c_j$ for $i \neq j$. We can easily show that the output becomes:

$$y^*{}_i(n) = c_i \sum_{\tau=-\infty}^{\infty} h_1(\tau)x(n-\tau) + c_i^2 \sum_{\tau_1=-\infty}^{\infty}\sum_{\tau_2=-\infty}^{\infty} h_2(\tau_1,\tau_2)x(n-\tau_1)x(n-\tau_2) + \ldots +$$

$$c_i^N \sum_{\tau_1=-\infty}^{\infty} \ldots \sum_{\tau_N=-\infty}^{\infty} h_N(\tau_1,\ldots,\tau_N)x(n-\tau_1)x(n-\tau_2)\ldots x(n-\tau_N) \tag{3}$$

$(i=1,2,\ldots,N-1)$

From Equ.1, Equ.2, and Equ.3, we have:

$$\begin{bmatrix} y^*(n) \\ y_1^*(n) \\ \vdots \\ y_{N-1}^*(n) \end{bmatrix} = \begin{bmatrix} 1 & 1 \ldots 1 \\ c_1 & c_1^2 \ldots c_1^N \\ \ldots \ldots \ldots \ldots \\ c_{N-1} & c_{N-1}^2 \ldots c_{N-1}^N \end{bmatrix} \begin{bmatrix} y_1(n) \\ y_2(n) \\ \vdots \\ y_N(n) \end{bmatrix} \tag{4}$$

The coefficient matrix is a transposed Vandermonde matrix, therefore $y_i(n)$ ($i=1,2,\ldots,N$) can be uniqely determined from output $y^*(n)$.

We have thus developed a method to identify each of the Volterra kernels $y_k(n)$ from the output signal by using a series of multilevel input, or an input scaled to several levels of magnitude, $c_i x(n)$. The unscaled input x(n) could be any signal, such as step, sinusoidal or noise, but they must be properly scaled to insure that Equ.4 can be solved with relative ease in numerical computation. In our case, only the first and second order (or third order for nonlinearity with odd symmetry) Volterra components are needed to identify the subsystems. However, in the actual measurement, higher order $y_k(n)$, for k>2, are used to improve the accuracy of computation. In Fig.1, if F(.) is equi-continuous or it is satisfied by Weierstrass Criteron, then:

$$| F(x_1) - F(x_2) | < M|x_1 - x_2| \tag{5}$$

where $x_1 \neq x_2$ and $x_1, x_2 \in I$. As M is a finite number, F(.) can be approximated by a polynomial within interval I.

$$F(.)=d_1(.)+d_2(.)^2+\ldots+d_N(.)^N \tag{6}$$

From Fig.1 and Equ.6, we can then express the relationship between the first order kernel and the linear subsystems as:

$$h_1(n)=d_1\sum_{i_1=0}^{n}g_1(i_1)g_2(n-i_1) \tag{7}$$

and the second order response $y_2(n)$:

$$y_2(n)=d_2\sum_{i_1=0}^{n}g_2(n-i_1)\left(\sum_{i_2=0}^{i_1}g_1(i_2)x(i_1-i_2)\right)^2 \tag{8}$$

where $h_1(n)$ is obtained from $y_1(n)$ by a deconvolution process and both $y_1(n)$ and $y_2(n)$ are obtained by solving Equ.4.

To obtain $g_1(n)$ and $g_2(n)$, from Equ.7 and Equ.8 we can see that there are $2(L+1)$ equations for $2(L+1)+2$ unknowns, including $g_1(n)$, $g_2(n)$, $n=0,1,\ldots$, L and d_1, d_2. However, we can let $g_1{}^*(n)=\beta_1 g_1(n)$, $g_2{}^*(n)=\beta_2 g_2(n)$ and $\beta_1\beta_2=d_1$, $\beta_1{}^2\beta_2=d_2$ and rewrite Equ.7 and Equ.8 as follows:

$$h_1(n)=\sum_{i_1=0}^{n}g_1{}^*(i_1)g_2{}^*(n-i_1) \tag{9}$$

$$y_2(n)=\sum_{i_1=0}^{n}g_2{}^*(n-i_1)\left(\sum_{i_2=0}^{i_1}g_1{}^*(i_2)x(i_1-i_2)\right)^2 \tag{10}$$

We can now proceed to solve $g_1{}^*(n)$ and $g_2{}^*(n)$ as will be discussed in following sections.

We have thus derived the equations to decouple the two linear subsystems from the nonlinear subsystem by the use of first and second order Volterra component of the output. This approach is slightly different to the method developed by either Korenberg or Billings and Fakhour, where they derived their equations by the use of first and second order cross-correlation functions. It is therefore conceivable that the input used in their method must be white Gaussian noise, although a somewhat colored white noise is allowable. The method we developed does not have any restrictions on the type of input signal, which could be step, sinusoidal or noise as long as they can be conveniently scaled and the algorithm we developed does not require a prior knowledge or assumption or the $g_1(t)$ and $g_2(t)$ functions and the computation is relatively simple.

COMPUTATION FOR g1*(t) AND g2*(t)

Let us define the original system as shown in Fig.2:

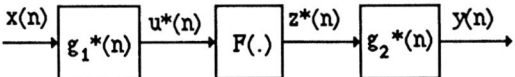

Fig.2. Normalized Block-Structured Model

and proceed to solve $g_1^*(t)$ and $g_2^*(t)$ as originally derived in Equ.9 and Equ.10.

We have developed a group nonliear programming to reduce the computation time and a predictor-corrector to compute $g_1^*(n)$ and $g_2^*(n)$ by Lagrange Polynomial approximation (Korn and Korn,1968) and the Flecher- Reeeves(F-R) (1964) Conjugate Gradient method of estimation to minimize a cost function.

(a). Cost Function

First, let us establish a cost function from Equ.9 and Equ.10 as follows:

$$f=\frac{1}{L+1}\sum_{i=-\infty}^{\infty}\left(\sum_{j=-\infty}^{\infty}g_1^*(j)g_2^*(i-j)-h_1(i)\right)^2 +$$

$$\frac{1}{L+1}\sum_{i=-\infty}^{\infty}\left(\sum_{i_2=-\infty}^{\infty}g_2^*(i-i_2)\left(\sum_{i_1=-\infty}^{\infty}g_1^*(i_1)x(i_2-i_1)\right)^2-y_2(i)\right)^2 \tag{11}$$

as f tends to zero, the solution represents the minimum mean square error. If $x(n)$ is causal, i.e. $x(n)=0$ for $n<0$ and both $g_1(n)$ and $g_2(n)$ are also causal, we can rewrite the above equation as:

$$f=f_0+f_1+..+f_N \tag{12}$$

where $N<L$.

To save computation time we can use the method of group nonliear programming. Let $(N+1)$ be exactly divisable by $(L+1)$ or

$$M=(L+1)/(N+1) \tag{13}$$

where M represents the number of element in each group. We can therefore write f_k, $k=0,1,2,...N$ as:

$$f_k=\frac{1}{M}\sum_{i=kM}^{(k+1)M-1}\left(\sum_{j=0}^{i}g_1^*(j)g_2^*(i-j)-h_1(i)\right)^2 +$$

$$\frac{1}{M}\sum_{i=kM}^{(k+1)M-1}\left(\sum_{i_2=0}^{i}g_2^*(i-i_2)\left(\sum_{i_1=0}^{i_2}g_1^*(i_1)x(i_2-i_1)\right)^2-y_2(i)\right)^2 \tag{14}$$

Examination of Equ.14 reveals that f_k does not depend on $g_1^*(i)$ and $g_2^*(i)$ for $i>(k+1)M-1$. Therefore we can minimize f by minimizing f_k in terms of variables $g_1^*(i)$, $g_2^*(i)$, where $kM\leq i\leq(k+1)M-1$, recursively for k varies from o to N. The computation time is thus reduced by a factor of M.

(b) Predictor-Corrector Algorithm

The general form of our predictor-corrector algorithm for kth cost function f_k has the following form:

Predictor:

$$G_k^{(0)} =$$

$$
\begin{bmatrix}
g_1^{*(0)}(kM) \\
g_1^{*(0)}(kM+1) \\
\vdots \\
g_1^{*(0)}((k+1)M-1) \\
g_2^{*(0)}(kM) \\
g_2^{*(0)}(kM+1) \\
\vdots \\
g_2^{*(0)}((k+1)M-1)
\end{bmatrix}
=
\begin{bmatrix}
\alpha_0^{(p_1)} .. \alpha_{p_1}^{(p_1)} \ 0 \ ..0 & \\
\alpha_0^{(p_1)} .. \alpha_{p_1}^{(p_1)} 0..0 & \\
.... & \\
\alpha_0^{(p_1)} .. \alpha_{p_1}^{(p_1)} & \\
& \alpha_0^{(p_2)} .. \alpha_{p_2}^{(p_2)} \ 0 ..0 \\
& \alpha_0^{(p_2)} .. \alpha_{p_2}^{(p_2)} 0..0 \\
& .... \\
& \alpha_0^{(p_2)} .. \alpha_{p_2}^{(p_2)}
\end{bmatrix}
\begin{bmatrix}
g_1^{*}(kM-p_1-1) \\
g_1^{*}(kM-1) \\
g_1^{*(0)}(kM) \\
\vdots \\
g_1^{*(0)}((k+1)M-2) \\
g_2^{*}(kM-p_2-1) \\
\vdots \\
g_2^{*}(kM-1) \\
g_2^{*(0)}(kM) \\
\vdots \\
g_2^{*(0)}((k+1)M-2)
\end{bmatrix}
\qquad (15)
$$

where p_1, p_2 are orders of predictors.
For k=0, then:

$$G_0^{(0)} = [g_1^*(0) \, g_1^*(0) .. g_1^*(0) \, g_2^*(0) .. g_2^*(0)]_{2M+1}^T \qquad (16)$$

Corrector:

$$G_k^T = \{G_k^T \mid f_k(G_k^T) = \min f_k; \text{ with inital guess } G_k^{(0)}\} \qquad (17)$$

In present study, Lagrange polynomial has been chosen as the predictor and Fletcher-Reeves Conjugate gradient method is used to minimize f_k. The Lagrange polynomial associated with $g_1^*(i_1)$, $g_2^*(i_2)$ are ($kM-p_1-1 \leq i_1 \leq kM-1$, $kM-p_2-1 \leq i_2 \leq kM-1$):

$$L_1(j) = \sum_{i_1=kM-p_1-1}^{kM-1} l_{i_1}^{(p_1)}(j) g_1^*(i_1) \qquad (18a)$$

$$l_2(j) = \sum_{i_2=kM-p_2-1}^{kM-1} l_{i_2}^{(p_2)}(j) g_2^*(i_2) \qquad (18b)$$

where

$$l_{i_1}^{(p_1)}(j) = \frac{(j-kM+p_1+1)(j-kM+p_1)...(j-i_1+1)(j-i_1-1)...(j-kM+1)}{(i_1-kM+p_1+1)(i_1-kM+p_1)...(1)(-1)...(i_1-kM+1)} \qquad (19a)$$

$$l_{i_2}^{(p_2)}(j) = \frac{(j-kM+p_2+1)(j-kM+p_2)...(j-i_2+1)(j-i_2-1)...(j-kM+1)}{(i_2-kM+p_2+1)(i_2-kM+p_2)...(1)(-1)...(i_2-kM+1)} \qquad (19b)$$

$j \neq i_1$ ($i_1 = kM-p_1-1$, $kM-p_1$, ..., $kM-1$); $l_0^{(0)}(j) = 1$
$j \neq i_2$ ($i_2 = kM-p_2-1$, $kM-p_2$, ..., $kM-1$).

Comparing Equ.15 and Equ.18, we have:

$$\alpha_i^{(p_1)} = l_i^{(p_1)}(kM) \qquad (i=0,...,p_1) \qquad (20)$$

and $\alpha_i^{(p_1)}$ does not change with k due to the uniform step size. By using this property, the redundant computation and computer memory can be obviated and changing the order of predictor turn out to be relatively uncomplicated.

To insure the stability of the interpolation polyomial we let the highest order of the predictor be limited to 6. $\alpha^{*(p)}$ can be derived from eq(19) and eq(20) as:

$$\alpha^* = (\alpha^{*(0)} \ \alpha^{*(1)} \ \alpha^{*(2)} \ \alpha^{*(3)} \ \alpha^{*(4)} \ \alpha^{*(5)} \alpha^{*(6)})$$

$$= \begin{bmatrix} 1 & -1 & 1 & -1 & 1 & -1 & 1 \\ 0 & 2 & -3 & 4 & -5 & 6 & -7 \\ 0 & 0 & 3 & -6 & 10 & -15 & 21 \\ 0 & 0 & 0 & 4 & -10 & 20 & -35 \\ 0 & 0 & 0 & 0 & 5 & -15 & 35 \\ 0 & 0 & 0 & 0 & 0 & 6 & -2 \\ 0 & 0 & 0 & 0 & 0 & 0 & 7 \end{bmatrix} \tag{21}$$

(c) An algorithm for changing the order of predictor

For uniform step size, there exists an optimum order of predictor such that the best estimation can be obtained. This would require an efficient computing method for estimating the local truncation error at each step.

The truncation error for predictor in Equ.18 at kM can be derived (see appendix I) as:

$$E_1^{(p_1)} = g_1^{(p_1+1)} (\S_1) \tag{22a}$$

$$E_2^{(p_2)} = g_2^{(p_2+1)} (\S_2) \tag{22b}$$

where $g_1^{(p_1+1)}$ stands for (p_1+1)th order derivitive of g_1 at $\S_1$ and $g_2^{(p_2+1)}$ stands for (p_2+1)th order derivitive of g_2 at $\S_2$, and $(kM-p_1-1 \leq \S_1 \leq kM, \ kM-p_2-1 \leq \S_2 \leq kM)$.

Under certain assumption, the $E_1^{(p_1)}$ and $E_2^{(p_2)}$ can be approximated by the following expressions (see Appendix II):

$$E_1^{(p_1)} \approx \nabla^{(p_1+1)} \ g_1^*(kM-1) \tag{23a}$$

$$E_2^{(p_2)} \approx \nabla^{(p_2+1)} \ g_2^*(kM-1) \tag{23b}$$

where $\nabla^{(p_1+1)} g_1^*(kM-1)$ and $\nabla^{(p_2+1)} g_2^*(kM-1)$ are (p_1+1)th and (p_2+1)th order backward difference respectively and can be computed recursively.

Suppose the present order is k, then the optimum order for next step can be determined by comparing $|E_1^{(k-1)}|$, $|E_1^{(k)}|$ and $|E_1^{(k+1)}|$.

$$E_1^{(p_1)} = \min (\ |E_1^{(k-1)}|, \quad |E_1^{(k)}| \ , |E_1^{(k+1)}|) \tag{24}$$

$p_1 \in \{k-1, \ k, \ k+1\}$

Same method can be used for $E_2^{(p_2)}$.

(d) Method to start the predictor-corrector algorithm

To compute $g_1^*(n)$ $g_2^*(n)$, we must first choose the initial points and then use the predictor-corrector algorithm to derive the additonal values beyond these initial points. Two cases can be chosen for the inital values:

(1) strictly proper function --- In this case $g_1^*(0)$ $g_2^*(0)$ are zeros, or:

$$g_1^*(1) = y_1(2) / [h_1(2) x_2^2(0)] \qquad (25a)$$
$$g_2^*(1) = h_1(2) / g_1(1) \qquad (25b)$$

We have chosen M=2 in this case. First order predictor is used here for predicting $g_1^*(1)$ and $g_2^*(2)$ and so on.The truncation error, as will be shown later, is then calculated, the optimum order is chosen. Since $g_1^*(0) = g_2^*(0) = 0$, one bit shift for all the algorithm in Equ.14 must be considered.

(2) non-strictly proper transfer function

In this case $g_1^*(0), g_2^*(0)$ are not equal to zero, or:

$$g_1^*(0) = y_2(0) / [h_1(0) x^2(0)] \qquad (26a)$$
$$g_2^*(0) = h_1(0) / g_1^*(0) \qquad (26b)$$

Zeroth order predictor is used here for $g_1^*(1)$ and $g_2^*(1)$ and so on. Automatically changing order algorithm is used after the second step.

(e) Calculating the gradient

We choose the Flecher-Reeves's method to minimize the cost function f_k. The gradient of cost function must be calculated during the iteration. For the kth cost function f_k, the gradient ∇f_k is:

$$\nabla f_k = \left(\frac{\partial f_k}{\partial g_1^*(kM)}, \cdots, \frac{\partial f_k}{\partial g_1^*((k+1)M-1)}, \frac{\partial f_k}{\partial g_2^*(kM)}, \cdots, \right.$$

$$\left. \cdots, \frac{\partial f_k}{\partial g_2^*((k+1)M-1)} \right)^T \qquad (27)$$

$$A_1 = \frac{2}{M} \sum_{i=kM}^{(k+1)M-1} \left(\sum_{j=0}^{i} g_1^*(j) g_2^*(i-j) - h_1(i) \right) g_2^*(i-1) \qquad (28a)$$
$$(i >= 1)$$

$$A_2 = \frac{4}{M} \sum_{i=kM}^{(k+1)M-1} \left(\sum_{i_2=0}^{i} g_2^*(i-i_2) \left(\sum_{i_1=0}^{i_2} g_1^*(i_1) x(i_2-i_1) \right)^2 - y_2(i) \right) \cdot$$

$$\left(\sum_{i_2=0}^{i}g_2{}^*(i-i_2)\left(\sum_{i_1=0}^{i_2}g_1{}^*(i_1)x(i_2-i_1)\right)x(i_2-l)\right) \qquad (i_2\geq l) \qquad (28b)$$

$$A_3=\frac{2}{M}\sum_{i=kM}^{(k+1)M-1}\left(\sum_{j=0}^{i}g_1{}^*(j)g_2{}^*(i-j)-h_1(i)\right)g_1{}^*(i-l) \quad (i\geq l) \qquad (28c)$$

Consider the schematic in Fig.3:

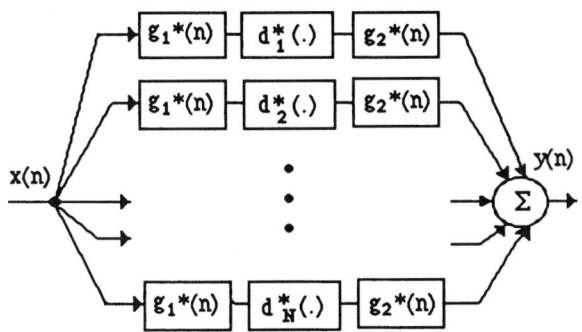

Fig.3. Realization of the Nonlinear System

From Fig2 and Equ.9 and Equ.10, we have
$h_1(n)=g_1{}^*(n)\cdot g_2{}^*(n)=d_1{}^*\ g_1{}^*(n)\cdot g_2{}^*(n)$
So, $d_1{}^*=1$. $\qquad\qquad\qquad\qquad\qquad\qquad\qquad\qquad (30)$
$y_2(n)=g_2{}^*(n)*(\ g_1{}^*(n)*x(n))^2=g_2{}^*(n)\ *d_2{}^*(g_1{}^*(n)\ *x(n))^2$
So, $d_2{}^*=1$. $\qquad\qquad\qquad\qquad\qquad\qquad\qquad\qquad (31)$
for $d_k{}^*$ $\quad(k=3,\ ...,\ N)$
$y_k(n)=g_2{}^*(n)\ *d_k{}^*(\ g_1{}^*(n)\ *x(n))^k \qquad\qquad\qquad (32)$

$d_k{}^*$ can be determined by the sampled average method:

$$d_k{}^*=Avg\left(\frac{y_k(n)}{g_2{}^*(n)*(g_1{}^*(n)*x(n))^k}\right)$$

$$=\frac{1}{L+1}\sum_{i=0}^{L}\left(\frac{y_k(i)}{\left(\sum_{j_1=0}^{i}g_2{}^*(i-j_1)\left(\sum_{j_2=0}^{j_1}g_1{}^*(j_2)x(j_1-j_2)\right)\right)^k}\right) \qquad (33)$$

COMPUTATION PROCEDURE FOR NONLINEAR ELEMENT WITH SMOOTH NONLINEARITY

To implement the above method, a brief outline of procedure is given below from the input $x(t)$ and output $y(t)$ data of the system:

1. A series of input signals $c_i x(n)$, scaled to various levels of

c_i, i=1, 2, ..., N-1, are applied to the system, and the corresponding number of output signals $y_i^*(n)$, i=1,2,..., N-1 are measured from the experiment. The input can be step, sinusoidal, or any signal that can be scaled to proper level for measurement purpose.

2. The Volterra components of the output signal, $y_1(n)$, $y_2(n)$, are computed from Equ.4. Higher order of $y_i^*(n)$, for i>2 are used here and N should be chosen to be higher than 2 to improve the accuracy.

3. Solve for $g_1^*(n) = \beta_1 g_1(n)$ and $g_2^*(n) = \beta_2 g_2(n)$ as follows:

(a). Obtain the cost functions, f_k, k=0, ..., N. Usually let M=2 for grouping purpose;

(b). Obtain the initial values from Equ.25 for strictly proper system and from Equ.26 for nonproper system. The optimum order of predictor can be determined by comparing the truncation error $|E_1^{(p-1)}|$, $|E_1^{(p)}|$ and $|E_1^{(p+1)}|$, as in Equ.24. The smallest error determined the order. Once the order is determined, the initial guess can be made.

(c). Use F-R method to minimize f_k to obtain $g_1^*(n)$, $g_2^*(n)$. A brief flowchart is shown in Fig.4.

4. The nonlinear elements can be obtained from Equ.33.

A program IDENT has been written in C-language on MASCOMP. The F-R method is included as part of the program and is shown in Appendix III.

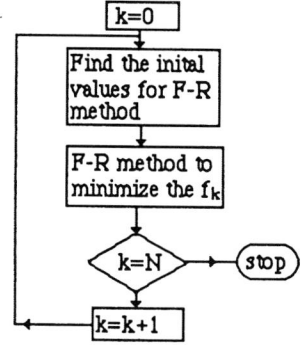

Fig.4. Flowchart to obtain $g_1^*(n)$ and $g_2^*(n)$.

We have thus presented a method to identify a block-structured model of LNL system based on the input-output relationship. The nonlinear element must contain second order Volterra kernel in its output. In addition to that it must be the non-memory smooth type that can be represented by polynomial approximation. However, in most practical cases, nonlinearity could very well be the abrupt type which may consist of a finite number of discontinuities. In this case, the best way to represent the nonlinear element is usually the piece wise approximation. In some extreme case, where the nonlinear element may consist of finite number of discontinuities of the first kind, or finite jump (saltus) at its point of discontinuity (such as on-off type), an approximation by Volterra-Fourier method has been derived and the result compared faverably well with orther method such as Hermite polynomial series approximation (Bartos, Sun, 1988).

ALGORITHM FOR NONLINEARITY WITH ODD SYMMETRY

If the static nonlinear element $F(\cdot)$ has the property of odd symmetry, then all the even components in the Volterra series of Equ.1 will be zero. This characteristic of nonlinearity can be easily detected by inspecting the solution of Equ.4, or $y_i(n)=0$ for $i=2,4,\ldots$. We shall then proceed to find $y_3(n)$ from Equ.4 and rewrite Equ.8 as follows:

$$y_3 = a_3 \sum_{i_1=0}^{n} g_2(n-i_1) \left(\sum_{i_2=0}^{i_1} g_1(i_2)x(i_1-i_2) \right)^3 \tag{34}$$

To eliminate a_1 and a_3, we let $\beta_1\beta_2 = a_1, \beta_1^3\beta_2 = a_3$, and rewrite Equ.34 as:

$$y_3 = \pm \sum_{i_1=0}^{n} g_2^*(n-i_1) \left(\sum_{i_2=0}^{i_1} g_1^*(i_2)x(i_1-i_2) \right)^3 \tag{35}$$

Since a_1 and a_3 may be negative, this means $\beta_1 = \text{sqroot}(a_3/a_1)$, and $\beta_2 = a_1 \text{sqroot}(a_1/a_3)$, becomes irrational if a_1 and a_3 have opposite signs. This will make the iteration process divergent. We should then take the negative sign of Equ.35 or reverse the sign of either a_1 or a_3 in order to continue the process and then reverse back the sign of a_1 or a_3 again after the model has been developed. The cost function and its derivatives function are modified as:

Cost Function:

$$f_k = \frac{1}{M} \sum_{i=kM}^{(k+1)M-1} \left(\sum_{j=0}^{i} g_1^*(j)g_2^*(i-j)-h_1(i) \right)^2 +$$

$$\frac{1}{M} \sum_{i=kM}^{(k+1)M-1} \left(\sum_{i_2=0}^{i} g_2^*(i-i_2) \left(\sum_{i_1=0}^{i_2} g_1^*(i_1)x(i_2-i_1) \right)^3 -y_3(i) \right)^2 \tag{36}$$

Gradient $-f_k$:
A_1 and A_3 are the same as Equ.28a and Equ.28c

$$A_2 = \frac{6}{M} \sum_{i=kM}^{(k+1)M-1} \left(\left(\sum_{i_2=0}^{i} g_2*(i_1-i_2) \left(\sum_{i_1=0}^{i_2} g_1*(i_1)x(i_2-i_1) \right) \right)^3 - y_3(i) \right) \cdot$$

$$\left(\sum_{i_2=0}^{i} g_2*(i-i_2) \left(\sum_{i_1=0}^{i_2} g_1*(i_1)x(i_2-i_1) \right) x(i_2-l) \right) \tag{37a}$$

$$A_4 = \frac{2}{M} \sum_{i=kM}^{(k+1)M-1} \left(\left(\sum_{i_2=0}^{i} g_2*(i_2) \left(\sum_{i_1=0}^{i-i_2} g_1*(i_1)x(i-i_2-i_1) \right) \right)^3 - y_3(i) \right) \cdot$$

$$\left(\sum_{i_1=0}^{i-l} g_1*(i_1)x(i-l-i_1) \right)^2 \tag{37b}$$

A computer program ODDIDENT is written for this purpose.

ALGORITHM FOR NONLINEARITY WITH PIECEWISE MODEL

From Fig.2, we can easily write the expression for $u*(n)$ or:

$$u*(n) = \sum_{i=0}^{n} g_1*(i)x(n-i) \tag{38}$$

Similarly, $z*(n)$ can be found from the following expression:

$$\begin{bmatrix} y(0) \\ y(1) \\ \vdots \\ y(n) \end{bmatrix} = \begin{bmatrix} g_2*(0) & 0 & 0 & & 0 \\ g_2*(1) & g_2*(0) & 0 & & 0 \\ & & \cdots\cdots\cdots\cdots & & \\ g_2*(n) & g_2*(n-1) & & & g_2*(0) \end{bmatrix} \begin{bmatrix} z*(0) \\ z*(1) \\ \vdots \\ z*(n) \end{bmatrix} \tag{39a}$$

for non-strictly proper transfer function and

$$\begin{bmatrix} y(1) \\ y(2) \\ \vdots \\ y(n+1) \end{bmatrix} = \begin{bmatrix} g_2*(1) & 0 & 0 & & 0 \\ g_2*(2) & g_2*(1) & 0 & & 0 \\ & & \cdots\cdots\cdots\cdots & & \\ g_2*(n+1) & g_2*(n) & & & g_2*(1) \end{bmatrix} \begin{bmatrix} z*(0) \\ z*(1) \\ \vdots \\ z*(n) \end{bmatrix} \tag{39b}$$

for strictly proper transfer function.

Let the nonlinear piecewise system be described by the input $u*(n)$ and output $z*(n)$ as follows:

$$z_i*(n) = \begin{cases} a_1 u_1*(n)+b_1 & \alpha_0 \le u_1*(n) < \alpha_1 \\ a_2 u_2*(n)+b_2 & \alpha_1 \le u_2*(n) < \alpha_2 \\ \cdots & \cdots \\ a_R u_R*(n)+b_R & \alpha_{R-1} \le u_R*(n) \le \alpha_R \end{cases} \tag{40}$$

where a_i, b_i, $i=1,\ldots R$, are the slope and intercept of the piecewise representation for each segment; α_i, $i=1,\ldots,$ R-1 be the break point or the boundary point of each segment with α_0 and α_R represent the initial and final values of the boundary or $\alpha_R - \alpha_0$ be the length of the total boundary of the nonlinear piecewise system.

The break point between two consecutive segments can be easily expressed as:

$$\alpha_i = (b_i - b_{i+1}) / (a_{i+1} - a_i) \qquad (41)$$
$$i = 1, 2, \ldots, R-1.$$

We shall use the χ^2 regression method to determine a_i and b_i. Let P be the Gaussian probability density function with mean $a_i u_i(n) + b_i$ and variance $s_i^2(n)$ for the $z_i(n)$ of each segment, as:

$$P[\, a_i\, u_i(n) + b_i,\ s_i^2(n)\,] \qquad (42)$$

and the χ^2 of each segment is defined as follows:

$$\chi^2 = \sum_{n=1}^{N_i} \frac{(z_i(n) - (a_i u_i(n) + b_i))^2}{\sigma_i^2(n)} \qquad (43)$$

where N_i represents the number of sampled points in the ith segment of the piecewise nonlinear system. The regressive line can be determined by minimizing χ^2 as follows:

Let:

$$\frac{\partial \chi^2}{\partial a_i}\Big|_{a_i = a_i{}^*; b_i = b_i{}^*} = 0 \qquad (44a)$$

$$\frac{\partial \chi^2}{\partial b_i}\Big|_{a_i = a_i{}^*; b_i = b_i{}^*} = 0 \qquad (44b)$$

We have thus:

$$\sum_{n=1}^{N_i} \frac{[z_i(n) - \{a_i{}^* u_i(n) + b_i{}^*\}]}{\sigma_i^2(n)} [u_i(n)] = 0 \qquad (45a)$$

$$\sum_{n=1}^{N_i} \frac{[z_i(n) - \{a_i{}^* u_i(n) + b_i{}^*\}]}{\sigma_i^2(n)} = 0 \qquad (45b)$$

Let us choose the same $s_i(n)$ within segment, then the expected values of $a_i{}^*$ and $b_i{}^*$ can be solved from Equ.55a and Equ.55b, respectively, with all the $s_i(n)$ cancelled each other, i.e.

$$a^*_i = \frac{N_i \sum_{n=1}^{N_i}(u_i(n)z_i(n)) - \left(\sum_{n=1}^{N_i}u_i(n)\right)\left(\sum_{n=1}^{N_i}z_i(n)\right)}{N_i\left(\sum_{n=1}^{N_i}u_i^2(n)\right) - \left(\sum_{n=1}^{N_i}u_i(n)\right)^2} \tag{46a}$$

$$b^*_i = \frac{\left(\sum_{n=1}^{N_i}u_i^2(n)\right)\left(\sum_{n=1}^{N_i}z_i(n)\right) - \left(\sum_{n=1}^{N_i}u_i(n)z(n)\right)\left(\sum_{n=1}^{N_i}u_i(n)\right)}{N_i\left(\sum_{n=1}^{N_i}u_i^2(n)\right) - \left(\sum_{n=1}^{N_i}u_i(n)\right)^2} \tag{46b}$$

We shall define the maximum error ϵ of each segment for the pair $u_i(n)$, $z_i(n)$ as:

$$\text{Max} \mid z_i(n) - [\, a_i^* u_i(n) + b_i^* \,] \mid \ \leq \ \epsilon \tag{47}$$

ϵ can also be defined as the tolerance and is set by the user. The smaller the ϵ is, the more segment will be resulted.

The procedure starts at point α_0 with a prescribed tolerance Œ and the number of points in the segment is increased by adding one pair at a time. The parameters a_i^*, b_i^* are then calculated from Equ.46. The segment is ended if the tolerance ϵ as defined by Equ.47 is no longer satisfied. The procedure is then repeated to find the next segment until α_R is reached. The break point of each segment is calculated from Equ.41. This completes the computation process. A computer program "P_lident" is written in C language for this purpose. It should be noted here that the piecewise nonlinear model obtained by the procedure above is valid only within interval I and I depends on the amplitude of the test signal.

NUMERICAL EXAMPLE

To illustrate the numerical computation, we shall present an example of nonlinear element with odd symmetry as:

Let the subsystems be defined as:

$g_1(n) = (0.8)^n$; $g_2(n) = (0.75)^n$
$F(x) = x - x^3 + 0.1x^5$

Input in this case is white noise. The result of the identified nonlinear subsystem by the use of ODDIDENT is tabulated as:

	d_1	d_2	d_3	d_4	d_5
original	1	0	-1	0	0.1
identified	1.00000	0.00000	-1.0000	0.00000	0.09999

$g_1(n)$

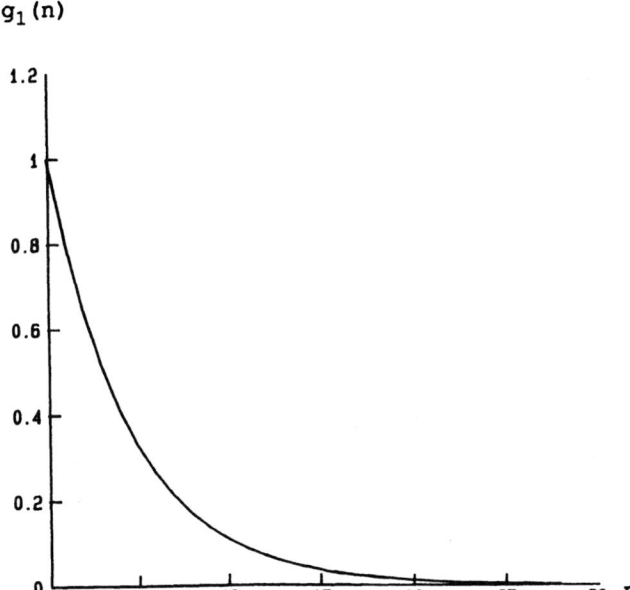

Fig.5a.   Identified $g_1^*(n)$ and original $g_1^*(n)$. Both curves are too close to separate.

$g_2(n)$

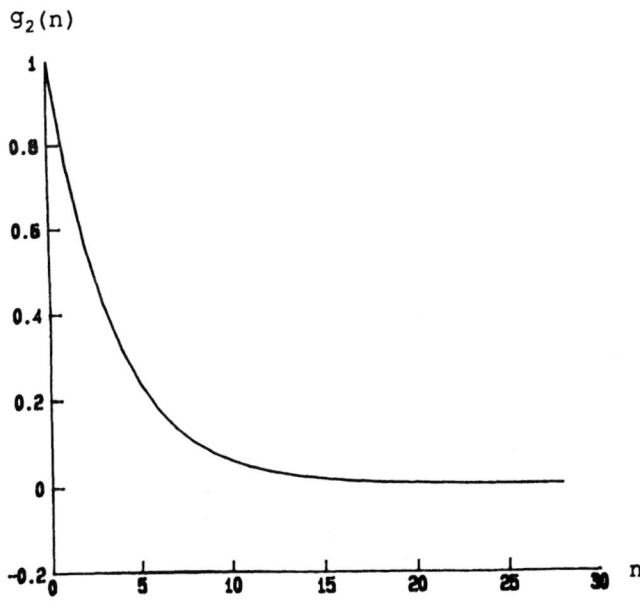

Fig.5b.   Identified $g_2^*(n)$ and original $g_2^*(n)$. Both curves are too close to separate.

CASCADE MODEL FOR BIOELECTRODE POLARIZATION

The interfacial polarization immittance between metal electrode and electrolyte solution is a very important phenomenon in the monitoring of bioelectric signals and stimulation application. The behavior of such interface is usually linear for small perterbations, but in most cases it is nonlinear for large signal usage. Schwan(1965) has defined the limit law of linearity for the onset of nonlinearity whose behavior can be explained by a fractional power factor.Onaral and Schwan(1982) have extended measurement capability to over six decade of frequency with sinusoidal perterbation at various amplitudes toward the nonlinear range.

Sun and Onaral(1983) have generalized the Davidson and Cole model into multiple fractional power poles for linear representation of the polarization impedance and used the positive real principle to prove its existence(Sun & et all,1984) over the six decades frequency range.

The nonlinear range of the polarization immittance was approximated by a cascaded model of a single fractional power pole and a modified saturation element using the Describing Function method(Sun, Onaral,1983). The system was further generalized as a fractal dimension system where the nonlinear element was modeled by a piece wise saturation type element and the same datawas reconstructed from this model with reasonable accuracy(Sun, 1988).

We shall now use the new algorithm method to develop a cascade model for the polarization immittance system and the result proves the existense of the piecewise saturation model which is very similar to the one developed before.

The Onaral-schwan data(1983) showed the dependance of Pt electrode overpotential on the increasing current densities with rectangular input pulses for six decades of time response as shown in Fig.6. Three input levels are chosen for our model, with J=46, 4.6 and 0.47mF/cm2. These three response curves(overpotentials) are first approximated by the following equations (converted to rectangualr scale) as:

$J = 46 mA/cm^2$:

$$V_{pol} = \begin{bmatrix} t^{0.8} & 0.01 \leq t < 0.1 \\ 0.42t^{0.425} & 0.1 \leq t < 10 \\ 0.99t^{0.053} & 10 \leq t \leq 7079.5 \end{bmatrix}$$

$J = 4.6 mA/cm^2$:

$$V_{pol} = \begin{bmatrix} 0.089t^{0.75} & 0.01 \leq t < 1 \\ 0.089t^{0.58} & 1 \leq t < 8.91 \\ 0.16t^{0.303} & 8.91 \leq t \leq 398.1 \end{bmatrix}$$

$J = 0.47 mA/cm^2$:

$$V_{pol} = \begin{bmatrix} 0.0079t^{0.58} & 0.01 \leq t < 0.2 \\ 0.0115t^{0.79} & 0.2 \leq t < 10 \\ 0.0275t^{0.41} & 10 \leq t < 70.79 \\ 0.0398t^{0.325} & 70.79 \leq t < 7079.5 \end{bmatrix}$$

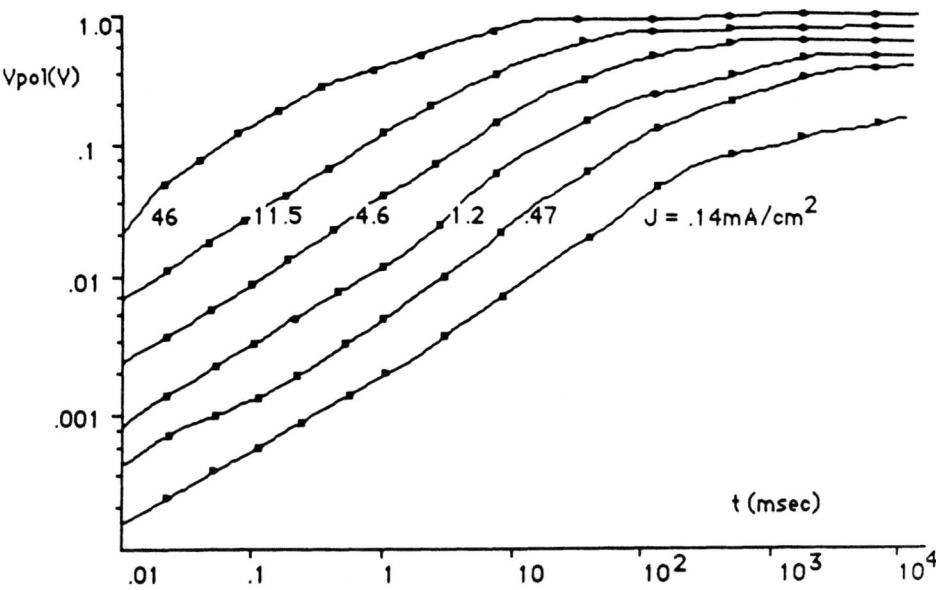

Fig.6.   Experimental data of cathodic overpotential versus
           time with current density as parameter (Onaral and
           Schwan, 1978) Pt electrode, parallel cell, area
           0.0855cm², 0.9% NACL, T:25±1°C.

Numerical computation of the Volterra components from equ.3 is carried out by replacing: $m \to n\Delta t$, $\tau \to i\Delta t$, $\tau_1 \to i_1\Delta t$, $\tau_2 \to i_2\Delta t$ etc and the limits of summation are from 0 to N-1. Similarly for Equ.4. To same computation time, two separate routines for two range of t were used:

N=50,	Δt=1ms	$0.01 \leq t \leq 50$ ms
N=100,	Δt=10ms	$1 \leq t \leq 10^3$ ms

There is a small overlap region which can be easily seen from the identified results. The result of $g_1(t)$ is plotted in Fig.7 and it is strikingly similar to the fractal model developed before (Wang,1986). $g_2(t)$ here is assumed to be unity, so the model is essentially a Wiener type. The nonlinear subsystem is obtained by using the piecewise program, P_LIDENT, and the result is plotted in Fig.8 which again is remarkably similar to the previous result(Sun et all,1988, Wang,1986). The original data are reproduced from the model in Fig.9. The slight discontinction in the curves from the model around t=10ms is mainly due to the change over in the range of t from one set to another.

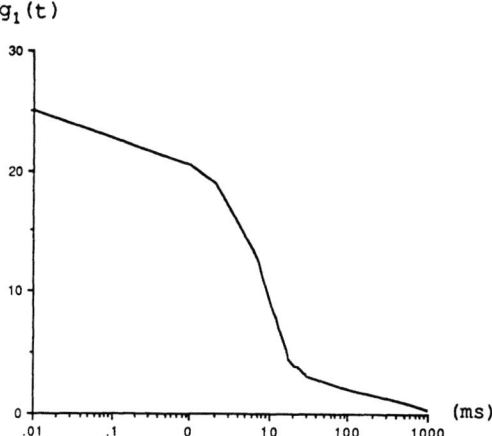

Fig.7.   The linear subsystem for the Pt electrode model
          with t in log scale.

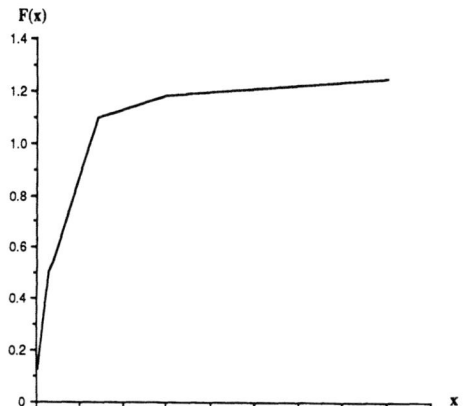

Fig.8.   Piece wise saturation model of the Pt-electrode
          inter-facial impedance.

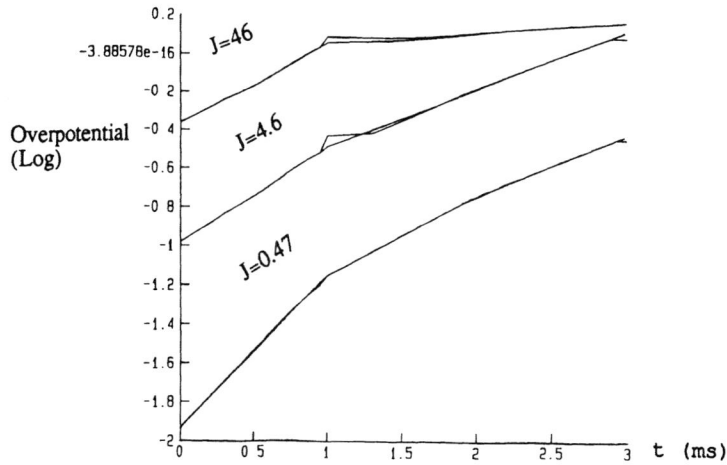

Fig.9.   Data reproduced from the model as compared with
          the original value; t is in log scale.

CONCLUSION

A procedure for the identification of nonliear system in LNL cascaded model is presented here and the method is based strictly on the input and output data and no prior knowledge of the system is required. The method presented is a different approach than the one developed by Korenberg and Billings in that it is based on the discrete data and the decoupling of the linear and nonlinear subsystem is based strictly on the multilevel of input signal instead of the first and second order crosscorrelation functions. Therefore, the input signal is not restricted to white noise or mixture of sinusoidal and for most practical purpose a step input with different scaling factor is all one needed for the measurement purpose. The computation of $g_1(n)$ and $g_2(n)$ are based on the minimization of cost function by using the predictor-corrector method. Three computer programs, namely, IDENT, ODDIDENT, P_LIDENT have been developed and our experience shows that they are comparatively simple and takes much less time than all the previous methods. The method is equally applicable to both smooth and abrupt type of nonlinear element such as piecewise as long as it satisfies the Weierstrass Criteron.

In the practical case of system identification, most of the informations we can obtain include only input and output signals and no prior knowledge of the system such as the order of the subsystem for $g_1(n)$ and $g_2(n)$ are available. Although white noise input have been used mostly for this purpose, there are cases where a pure white noise signal is not available and also the system may not be able to sustain the long duration required for white noise input in order to obtain meaningful result. For most of the measurement purpose, a step or single sinusiodal signal might be the only signal available and it is for these reasons that the method presented here have a very unique advantage over the other methods.

The K-B model described above has been widely used for system identification in biological system and to some extent in structured or mechanical system. It is not applicable to systems with feedback, multiloops, or system with cogutive or adaptive nature. Palm(1979) has suggested a LNL sandwich type but no algorithm was provided for identification purpose. Kolmogoroff(1957) has shown that any continuous system can be represented exactly as a double sandwich of NLNL and the function in first nonlinearity could be quite "wide", not diffrentiable or abrupt type. We have shown(Shi, 1988) that a number of parallel array of K-B structure can be used for general nonlinear system identification. This means that an iterative procedure can be denied to approximate a large class of nonlinear function in discrete domain. We have also shown that the number of parallel arraies can be reduced if the system under study have weak nonlinearity. In case where the system exhibit low pass and low Q frequency resonant phenomena, a single array representation will be sufficient for approximation with reasonable accuracy. Therefore, the result as presented can be easily extended to a more general type of system identification.

ACKNOWLEDGEMENT

The authors wish to express deep appreciation to the Office of Naval Research, N00014-87-0210 and GE-RCA Moorestown N.J. for their general support of this research project.

REFERENCES

Bartos J. and Sun H. H., 1988, Characterization of Abrupt Nonliearity by Volterra-Fourier's Method, _J Franklin Inst_, 325:465 -484.

Billings SA and Fakhour SY, 1978, Identification of a Class of Nonlinear Systems Using Correlation Analysis, _Proc. IEE_, 125:691-697.

Billings SA and Fakhouri SY, 1982, Identification of Systems Containing Linear Dynamic and Static Nonlinear Elements, _Automatica_, 18:15-26.

Flecher R. and Reeves CM, 1964, Function Minimization by Conjugate Gradients, _Computer J._ 7:149-153.

Gardner AB, 1973, Identification of Processes Containing Single Valued Nonlinearities, _Int. J. control_, 18:1029-1039.

Korenberg M.J., 1985, Orthogonal Identification of Nonlinear Difference Equation Method, _Proc. 28th Midwest Sym on CAS_, 90-95.

Korenberg M.J., 1988, Identifying Nonlinear Difference Equation and Functional Expansion Representation: The Fast Orthogonal Algorithem, _Annals of Biomedical Eng._16:123-142 (Special Issue on Nonlinear Modeling of Physiological Systems).

Kolmogoroff AN, 1957, On the Representiation of Continuous Function of Several Variables by Supperpostion of Contiuous Function of One Variable and Addition, _Dokl Akad Nawk USSR_, 114:953-954. AMS Translation,2:55-59.

Korenberg MJ, and Hunter IW, 1986, The Identification of Nonliear Biological Systems: LNL Cascaded Models, _Biol. Cybern_. 55:125-134.

Korn GA and Korn TM, 1968, "Mathmatical Handbook for Scienticsts and Engineers", McGraw-Hill Corp NY.

Korenberg MJ, 1973a, Identification Biological Cascades of Linear and Static Nonlinear Systems, _Proc. 16th Midwest Symp Circuit Theory_ 18.2:1-9.

Korenberg MJ,1973b, Cross-correlation of Neural Cascades, _10th Rocky Mountain Bio Eng. Symp_. 47-51.

Onaral B, Schwan HP, 1983, Linear and Nonlinear Properties of Platinum Electrode Polarization II: Time Domain Analysis, _Mde & Biol Eng & Comp_ ,21:210-216.

Onaral B, Schwan HP, 1982, Linear and Nonlinear Properties of Platinum Electrode Polarization I:Frequency Domain Analysis, _Med & Biol Eng & Comp_, 20:299-306.

Schwan HP, Maczuk JG, 1965, Electrode Polarization Impedance: Limits of Linearity, _Proc 18 Ann Conf EMB_, 24.

Sun HH, Onaral B, 1983, A Unified Approah to Represent Metal Electrode Polarization, _IEEE Trans on BME_, 21:399-406.

Sun HH, Onaral B and Tsao YY, 1984, Application of the Positive Reality Principle to Metal Linear Polarization Phenomena, _IEEE Trans on BME_, 664-674.

Sun HH, Wang X and Onaral B, 1988, Onset of Nonlinearity in Fractal Dimension System: An Application to Polarized Bioelectrode Interface, _Annals of Biomed Eng._ 111-121.

Schetzen M, 1980, "The Volterra and Wiener Theories of Nonlinear Systems", John Wiley & Son NY.

Tuis L, 1978, "Identification of Nonlinear Systems by Means of Multilevel Pseudo-Random Signals Applied to a Water Turbine Unit", Identification and System Parameter Estimation, Rajbman, Editor, North-Holland Publishing Co. 2027-2036.

Wang X, 1986, "Phase Plane Analysis in the Computer Simulation Processes on Bioelectrode Polarization Phenomena", M.S. thesis, Drexel University lib.

APPENDIX I   PROOF OF Equ.22

$$E^{(P_1)}(kM) = g_1*(kM) - L_1(kM) \tag{A-1}$$

Consider a auxilary function $f(t)$:

$$f(t) = g_1*(t) - L_1(t) - C(t-kM+p_1+1)(t-kM+p_1)\ldots(t-kM+1) \tag{A-2}$$

C is a constant such that $f(t) = 0$ at $kM+p_1-1, kM-p_1, \ldots, kM$.
Repeatedly using Rolle theorem, we have:

$$f^{(n+1)}(t) = g_1*^{(n+1)}(t) - C(p_1+1) \tag{A-3}$$

Using Rolle theorem again, we have $f^{(n+1)}(\S_1) = 0$
($kM+p_1-1 \leq \S_1 \leq kM$)
So, $C = g_1*^{(n+1)}(\S_1)/(p+1)!$ \hfill (A-4)

Substituting (A-4) into (A-2), we have:

$f(kM) = g_1*(kM) - L_1(kM) - g_1*^{(n+1)}(\S_1) = 0$.
From (A-1)
$$E^{(P_1)}(kM) = g_1*^{(n+1)}(\S_1) \tag{A-5}$$
Same proof for $E^{(P_2)}(kM)$.
        Q.E.D

APPENDIX II   PROOF OF Equ.23

Assume the sample interval is small, then:

$$g_1*^{(n+1)}(\S_1) \approx (p_1+1)! \, g_1*[km-p-2, kM-p-1, \ldots, kM-1] \tag{B-1}$$
where $g_1*[kM-p-2, kM-p-1, \ldots, kM-1]$ is (p_1+1)th order difference quotient.

According to the relation between difference quotient and difference, i.e.

$$\nabla^{(P_1+1)} g_1*(kM-1) = h^{(P_1+1)}(p1+1)! \, g_1*[kM-p-2, kM-p-1, \ldots, kM-1] \tag{B-2}$$
where h is sample interval.

For uniform normalized sample interval, we have:

$$g_1*^{(n+1)}(\S_1) \approx \nabla^{(P_1+1)} g_1*(kM-1) \tag{B-3}$$
From (A-5)

$$E^{(P_1)}(kM) \approx \nabla^{(P_1+1)} g_1*(kM-1) \tag{B-4}$$
Same proof for $E^{(P_2)}(kM)$.
        Q.E.D.

NONLINEAR FILTERS FOR TRACKING CHAOS

IN NEUROBIOLOGICAL TIME SERIES

Bernard Saltzberg and William D. Burton, Jr.

Department of Psychiatry and Behavioral Science
University of Texas Health Science Center at Houston
Medical School
Houston, Texas 77030

Epilepsy Research Center
Baylor College of Medicine
Houston, Texas 77030

SUMMARY

A nonlinear digital filtering approach to tracking the nonstationary
dynamical behavior of neurobiological time series is described and
evaluated. Nonlinear digital filters were designed with the objective of
achieving high resolution tracking of the changing properties of complex
nonstationary time series. Three simple nonlinear autoregressive models;
the logistic, quadratic, and cubic models, were chosen because they each
exhibit a rich variety of behavior including simple periodic, complex
periodic, and chaotic behavior depending upon the value of a critical
model parameter (the bifurcation parameter). The sensitivity of the time
series to the value of the critical bifurcation parameter in each model is
shown in the bifurcation diagrams Figs. 1a, 1b, and 1c.

Digital filters based on the three models were derived based on the
least-squares estimate of the critical parameter. The tracking capability
of these filters was evaluated using nonlinear, nonstationary test data.
The nonstationarity was introduced by continuous time modulation of the
bifurcation parameter in each model. The parameter was modulated over a
range which results in the production of both periodic and chaotic
behavior.

The test results obtained with these filter models demonstrate that
accurate high resolution tracking of the critical bifurcation parameters
can be achieved in essentially real-time using a standard laboratory
computer. The high resolution capability of the filter is particularly
useful for identifying chaotically stationary data epochs, an essential
condition for estimating other chaotic features such as the Hausdorff
dimension, the Lyapunov exponent or the attractor topography. Since the
critical bifurcation parameter values increase with data complexity (i.e.,
increasing when data in the filter window changes from periodic to

(a) LOGISTIC Model

$X(t+\Delta t) = AX(t) - BX^2(t)$

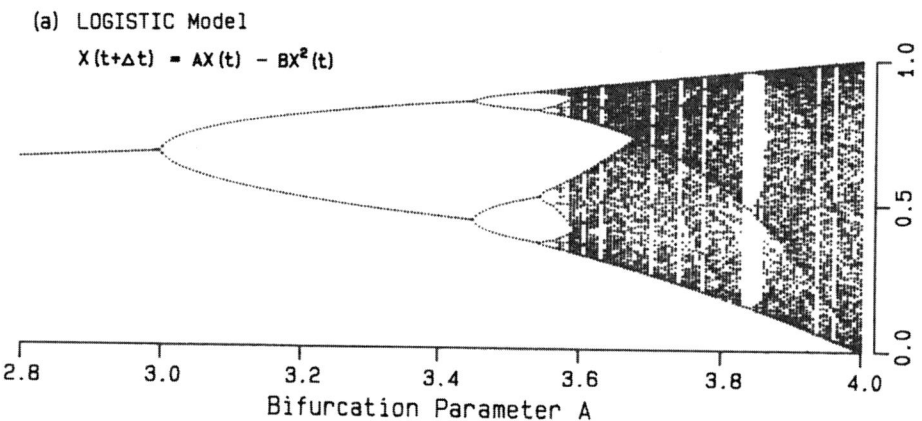

Bifurcation Parameter A

(b) QUADRATIC Model

$X(t+\Delta t) = C - X^2(t)$

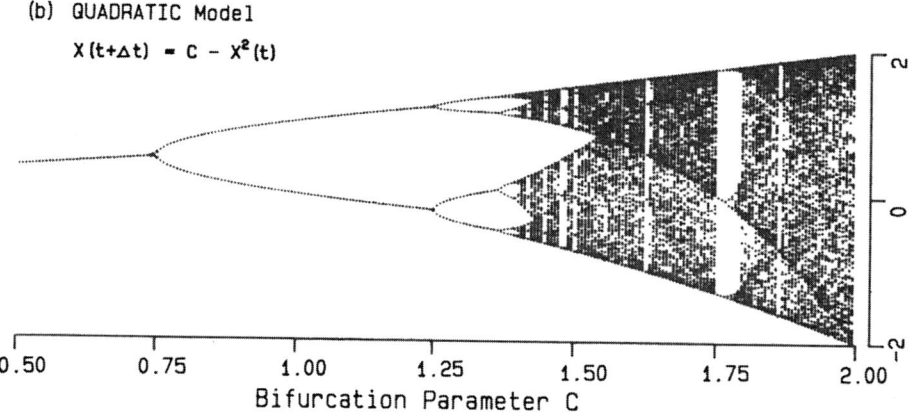

Bifurcation Parameter C

(c) CUBIC Model

$X(t+\Delta t) = aX(t) - bX^3(t)$

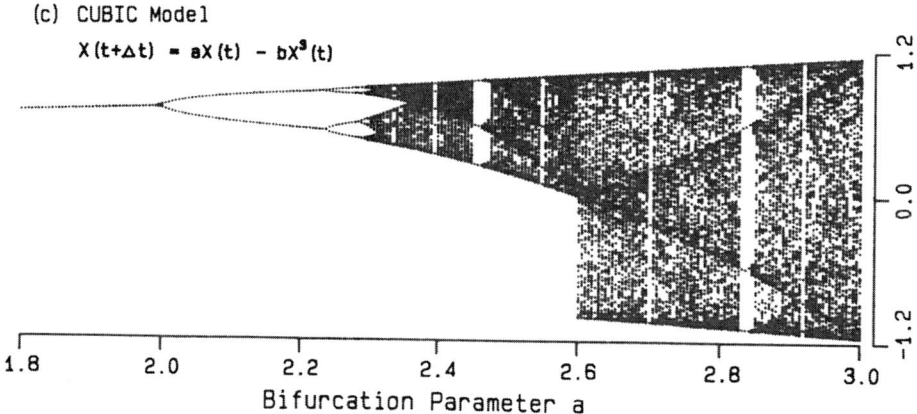

Bifurcation Parameter a

Figure 1. Bifurcation diagrams for three nonlinear models.

chaotic), such filters can provide an effective economical means of monitoring the nonlinear and nonstationary chaotic properties of a long time series.

INTRODUCTION

Several difficulties have been encountered in studies of neurobiological time series (Layne et al., 1986; Babloyantz and Destexhe, 1986) which estimate the Chaotic or Hausdorff Dimension (HD) as a basis for discriminating neurophysiological change of state. One difficulty is due to nonstationarities which can produce changes in the Hausdorff Dimension of the time series and thereby greatly compromise the accuracy of the estimate for lack of a stationary sample of sufficient length. Another difficulty involves the immense computational requirements associated with estimating the Hausdorff Dimension of long duration time series. For example, derivation of the Hausdorff Dimension (Babloyantz et. al., 1985) of a time series based on 2,000 samples can require the computation of 20 million vector distances between points in vector spaces determined from combined samples of the time series. This severely limits the amount of data that can reasonably be subjected to HD analysis even with a supercomputer.

A parametric modeling approach which is not burdened by the above difficulties is proposed in this paper as an alternative to HD estimation. The approach is based on the use of digital filters which estimate the parameters of nonlinear autoregressive models that are used to represent a time series. This form of representation of complex nonstationary neurobiological signals was suggested by the observation that certain simple nonlinear models exhibit a variety of behavior similar to the nonstationary signals encountered in electroencephalography.

FILTER DERIVATION

The filters corresponding to three nonlinear models; the logistic model, the quadratic model, and the cubic model are derived in this section.

LOGISTIC MODEL

The conventional form of the logistic model is:

$$X(t+\Delta t) = \alpha X(t) [1-X(t)] \tag{1}$$

To avoid amplitude scaling constraints, Eq. (1) can be rewritten as:

$$X(t+\Delta t) = AX(t) - BX^2(t) \tag{2}$$

$$\text{with } 0 < A \leq 4 \text{ and } B > 0$$

It can be shown that the chaotic behavior of the logistic model is dependent on the parameter A alone in Eq. 2 in a manner equivalent to the parameter α in Eq. 1. Therefore Eq. 2 was used to eliminate scaling dependence in obtaining a least square estimate for A. The behavior of the output of this model is critically dependent on the value of A as illustrated in the bifurcation map of Fig. 1(a). For example, over the range $0 < A < 3$ the output of (2) converges to a constant value. In the range $3 < A \leq 4$ the output changes from periodic to chaotic when A exceeds approximately 3.57. For $A > 4$ the output increases without bound with increasing time.

The minimum least squares estimate (MLE) for A, based on Eq. (2) is obtained from the squared error function as follows:

$$\epsilon^2 = \sum_{k=0}^{k=N} \left[X_{(t+\{k+1\}\Delta t)} - AX_{(t+k\Delta t)} + BX^2_{(t+k\Delta t)} \right]^2 \tag{3}$$

where ϵ^2 is the squared error of estimation.

The value of A (denoted as Â) which minimizes the error is determined by setting the partial derivatives with respect to A and B equal to zero and solving the resulting simultaneous equations for Â. This gives an Â filter estimate of the form:

$$\hat{A}(t) = \frac{\sum_{k=0}^{k=N} X^3_{(t+k\Delta t)} \sum_{k=0}^{k=N} X^2_{(t+k\Delta t)} X_{(t+\{k+1\}\Delta t)} - \sum_{k=0}^{k=N} X^4_{(t+k\Delta t)} \sum_{k=0}^{k=N} X_{(t+k\Delta t)} X_{(t+\{k+1\}\Delta t)}}{\left\{ \sum_{k=0}^{k=N} X^3_{(t+k\Delta t)} \right\}^2 - \sum_{k=0}^{k=N} X^4_{(t+k\Delta t)} \sum_{k=0}^{k=N} X^2_{(t+k\Delta t)}} \tag{4}$$

QUADRATIC MODEL

The quadratic model is given by:

$$X(t+\Delta t) = C - X^2(t) \tag{5}$$

$$\text{with} \quad 0 < C \leq 2$$

The nature of the output of this model is critically dependent on the value of C as illustrated in the bifurcation map of Fig. 1(b). For example, over the range $0 < C \leq 3/4$, the output of (5) converges to a constant value. In the range $3/4 < C < 2$ the output changes from periodic to chaotic when C exceeds approximately 1.43. For $C > 2$ the output tends to negative infinity with increasing time. Proceeding as before, we obtain the MLE for C:

$$\hat{C} = \frac{\sum_{k=0}^{k=N} X^2_{(t+k\Delta t)} + \sum_{k=0}^{k=N} X_{(t+\{k+1\}\Delta t)}}{N + 1} \tag{6}$$

CUBIC MODEL

In order to study complex, zero mean, time series such as those encountered in EEG, a cubic model is introduced here which produces zero mean chaotic behavior over a significant range of its critical parameter. The conventional form of the cubic model is given by:

$$X(t+\Delta t) = \alpha X(t) [1-X^2(t)] \tag{7}$$

To avoid the scaling constraint, Eq. (7) can be rewritten as:

$$X(t+\Delta t) = aX(t) - bX^3(t) \tag{8}$$

The parameter a alone in Eq. (8) determines the chaotic behavior of this model in a manner equivalent to the parameter α in Eq. (7). It can be noted from the bifurcation map of this model (Figure 1(c)) that for $0 < a < 2$ the output of Eq. (8) converges to a constant value. In the range $2 < a \leq 3$ the output changes from periodic to chaotic when a exceeds approximately 2.3. It can also be noted from Fig. 1(c) that zero mean chaotic bahavior occurs when $a \geq \dfrac{3\sqrt{3}}{2} = 2.598$. For $a > 3$ the amplitude of the chaotic behavior grows without bound. The time domain filter for tracking the critical parameter a, is given by:

$$\hat{a}(t) = \frac{\sum_{k=0}^{k=N} X^4(t+k\Delta t) \sum_{k=0}^{k=N} X^3(t+k\Delta t)\,X(t+[k+1]\Delta t) - \sum_{k=0}^{k=N} X^6(t+k\Delta t) \sum_{k=0}^{k=N} X(t+k\Delta t)\,X(t+[k+1]\Delta t)}{\left\{ \sum_{k=0}^{k=N} X^4(t+k\Delta t) \right\}^2 - \sum_{k=0}^{k=N} X^6(t+k\Delta t) \sum_{k=0}^{k=N} X^2(t+k\Delta t)} \tag{9}$$

RESULTS: FILTER TRACKING PERFORMANCE

The foregoing filter equations (4) (6) and (9) have been applied to test data to determine the number of data points that are required in the filter window in order to obtain an accurate estimate of the parameters associated with each model.

Examples of test data for the logistic model, the quadratic model, and the cubic model are shown in Fig. 2. For these test data estimates of the bifurcation parameters were computed using the corresponding filter equations (4), (6), (9) for several values of N, the number of time series points used. The results are shown in Table 1. In this test the filters estimated the bifurcation parameters to essentially the machine arithmetic precision. This result implies that short-term stationarity of the bifurcation parameters can be accurately estimated with these nonlinear filters.

A test of the ability of the filters to track nonstationarities in the bifurcation parameter was carried out by modulating the bifurcation parameter so that transitions between chaotic and non-chaotic behavior would occur. Tests of different sample sizes in the filter window were conducted to determine tracking accuracy as a function of sample size. Figs. 3a, b, and c represent test data in which the critical parameters have been sinusoidally modulated. Figs. 4a, b and c present the output of the filters and a superimposed plot of the actual modulation for comparison in order to demonstrate filter performance. As Figure 4 and Table 2 illustrate the trade-offs among the number of samples in the filter window, tracking accuracy, and tracking resolution for various rates of nonstationary change can be obtained from this testing procedure.

Since in studying neurophysiological time series the applicable filter model is not known tests were performed to determine whether the specific choice of nonlinear model is critical to filter performance. These tests evaluated the capability of the logistic filter for tracking nonstationary chaotic data generated by the quadratic model, and vice versa. The results shown in Figs. 5a and 5b demonstrate that the two filters are essentially interchangeable, indicating that the choice of second degree nonlinear filter models is not critical in tracking second degree nonlinear data.

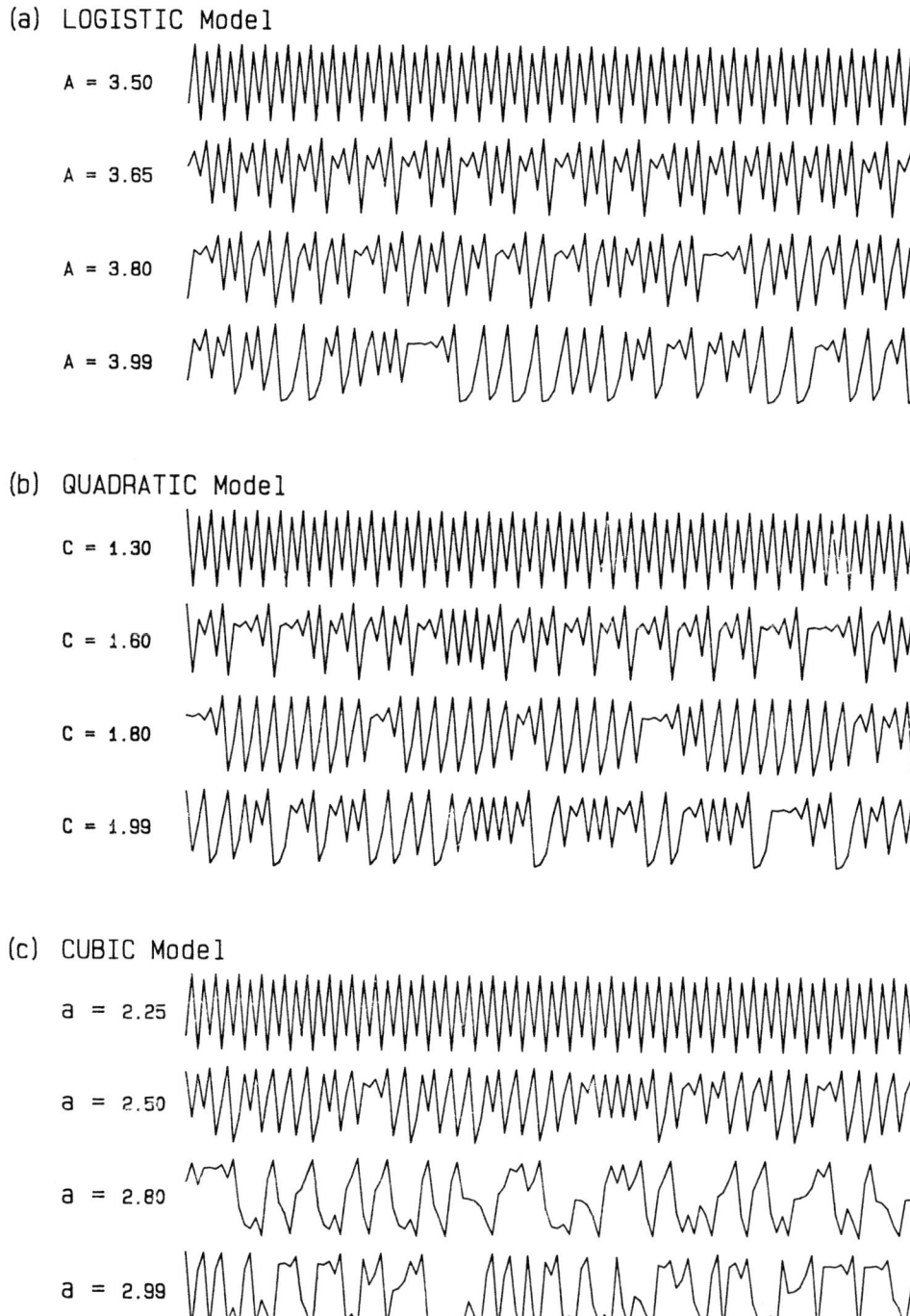

(a) LOGISTIC Model

A = 3.50

A = 3.65

A = 3.80

A = 3.99

(b) QUADRATIC Model

C = 1.30

C = 1.60

C = 1.80

C = 1.99

(c) CUBIC Model

a = 2.25

a = 2.50

a = 2.80

a = 2.99

Figure 2.   Time series generated by each model for various constant values
of critical parameter.

Table 1. Filter Estimates of Bifurcation Parameters Averaged Over 125 Data Points

LOGISTIC MODEL

True A	Data Points in Filter Window	Estimates for A Avg	Std. Dev.	Max. Absolute Deviation
3.50	5	3.50	0	0
	11	3.50	0	0
	23	3.50	0	0
	47	3.50	0	0
3.65	5	3.65	2.7×10^{-7}	7.1×10^{-7}
	11	3.65	1.9×10^{-7}	4.8×10^{-7}
	23	3.65	1.7×10^{-7}	2.4×10^{-7}
	47	3.65	1.7×10^{-7}	2.4×10^{-7}
3.80	5	3.80	4.6×10^{-7}	2.9×10^{-6}
	11	3.80	2.5×10^{-7}	4.8×10^{-7}
	23	3.80	2.2×10^{-7}	2.4×10^{-7}
	47	3.80	2.3×10^{-7}	2.4×10^{-7}
3.99	5	3.99	2.6×10^{-7}	7.1×10^{-7}
	11	3.99	1.6×10^{-7}	4.8×10^{-7}
	23	3.99	9.5×10^{-8}	2.4×10^{-7}
	47	3.99	2.1×10^{-8}	2.4×10^{-7}

QUADRATIC MODEL

True C	Data Points in Filter Window	Estimates for C Avg	Std. Dev.	Max. Absolute Deviation
1.30	5	1.30	5.6×10^{-7}	6.8×10^{-6}
	11	1.30	2.6×10^{-7}	6.2×10^{-6}
	23	1.30	1.6×10^{-7}	6.1×10^{-6}
	47	1.30	1.0×10^{-7}	6.0×10^{-6}
1.60	5	1.60	2.6×10^{-6}	6.1×10^{-6}
	11	1.60	1.5×10^{-6}	4.2×10^{-6}
	23	1.60	1.0×10^{-6}	2.4×10^{-6}
	47	1.60	9.9×10^{-7}	1.5×10^{-6}
1.80	5	1.80	3.2×10^{-6}	9.4×10^{-6}
	11	1.80	2.3×10^{-6}	6.4×10^{-6}
	23	1.80	1.4×10^{-6}	4.6×10^{-6}
	47	1.80	8.8×10^{-7}	3.6×10^{-6}
1.99	5	1.99	3.5×10^{-6}	8.2×10^{-6}
	11	1.99	2.6×10^{-6}	6.7×10^{-6}
	23	1.99	2.6×10^{-6}	4.8×10^{-6}
	47	1.99	2.6×10^{-6}	3.8×10^{-6}

Table 1. Filter Estimates of Bifurcation Parameters Averaged Over 125
          Data Points (Cont'd)

CUBIC MODEL

True a	Data Points in Filter Window	Estimates for a Avg	Std. Dev.	Max. Absolute Deviation
2.25	5	2.25	5.0×10^{-7}	1.7×10^{-6}
	11	2.25	2.6×10^{-7}	1.2×10^{-6}
	23	2.25	2.0×10^{-7}	9.5×10^{-7}
	47	2.25	1.7×10^{-7}	9.5×10^{-7}
2.50	5	2.50	9.8×10^{-7}	2.6×10^{-6}
	11	2.50	5.9×10^{-7}	1.7×10^{-6}
	23	2.50	4.0×10^{-7}	1.2×10^{-6}
	47	2.50	5.7×10^{-7}	1.2×10^{-6}
2.80	5	2.80	5.3×10^{-6}	2.5×10^{-5}
	11	2.80	4.9×10^{-6}	1.0×10^{-5}
	23	2.80	3.8×10^{-6}	6.0×10^{-6}
	47	2.80	2.4×10^{-6}	3.6×10^{-6}
2.99	5	2.99	1.3×10^{-5}	4.9×10^{-5}
	11	2.99	9.0×10^{-6}	2.5×10^{-5}
	23	2.99	6.8×10^{-6}	1.6×10^{-5}
	47	2.99	3.5×10^{-6}	9.3×10^{-6}

(a) LOGISTIC Model

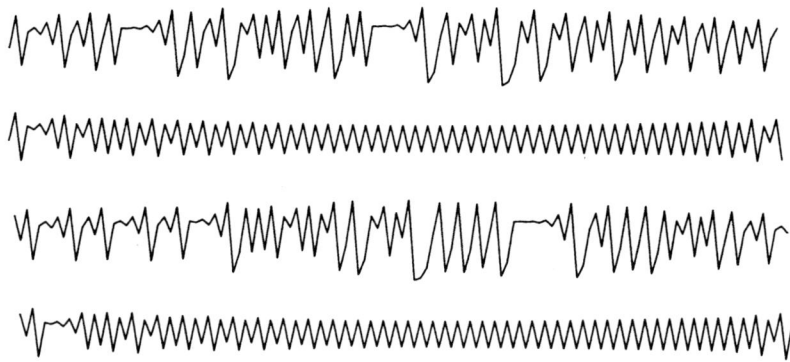

(b) QUADRATIC Model

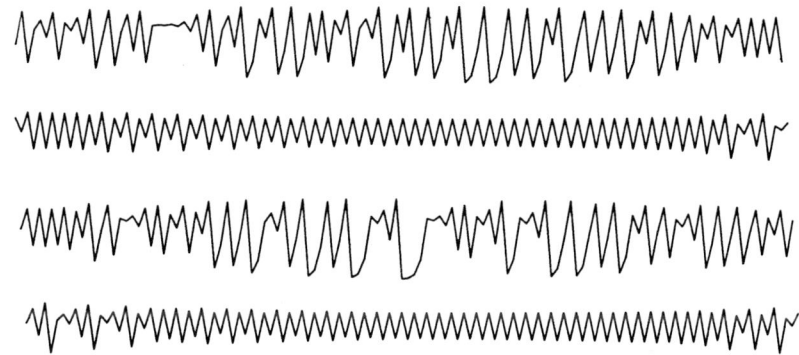

(c) CUBIC Model

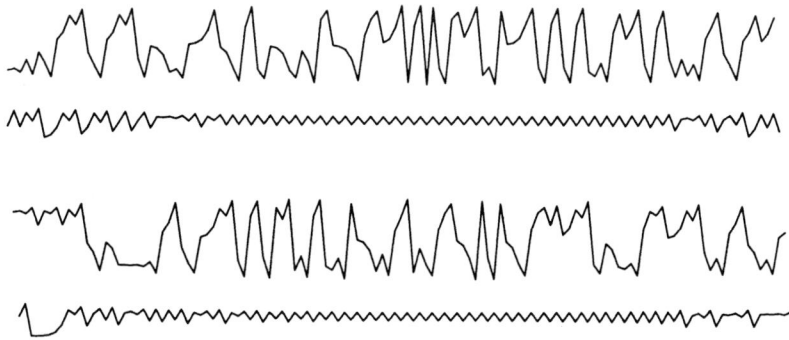

Figure 3. Time series generated by each model with sinusoidally varying critical parameter.

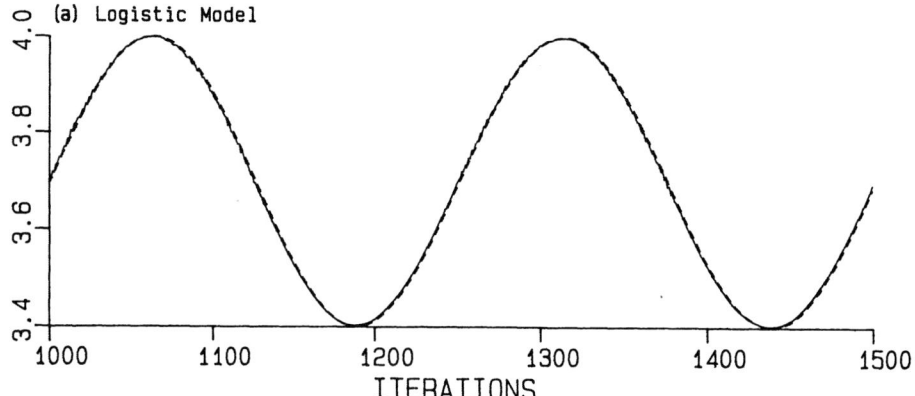

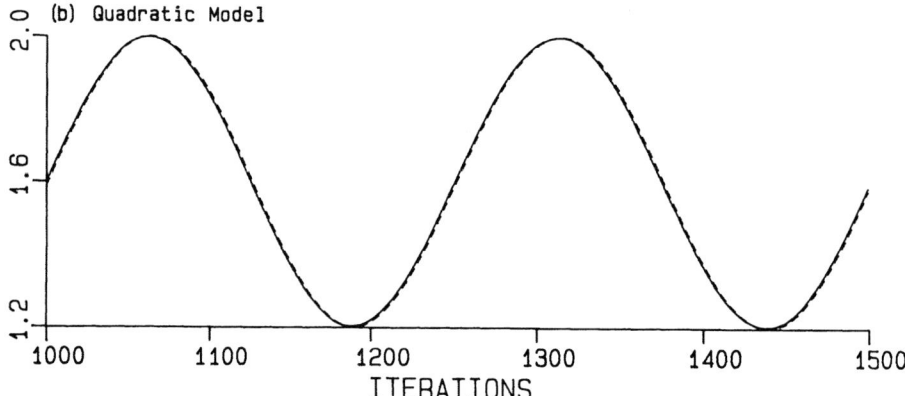

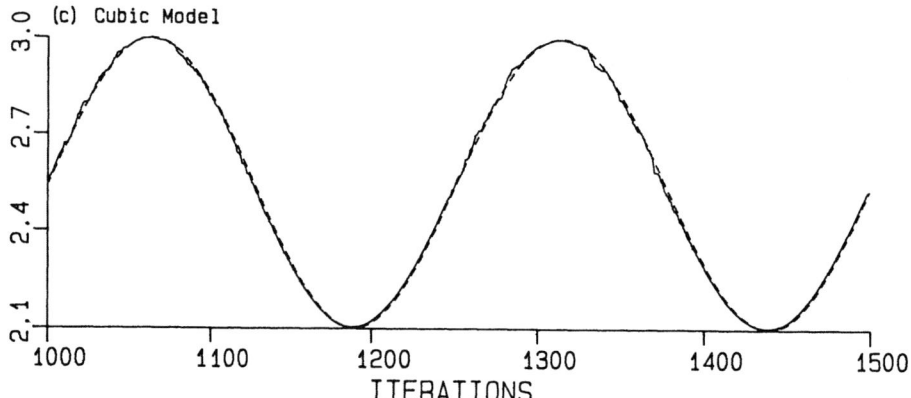

Figure 4. Tracking of modulated critical parameter.
Dotted line, true parameter values;
Solid line, estimated parameter values (11 points in window).

Table 2. Error tracking modulated chaotic parameter. (Average of 500 points, 2 cycles of sinusoidal modulation)

CUBIC MODEL

Data Points in Filter	Maximum Absolute Error	Mean Absolute Error	Maximum Relative Error	Mean Relative Error
5	0.0205	0.0027	0.0054	0.0007
11	0.0109	0.0026	0.0029	0.0007
23	0.0137	0.0026	0.0037	0.0007
47	0.0179	0.0039	0.0049	0.0011

QUADRATIC MODEL

5	0.0050	0.0032	0.0033	0.0020
11	0.0050	0.0032	0.0033	0.0020
23	0.0050	0.0032	0.0033	0.0020
47	0.0048	0.0031	0.0033	0.0019

LOGISTIC MODEL

5	0.0256	0.0043	0.0097	0.0017
11	0.0293	0.0047	0.0112	0.0018
23	0.0397	0.0055	0.0156	0.0021
47	0.0544	0.0118	0.0225	0.0047

Notes:

(1) Absolute Error = absolute value of (Estimate - Average True parameter value)

(2) True parameter value = Average of actual parameter values in filter window

(3) Mean Absolute Error = Average of Absolute errors over 500 points.

(4) Relative Error = Absolute Error divided by Average True Parameter

(a) Tracking of logistic parameter using quadratic filter

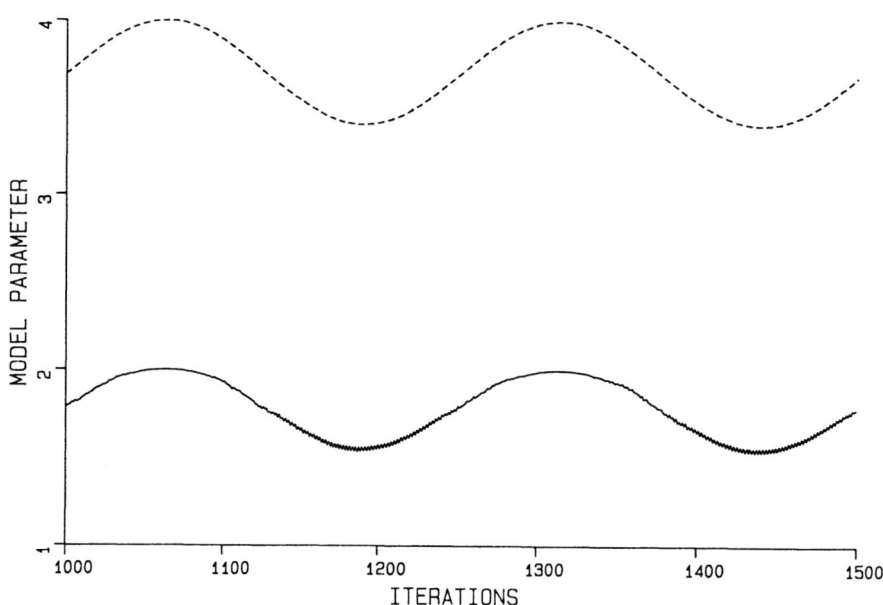

(b) Tracking of quadratic parameter using logistic filter

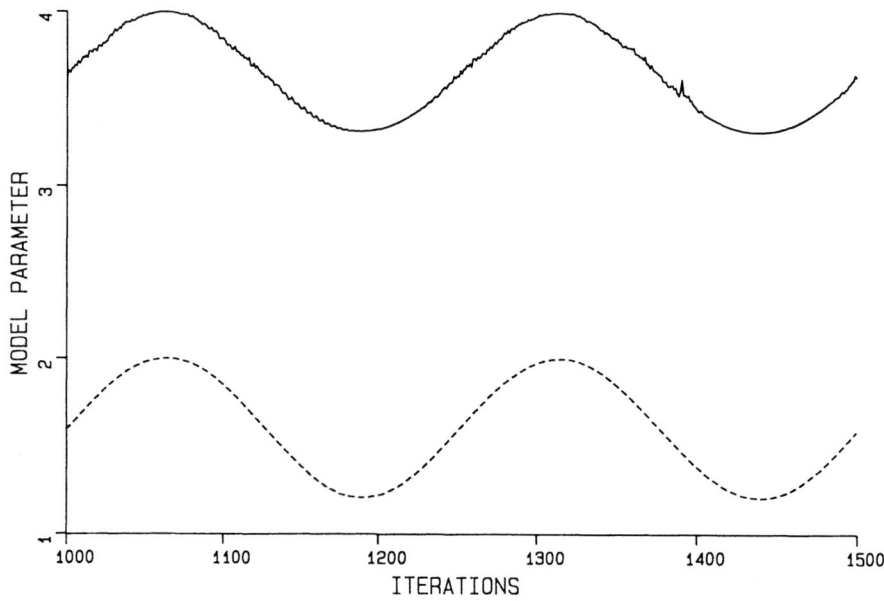

Figure 5. Tracking of modulated critical parameter using another model's
filter.
Dotted line, true parameter values;
Solid line, estimated parameter values (11 points in window).

CONCLUDING DISCUSSION

The test results indicate that each of the nonlinear autoregressive filters investigated can effectively track nonstationarities in the bifurcation parameters with high time resolution. In each test the filters tracked the imposed nonstationarity and produced parameter estimates lying in the approximately correct regions of the bifurcation diagram independent of the model source used for generating the test data. This suggests that an underlying chaotic similarity principle may be involved in applying the parameter filter associated with a specific model to the data generated by another model. The hypothesized principal is that a filter designed for tracking the sensitive parameter of a chaotic nonlinear system can also track the nonstationary chaotic behavior of higher order nonlinear systems. In other words, it is hypothesized that changing levels of chaos in time series can be effectively monitored by filter estimates of the sensitive parameter of a particular chaotic model. A confirmation of this hypothesis in signal analysis applications in electrophysiology would offer two important advantages: 1) a significant reduction in computer requirements as compared to Hausdorff Dimension estimation; and 2) provide a high resolution tracking capability for lengthy nonstationary and nonlinear dynamical data sources.

This research has been supported in part by the following grants from the National Institutes of Health: MH-40715, HD-21888, NS-11535 and NS-25368.

REFERENCES

Layne, S. P., Mayer-Kress, G., and Holzfuss, J., 1986, Problems associated with dimensional analysis of EEG data. in: "Dimensions and Entropies in Chaotic Systems," G. Mayer-Kress, ed., Springer.

Babloyantz, A., and Destexhe, A., 1986, Low-dimensional chaos in an instance of epilepsy. Proceedings of the National Academy of Sciences of the U.S.A. 83:3513-17.

Freeman, W. J., 1987, Simulation of chaotic EEG patterns with a dynamic model of the olfactory system. Biological Cybernetics 56:138-50.

Babyloyantz, A., Nicolis, C. and Salazar, M., 1985, Evidence of chaotic dynamics of brain activity during the sleep cycle. Physics Letters 111A:152.

CHAOTIC HEART RATE DYNAMICS IN ISOLATED PERFUSED RAT HEARTS

Joseph P. Zbilut*, Gottfried Mayer-Kress**, Paul A. Sobotka***, Michael O'Toole*** and John X. Thomas, Jr.****

*Surgical Nursing and Physiology, Rush Univ., Chicago, IL; **Center for Nonlinear Studies, Los Alamos Nat'l Lab., Los Alamos, NM; ***Cardiology and Physiology, and ****Physiology, Loyola Univ. Medical Center, Maywood, IL

ABSTRACT

The application of the theory of chaotic dynamical systems has gradually evolved from computer simulations to assessment of erratic behavior of physical, chemical, and biological systems (Rössler, 1976; Swinney, 1983; Glass, et al., 1983; West, et al., 1986, Zbilut, et al., 1988). Whereas physical and chemical systems lend themselves to fairly good experimental control, biological systems are limited in this respect. Indeed, investigations in this area are often restricted by constrained experimental designs necessary for laboratory control, or ethical concerns regarding human subjects (Koslow, et al., 1987). Moreover the transitions from the behavior of discrete units such as cardiac cells to synergistic cooperation and self-organization as in a heart have not yet been fully explored (Haken, 1982). To overcome some of these lacunae, we studied the heart beat intervals (R wave to R wave of the electrocardiogram) of isolated, perfused rat hearts and their response to a variety of external perturbations. We have chosen the interbeat interval as a primary observable because we are interested in the long time dynamics of the heart as opposed to the temporal pattern of individual heart beats, which have been the subject of clinical studies for some time. The results indicate bifurcations between complex patterns, states with positive dynamical entropies and low values of fractal dimensions frequently seen in physical, chemical and cellular systems. The fact that these chaotic dynamical states can be elicited by global forcing situations and external perturbations suggest a hierarchical self-similarity in the organization of the heart.

Seemingly irregular temporal behavior has been found to be the result of deterministic processes in many natural systems. A key feature of these dynamics has been the geometrical representation of these processes by a reconstructed state space in which chaotic trajectories asymptotically approach a fractal "strange attractor." The complexity or degree of chaos present in the system can be quantified with the help of the fractal dimension and dynamical entropy. For dimensions and entropies there exist an infinity of possible variants ($-\inf < D_q < +\inf$; $-\inf < K_q < +\inf$),

RATE OF ISOLATED RAT HEARTS

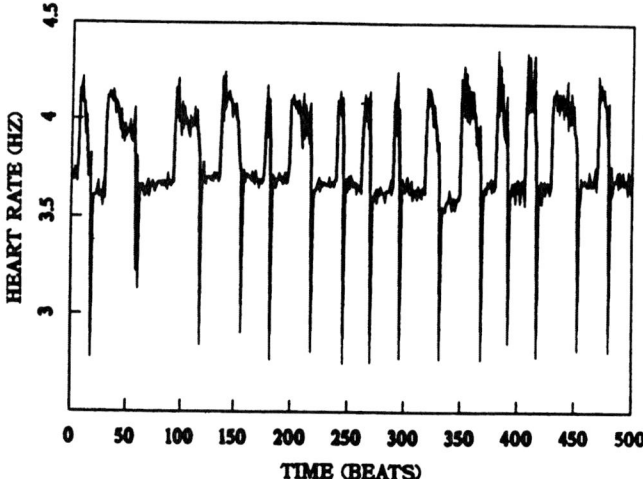

Fig. 1. Time series of R-R intervals obtained in a heart after
prolonged perfusion (approximately 5 hr). Note the very
short refractory period with a low heart rate (typically
only one beat long). <u>Methods</u>. Male Sprague-Dawley rats,
weighing 275-325 grams were anesthetized by intraperitoneal
injection of 55 mg/kg sodium pentobarbital. Following
intravenous injection of 2,000 IU of sodium heparin, hearts
were removed from the chest cavity and quickly immersed in
(4 C°) cold Krebs/Henseleit bicarbonate (KHB) buffer (in mM:
NaCl 118; KCl 4.8; CaCl₂ 2.5; MgSO₄ 1.2; NaHCO₃ 24; KH₂PO₄
1.2; and glucose 11). Extraneous tissue was removed from
the heart and the aorta was cannulated on the perfusion
apparatus. The apparatus was a triple reservoir,
non-recirculating gravity flow Langendorff apparatus
operated at 80-90 mm Hg perfusion pressure. Once mounted on
the apparatus, the heart was submerged in a thermally
insulated jacket filled with KHB to minimize edema
formation, and the ventricular temperature was monitored
with a thermistor probe inserted through the mitral valve
into the left ventricle. Temperature was maintained at 37 ±
5 C° with a Yellow Springs controller (Yellow Springs
Instrument Company, Yellow Springs, Ohio, USA), connected to
the ventricular thermistor probe. Sensing bipolar electrodes
were implanted on the left mid-ventricle, and the signal was
filtered and amplified (Gould Universal) to obtain a clear,
noise-free R wave as judged by inspection of the
oscilloscope tracing (Hewlett-Packard), and electrostatic
recording (Gould 2400). The signal was digitized through a
computer board (Data Translation, DT 2821), and the
interbeat intervals were calculated on an R-to-R basis by
compiled computer routines (ASYST) with millisecond accuracy
for off-line analysis. Typically 520 intervals were
measured per test condition. In our laboratory we have found
that perfused hearts become edematous due to lack of oncotic
pressure in the KHB, yet the gain in net weight is
negligible after 105 minutes total perfusion. During this
time period the preparation is stable with respect to
end-diastolic pressure, peak systolic pressure, and coronary
flow rate. Beyond this time, there is significant
deterioration in all these variables.

each describing slightly different static and dynamic invariants of the system respectively. Most of them are very difficult to obtain numerically for values of $q \neq 2$. Therefore especially in the context of biological experiments, results for the "correlation" or "Grassberger-Procaccia" dimensions are quoted in the literature with the corresponding entropy K_2 (Mayer-Kress, 1986). The disadvantage of this method is that many of the relevant dynamical features of the attractor are averaged out, and therefore the descriptive and analytical results obtained in this way have large error bars and do not discriminate well between data series of different complexity. Therefore we also use a modification of the fractal dimension algorithms in which we estimate dimensional complexity values as a time dependent quantity which gives us a better statistical and dynamical picture of the typically non-stationary time series. This result can be compared to the sliding window method used in the spectral monitoring of clinical data (Mayer-Kress, et al., 1988). (Fig. 1-3)

To elicit chaotic dynamics of heart beat intervals we used isolated rat hearts perfused by the Langendorff method (Langendorff, 1895). Several kinds of external perturbations or modes of driving were used to induce changes in the R-R dynamics. These included pacing, ischemia, reperfusion, and prolonged perfusion. Each of these perturbations produced bifurcations to a small class of distinguished patterns. The most complex and sustained dynamical patterns, and for which sufficient data was available for dimensional analysis, were observed with prolonged perfusion. Specifically, a 3-state oscillation suggestive of the chemical oscillations commonly observed in the Belousov-Zhabotinsky reaction was noted (Uppal and Ray, 1974; Othmer, 1975). An attempt was therefore made to model a class of observed interval patterns with a discrete mapping obtained through experimental reconstruction from this reaction (Hirsch, et al., 1982; Bagley, et al., 1986).

In the Belousov-Zhabotinsky reaction regular (periodic) and chaotic oscillations have been observed, and for the case of chaotic oscillations a "strange attractor" has been previously identified. The Poincare return map of this attractor shows a very small noise level and has the form of a unimodal one-dimensional function. For a set of different parameter configurations, Maselko, Swinney and others have observed bifurcation sequences which could be reproduced by a simple family of one-dimensional maps in which the different experimental conditions correspond to the modification of a relatively small set of parameters in the map (Roux, et al., 1982; Bayer, et al., 1986; Maselko and Swinney, 1985). Thus a range of

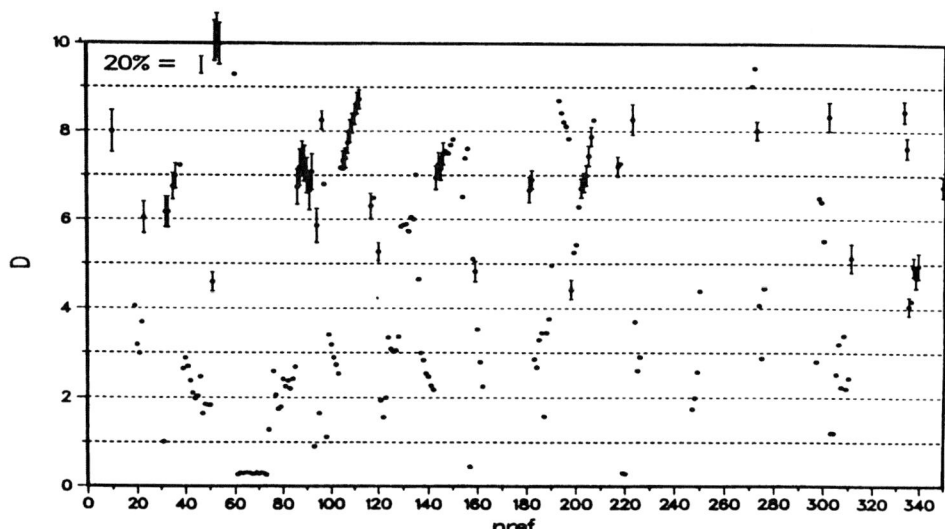

Fig. 2. Sequence of pointwise dimensional complexity values (see
Mayer-Kress, 1987 for details) of the experimentally
observed heart rate data of Fig. 1. n_{ref} indicates the
time-index of the reference points for which the dimension
was estimated for an embedding dimension of $d_{emb} = 20$. Note
the large fluctuations which reflect the strong
nonuniformity of the data.

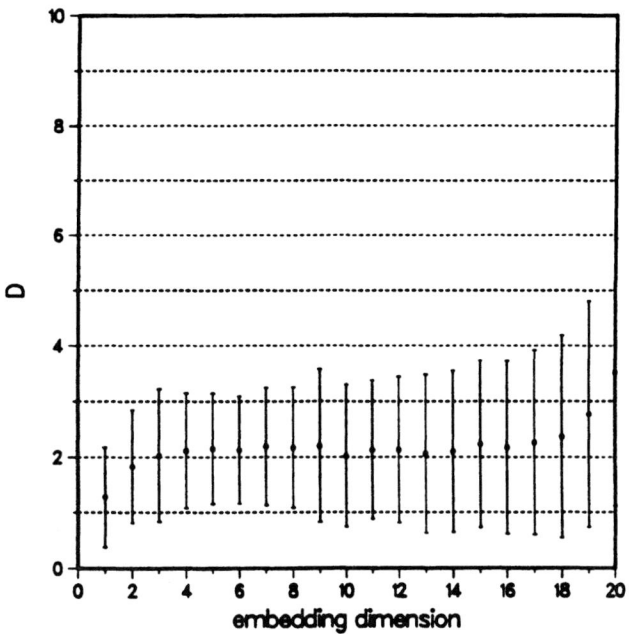

Fig. 3. Averaged pointwise dimension of the data of Fig. 1. Note the low value which indicates chaotic behavior of the heart rate. The corresponding correlation dimension is also below $d_2 = 3$ and the corresponding dynamical entropy is positive ($k_2 \sim 0.1$).

phenomena can be reproduced by changing very few parameters in the same basic family of one-dimensional maps.

In the dynamics of our perfused rat hearts we also found several parameter regions, for which the oscillatory behavior of the instantaneous heart rate (= $1/T_{RR}$ which includes not only SA node generated beats--but ectopic ones as well) can be grouped into oscillations between two distinct values (excitatory/quiescent) of R-R interval length. This pattern is very similar to the one described by Maselko and Swinney for the Belousov-Zhabotinsky reaction and by Bagley and colleagues (1986) for the one dimensional map model. In addition to these dynamical patterns, we observe a new variant of the same pattern in which a sequence of fast heart beats (high heart rate) is followed by exactly one extra slow beat (extra low heart rate) such that we have a regular alternation between three different states: quiescent -> excitatory -> refractory. This transition to an extra state of a large interbeat interval could be interpreted as the occurrence of a new refractory state during which an excitation to a fast heart rate is supressed (Fig. 1). The introduction

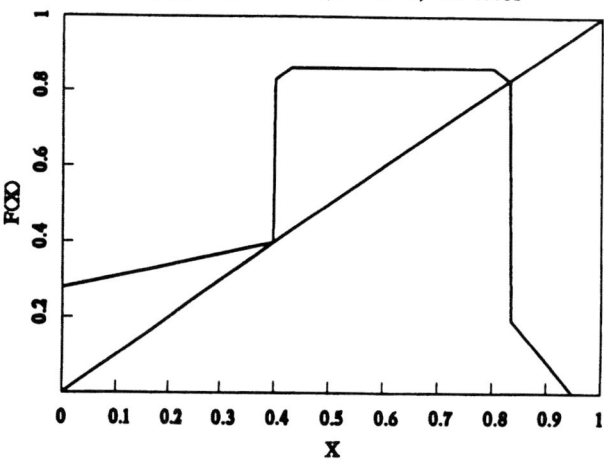

PIECEWISE LINEAR 1-D MAP
A=.3, B=1.1, E1=-.001, E2=1.E-5, SIG=.0031

Fig. 4. Graph of modification of one-dimensional map of Bagley and colleagues (1986). The function was constructed through linear sections which reproduce the features of the experimental data of Fig. 1: Section I ($0 < x < z_0$, $z_0 = 0.4$) produces the intermediate amplitude heart rates which we refer to as quiescent. The function in I is given by: $f_I(x) = a_0 + a_1 * x$, where $a_0 = z_0 + e_1 - a_1 * z_0$, $a_1 = 0.3$, and $e_1 = -0.001$. The value of e_1 determines the length of time the system remains in the "quiescent" state. The fact that we have chosen it to be negative would mean in the noise free case that the quiescent state would be stable for arbitrary long periods of time. Therefore we have added noise of an amplitude of sigma = 0.0031 in order to induce transitions from the quiescent state to the excitatory state at a rate which roughly corresponds to the one of the experimental data of Fig. 1. For a positive value of e_1 we can obtain a signal with identical statistical behavior but which is purely periodic. The second section of the map (II) ($z_0 < x < z_1$, $z_1 = 0.43$) is given by: $f_{II}(x) = b_1 * (z - z_0) + z_5 + e_2$ where $b_1 = 1.1$, $z_5 = (z_1 + b_1 * z_2) / (1 + b_1)$, $z_1 = b_1 * (z_1 - z_0) + z_3 + e_2$, $z_2 = 0.8$, $z_3 = 0.83$, $e_2 = 0.00001$. This part determines how the transition from quiescent to excitatory state takes place. The next section (III) ($z_1 < x < z_2$), $f_{III}(x) = z_1$ is just constant and not relevant for the attractor. In section IV ($z_2 < x < z_4$, $z_4 = z_3 + 0.1 * (z_3 - z_2)$) we have: $f_{IV}(x) = z_1 - b_1 * (z - z_2)$. This part determines the "excitatory" state. Finally we have section V ($z_4 < x$) which determines the transition from "excitatory" state to "refractory", in which the system only stays for one single time step. This has been introduced to mimic the experimental data but of course it is no problem to modify the map such that it generates multiple refractory time steps.

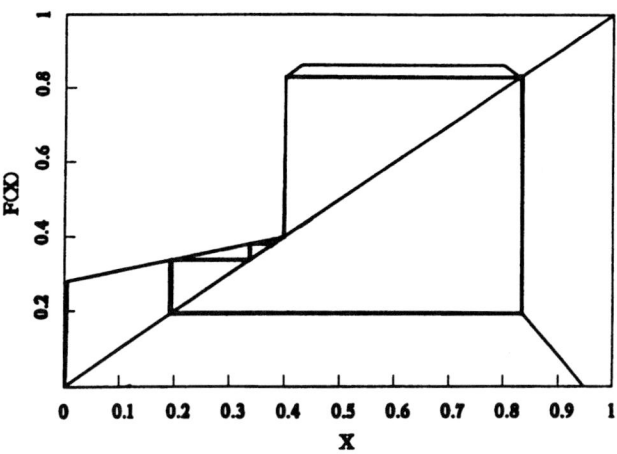

Fig. 5. For clarification, we have included the orbit of the system for a set of several hundred time steps.

of this refractory state is also necessary for the formation of traveling waves as observed both in the chemical reaction and in fibrillating hearts. By refractory state we do not mean the repolarization phase of action potentials; rather, here we refer to higher order level of organization subsuming perhaps multiple ionically driven nonlinear relaxation oscillators "percolating" through the excitable medium of heart tissue which is inherently heterogeneous and can be made more so pathologically, as well as through the naturally occurring mechanical tensions attendant to contraction (Mandell, 1983; Winfree, 1981; Kaplan, et al., 1988).

Inspection of the map introduced in the reference of Bagley, et al. (1986) shows that the change of a single parameter again can produce a transition from oscillations between two states to those where an extra state of oscillation is formed. (Compare Fig. 4-6 with Fig. 1, a-d of the Bagley map.) The only relevant change in the model lies in the larger function values in the region of small argument values (region I). We find this fact fairly surprising and think that this motivates further studies into more realistic models of the heart dynamics, which take these transitions into account.

The reason why the oscillations observed in the rat heart are not purely periodic as in the chemical reaction and in the corresponding model can have several reasons which are well discussed in the nonlinear dynamics

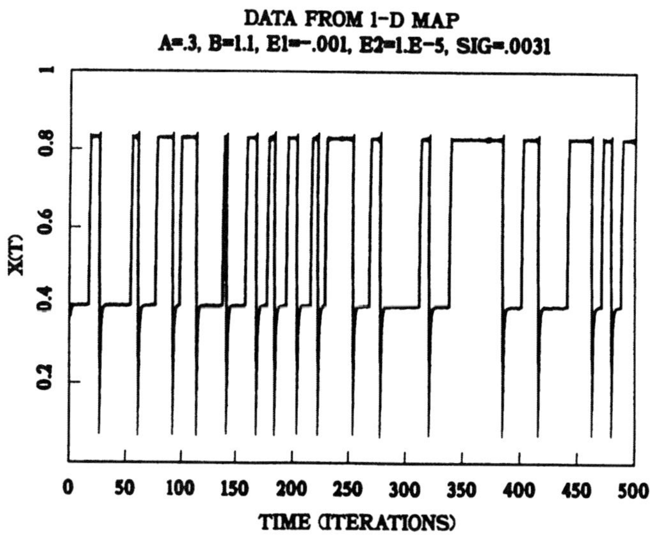

DATA FROM 1-D MAP
A=.3, B=1.1, E1=-.001, E2=1.E-5, SIG=.0031

Fig. 6. Time series generated from the map of Fig. 1. Note the intermittent occurrence of transitions between state ($x = 0.4$), excitatory state ($x = 0.83$), and a single refractory state ($x = 0.18$). Since the unperturbed map would remain in the quiescent state forever, we have added noise to the system. Thus we can interpret the observed signal as showing "noise induced intermittency."

literature. First of all, the conditions for the chemical experiments are very well controlled and kept stationary. This is a priori different from conditions found in a biological environment. As has been shown in a different context, external small scale fluctuations can destroy stable periodic attractors and produce "noise induced intermittency" (Mayer-Kress and Haken, 1981). The periodic pattern thereby is replaced by patterns which appear periodic for some time but undergo a "chaotic burst" and then are reinjected into a similar apparently periodic phase. Besides noise induced intermittency, another interpretation of the irregular sequence of the oscillatory patterns consists in a different perturbation of the original one-dimensional map in which the shape of the map is the same, but now the local slope of the map is such that regions with local instability (absolute value of the slope is larger than one) increase and produce deterministic intermittent behavior of the mapping. Universal scaling laws for both of these cases have been given by Hirsch et al. (1982).

In conclusion, we have demonstrated that bifurcations and chaotic dynamics can occur in cardiac tissue on the organ level of complexity. Furthermore, we have demonstrated the possibility of reproducing a number of dynamical patterns of the instantaneous heart rate with a very simple non-linear dynamical model, and also embedding it into a larger class of oscillatory modes of chemical oscillators. What is required at this stage is a sustained effort to determine what parameters are relevant to the initiation of such dynamics. We think that our simple model will be helpful in the classification and prediction of transitions and bifurcations to previously unobserved modes of oscillation. Clearly, the significance for the understanding of human cardiac dysrhythmias cannot be overstated (Mandell, 1988).

ACKNOWLEDGEMENTS

One of us (GMK) would like to thank K. Kaneko for discussions and F.E. Yates for his encouragement. JPZ would like to thank A. Mandell, A. Goldberger and J. Keithley for their support; and R. Nemicas for technical assistance.

REFERENCES

Bagley, R.J., Mayer-Kress G., and Farmer, J.D., 1986, Mode locking, the Belousov-Zhabotinsky reaction, and one-dimensional mappings, Phys Lett 114A: 419.

Bayer, W.A., Mauldin, R.D., and Stein, P.R., 1986, Shift-maximal sequences in function iteration: existence, uniqueness, and multiplicity, J Math Analysis and Applications 115:305.

Glass, L., Guevara, M.R., Shrier, A., and Perez, R., 1983, Bifurcation and chaos in a periodically stimulated cardiac oscillator, Physica 7D: 1983.

Hirsch, J.E., Huberman, B.A., Scalapino, D.J., 1982, Theory of intermittency, Phys Rev 25A: 519.

Kaplan, D.T., Smith, J.M., Saxberg, B.E.H., and Cohen, R.J., 1988, Nonlinear dynamics in cardiac conduction, in: "Nonlinearity in Biology and Medicine" [Vol 19 (1-2) of Math Biosciences], A.S. Perelson, B.

Koslow, S.H., Mandell, A.J., Shlesinger, M.F., eds., 1987, "Perspectives in Biological Dynamics and Theoretical Medicine," Annals of the New York Academy of Sciences, Vol. 584, New York.

Langendorff, O., 1895, Untersuchungen am überlebenden Säugetierherzen, Pflugers Arch 61: 291.

Mandell, A.J., 1983, From intermittency to transitivity in neuropsychobiological flows, Am J Physiol 245:R484.

Mandell, A.J., 1987, Dynamical complexity and pathologic order in the cardiac monitoring problem, Physica 27D:235.

Maselko, J., and Swinney, H.L., 1985, A complex transition sequence in the Belousov-Zhabotinskii reaction, Phys Scr T9:35.

Mayer-Kress, G., 1987, Application of dimension algorithms to experimental chaos, in "Directions in Chaos," H. Bai-Lin, ed., World Scientfic, Singapore.

Mayer-Kress, G., and Haken, H., 1981, Intermittent behavior of the logistic system, Phys Lett 82A: 151.

Mayer-Kress, G., and Layne, S.P., 1987, Dimensionality of the human encephalogram, in Koslow, et al., 1987.

Mayer-Kress, G., Yates, F.E., Benton, L., Keidel, M., Tirsch, W., Pöppl, S.J., and Geist, K., 1988, Dimensional analysis of nonlinear oscillations in brain, heart, and muscle, in: "Nonlinearity in Biology and Medicine" [Vol 19 (1-2) of Math Biosciences], A.S. Perelson, B. Goldstein, M. Dembo, and J.A. Jacquez, eds., Elsevier, New York.

Othmer, H.G., 1975, On the temporal characteristics of a model for the Zhabotinskii-Belousov reaction, Math Biosciences 24:205.

Rössler, O.E., 1976, Chaotic behavior in simple reaction systems, Z Naturforsch 31a:259.

Roux, J.C., Turner, J.S.. McCormick, W.D., and Swinney, H.L., 1982, Empirical observations of complex dynamics in a chemical reaction, in: "Nonlinear Problems: Present and Future," A. Bishop, D. Campbell, and B. Nicolaenko, eds., North Holland, Amsterdam.

Uppal, A., Ray, W.H., and Poore, A.B., 1974, On the dynamic behavior of continuous stirred tank reactors, Chemical Engineering Science 29: 967.

West, B.J., Bhargava, V., and Goldberger, A.L., 1986, Beyond the principle of similitude: Renormalization in the bronchial tree, J Appl Physiol 60:189.

Winfree, A.T., 1981, Peculiarities in the impulse response of pacemaker neurons, Lect Appl Math 19:265.

Zbilut, J., Mayer-Kress, G., and Geist, K., 1988, Dimensional analysis of heart rate variability in heart transplant recipients, in: "Nonlinearity in Biology and Medicine," [Vol 90 (1-2) of Math Biosciences] A.S. Perelson, B. Goldstein, M. Dembo, and J.A. Jacquez, eds., Elsevier, New York.

INTRODUCTION TO THE PROPERTIES AND ANALYSIS OF

FRACTAL OBJECTS, PROCESSES, AND DATA

Larry S. Liebovitch

Department of Ophthalmology
Columbia University
630 West 168th Street, New York, NY 10032

SUMMARY

   Three important properties that fractal systems have are:
1) self-similarity, 2) scaling relationships that connect
different spatial or temporal regimes, and 3) "monstrous"
properties such as zero or non-finite values of the moments
such as the mean or variance. A useful parameter to
characterize such systems is the fractal dimension. We provide
both qualitative and more formal definitions of the dimension
and show how it can be evaluated from experimental data. We
also show how a time series can be turned into a geometric
object. The topological properties of this object, including
the dimension can then be determined. If the dimension is low,
then seemingly "random" noise is actually "chaos" produced by a
deterministic non-linear system.

PROPERTIES OF FRACTAL SYSTEMS

   Objects, processes, or systems that are statistically
self-similar in time or space have three important properties:
1) self-similarity, 2) scaling, and 3) zero or non-finite
moments.

1. Self-Similarity

   The coastline shown in Fig. 1 looks jagged. One would
expect that if the scale of the map were increased that its
details would be fully resolved and then appear to be smooth.
This is not the case. No matter how much the coastline is
magnified in still looks equally jagged. This is statistical
self-similarity. An object is exactly self-similar if its
shape measured at scale x is proportional to its shape measured
at a finer scale $x' = \alpha x$. An object, such as the coastline, is
statistically self-similar if a property, such as its length,
measured at scale x is proportional to that measured at finer
scale $x' = \alpha x$. If the coordinates of an object have different
constants of proportionality, that is, if the object or its

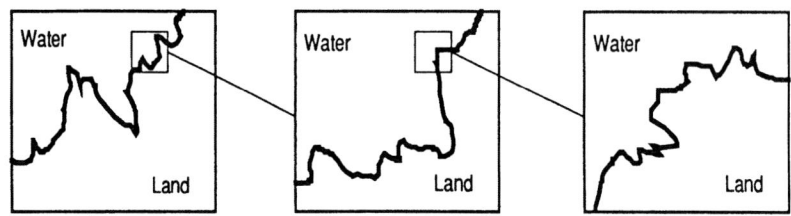

Fig. 1. A coastline is statistically self-similar, it looks
equally jagged at all scales of magnification.

measured properties are similar when x' = αx and y' = βy where
$\alpha \neq \beta$, then it is called self-affine.

For example, the Koch curve (Mandelbrot, 1983, p.42-44) is
exactly self-similar. It is constructed by starting with an
equilateral triangle. Then, an equilateral triangle of 1/3 the
size of the original is added to the middle third of each side,
and this process repeated forever. Blood vessels in the
circulatory system and bronchial tubes in the lung have
branches that have smaller branches that have smaller branches,
etc. and are statistically self-similar (Mandelbrot, 1983; West
and Goldberger, 1987). Brownian motion is self-affine (Voss,
1988). In 1909 Perrin recorded the position of a micron-sized
pollen grain at intervals Δt = 30 seconds. If the positions
had been recorded at finer temporal scale $(\Delta t)' = \alpha(\Delta t)$ then
the positions recorded would change by $(\Delta x)' = \alpha^{1/2}(\Delta x)$. This
is well illustrated by Lavenda (1985) where Perrin's data is
shown along with a modern simulation where Δt = 0.3 seconds.
Since there is no general term that includes exact
self-similarity, statistical self-similar, and self-affinity,
we will simply use the term self-similarity to describe all
three.

Another important biological example of self-similarity is
found in the ionic currents that pass through ion channels in
the cell membrane. Such channels are ubiquitous in cells
(Sakmann and Neher, 1983; Hille, 1984). In nerve they generate
and propagate the action potential, in muscle they control
calcium levels that initiate contraction, in epithelia they are
involved in net ion transport across the cell layer, in sensory
cells such as photoreceptors they are involved in signal
transduction, and in organelles such as mitochondria they
control ionic concentrations. Using the patch clamp technique
developed by Neher and Sakmann and their collaborators (Hamill
et al., 1981) small patches of cell membrane, approximately 1
μm in size are sealed within a micropipette. Such a patch may
contain only one ion channel. Thermal fluctuations provide the
energy for the channel to spontaneously change its conformation
from one that is open to the flow of ions to one that is
closed. Each time the channel is open, the picoamp currents
flowing through it can be resolved as shown in Fig. 2. Thus,
the kinetics, the switching between the open and closed
conformational states, can be studied in a single ion channel
molecule.

226

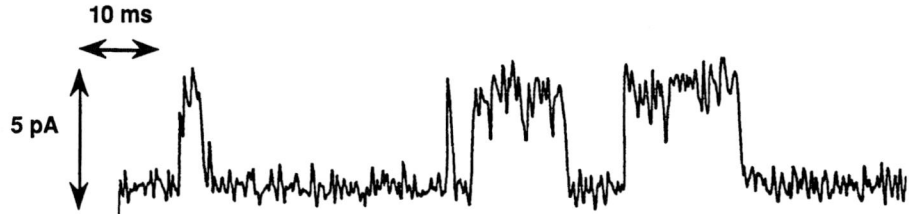

Fig. 2. Current measured through a single ion channel in the
apical membrane of a cell in the corneal
endothelium. The current is high when the channel
is open and low when it is closed.

We measured single channel currents from ion channels in the
corneal endothelium, the layer of cells that line the inside of
the cornea. We found that the kinetics of these channels are
self-similar (Liebovitch et al., 1987a; 1987b). That is, there
are bursts within bursts within bursts of opening and closing.
The same data, recorded on FM tape can be studied at different
time resolutions by using a computer to sample it at different
analog/digital sampling rates. We found that the frequency
histograms of closed time durations measured at different
analog/digital sampling rates are proportional to each other as
shown in Fig. 3. When observed over any limited time scale,
each of these histograms is approximately exponential, with a
time constant that depends on the time scale used for the
measurement. When observed over many time scales, these
histograms have a form known as a stretched exponential
(Liebovitch et al., 1987a; 1987b).

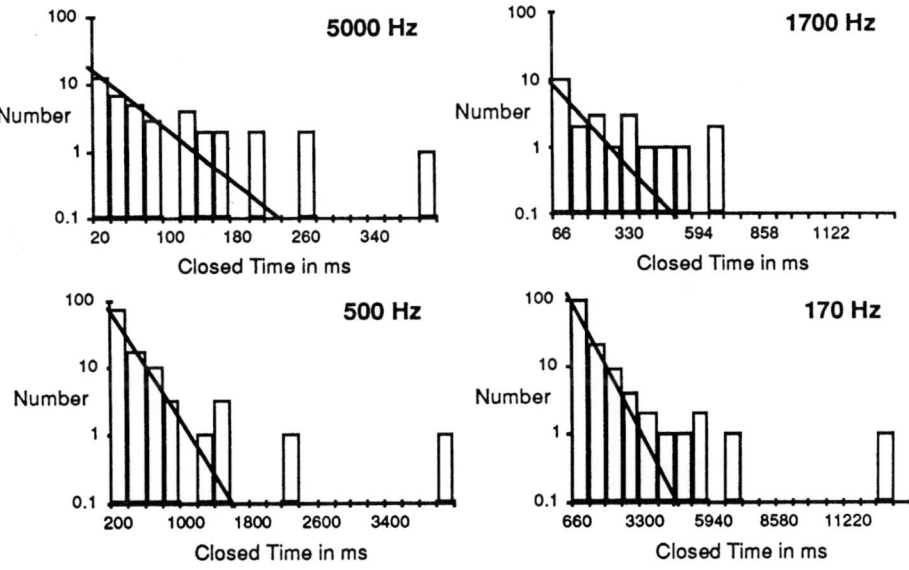

Fig. 3. The frequency histograms of the closed time
durations of the channel recorded from the corneal
endothelium are self-similar when the same data is
sampled at different analog/digital conversion
rates.

2. Scaling

Self-similarity means that structures at one spatial or temporal scale are related to structures at other spatial or temporal scales. Thus, there is an overall scaling connecting structures at different scales. Self-similarity means that a property L(x) measured at scale x is proportional to the same property L(αx) measured at scale αx, namely,

$$L(x) = k L(\alpha x), \quad \text{where } k = \text{constant}. \tag{1}$$

This will satisfied if L(x) is a power law

$$L(x) = A x^d \qquad \text{where } A, d = \text{constant}, \tag{2}$$

or more generally, if L(x) is the product of a power law and an oscillatory function f(1 + y) = f(y) such that

$$L(x) = A x^d f(\ln x / \ln \alpha). \tag{3}$$

This can be verified by direct substitution of either Eq. (2) or Eq. (3) into Eq. (1) to find that $k = \alpha^{-d}$.

Thus, a powerful method to detect fractals is to plot the logarithm of a measured property against the logarithm of the scale at which it is measured. On such a plot, fractals will display a straight line (or a straight line with an oscillatory component). For example, as the length of a coastline is measured at finer and finer scale, more and more bays and peninsulas are detected and so the length measured for the coastline increases as a power law of the scale of the map. As

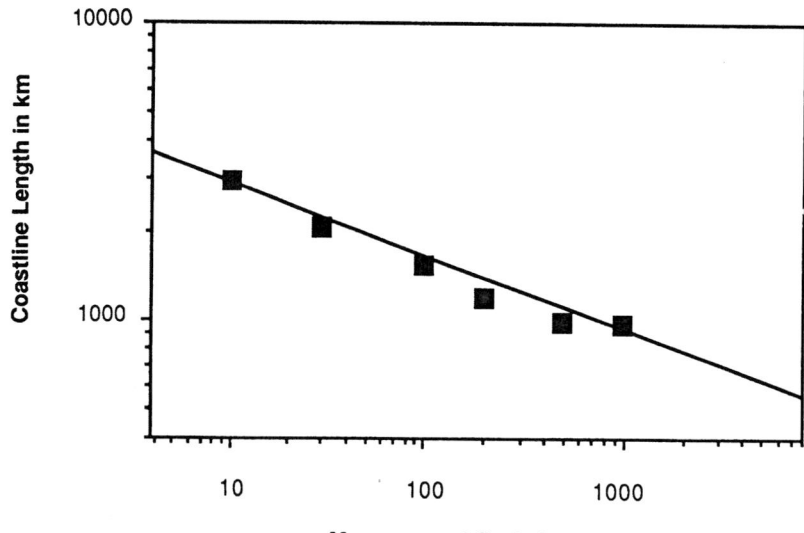

Fig. 4. The length of the western shore of Great Britain is a power law of the scale at which it is measured. After Richardson (1961) and Mandelbrot (1967).

shown in Fig. 4, Richardson (1961) and Mandelbrot (1967) found that for the western shore of Great Britain d = -0.25. Many such fractal-like scaling relationships have been found including the perimeters of clouds (Lovejoy, 1982) and soot particles (Sander, 1986); the surface areas of proteins (Lewis and Rees, 1985), minerals (Avnir et al., 1984), and cell membranes (Paumgartner et al., 1981); the distribution of the intensity of earthquakes (Kagan and Knopoff, 1981); the time course of the dielectric relaxation of glasses (Williams and Watts, 1970; Shlesinger, 1984); and the reaction rates of diffusion limited chemical reactions (Kopelman, 1988).

Ion channel kinetics can also be analyzed in a similar way. The kinetic rate constant is the probability per second that the channel will change states, for example, from closed to open. To observe the channel being closed at time scale t_{eff} requires that the channel remain closed long enough to be detected. Thus, we determine the effective kinetic rate constant k_{eff} at time scale t_{eff} as the probability per second that a closed channel will open given that it has already remained closed for at least a time t_{eff}. In renewal theory k_{eff}, is called the "age-specific failure rate" (Cox, 1962). It can be evaluated by evaluating the effective kinetic rate constant at time scales set by the analog/digital sampling rate, the bin size used in the closed time histograms, or the correlation time scale used in the correlation functions of the signal (Liebovitch et al., 1987a, 1987b, 1988a).

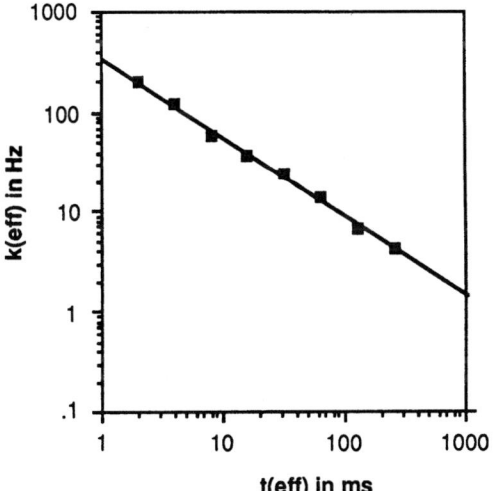

Fig. 5. The effective kinetic rate constant k(eff) for switching from the closed to open state measured for ion channels in the corneal endothelium is a power law of the effective time scale t(eff) at which the measurement is done. From Liebovitch et al. 1987a; 1987b.

As shown in Fig. 5 we found that the kinetics of the ion channels in the corneal endothelium have such a power law scaling characteristic of fractals. We also found similar results for ion channels in cultured hippocampal neurons (Liebovitch and Sullivan, 1987). Figure 5 can be interpreted in different ways. In terms of the effective kinetic rate constant, it says that the faster we can observe the channel, the faster we will see it open and close. In terms of the age specific failure rate, it says that the longer the channel remains closed, the less the probability per second that it will open. We have also shown how this plot can be interpreted in terms of the energy levels needed to produce such time behavior (Liebovitch, 1988b). Note, that such scaling relationships will extend over a finite range and no longer hold when new physical mechanisms become important. For the measurement of ion channel kinetics, the practical limit on the short time scale is set by the resolution of the electronics, which is approximately 100 µs. There is also a theoretical limit, which is the transit time of an ion through the channel, which is approximately 1 µs. At times faster than this, the discrete nature of the electrical current will be detected and the signal will not resemble that shown in Fig. 2. The long time scale limit is set by the length of time the patch can be maintained in the pipette, which can vary from minutes to hours.

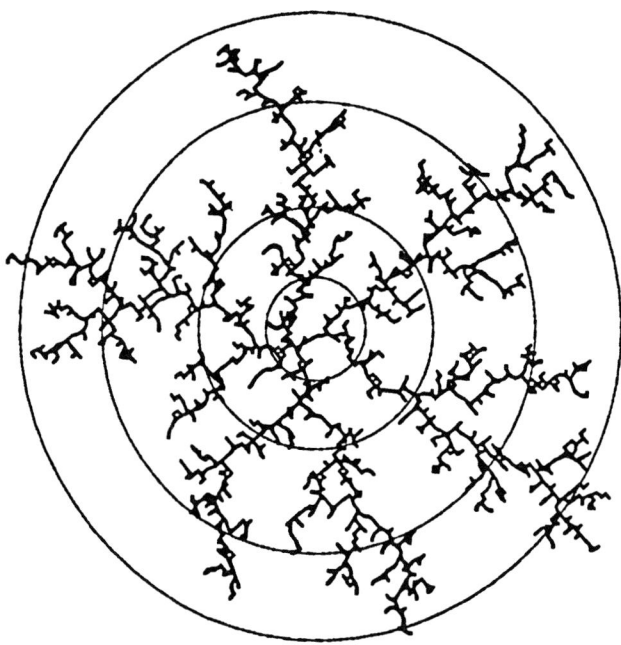

Fig. 6. A simulation of a diffusion limited aggregation. Note that at larger scales (circles) there is more and more space within the arms of this fractal, and thus the average density decreases with increasing radius.

3. Zero or Non-Finite Moments

Fractals can have such unusual properties that mathematicians called them "monsters." We now describe such properties. Shown in Fig. 6 is a fractal produced when one particle at a time is released and is allowed to random walk until it hits the existing structure, where it sticks, and then the next particle is released. This diffusion limited aggregation is self-similar. That is, there are small spaces within the structure at small scales and larger spaces at larger scales. Thus, as the average density is measured at larger scales, more of the larger spaces are included and thus the average density decreases. Meakin (1986) found that the density $\rho(r)$ measured within radius r has the form

$$\rho(r) = k \ r^{-d} \quad \text{where k = constant and d = 0.29.} \quad (4)$$

In other words, in the limit as $r \rightarrow \infty$ the density $\rho \rightarrow 0$. That is, this fractal object does not have a finite, non-zero density! The value obtained for the density will depend entirely on the scale at which it is measured. This is strikingly different from non-fractal objects where the density approaches a finite, non-zero value (Sander, 1987).

Moreover, the spatial autocorrelation function C(r) for Fig. 6 is also a power law (Meakin, 1986)

$$C(r) = k \ r^{-d} \quad \text{where k = constant and d = 0.29.} \quad (5)$$

This differs from non-fractal objects which (over some regime) will have a fixed length scale r_0 and thus

$$C(r) = k \ \exp(-r/r_0). \quad (6)$$

The fractal shown in Fig. 6 has no such fixed length scale r_0. Rather it has correlations at all length scales, as indicated by the power law form for the autocorrelation function C(r).

In Fig. 6 self-similarity causes the density, the first moment to be zero. Self-similarity can also cause the second moment, the variance to be infinite. Consider a spatial or time series with self-similar fluctuations. At larger scales there will be larger versions of the fluctuations at small scales. Thus, as longer data records, or data records at finer resolution, are analyzed, the variance will increase and approach infinity. This is observed for the variance of cotton prices (Mandelbrot, 1983), the roughness of rocks (Brown and Scholz, 1985), and the displacements of atoms in molecular dynamic simulations of globular proteins (Morgan et al., 1983). Bassingthwaighte (1988) found a physiological important example of such fractal behavior in the variance of the blood flow measured in the vessels of the heart. As shown in Fig. 7, the variance measured for the blood flow increased as the heart is divided into finer pieces. Van Beek, Bassingthwaighte, and King (1988) have shown that slightly asymmetric branching of the blood vessels in the heart will lead to such a distribution of variance.

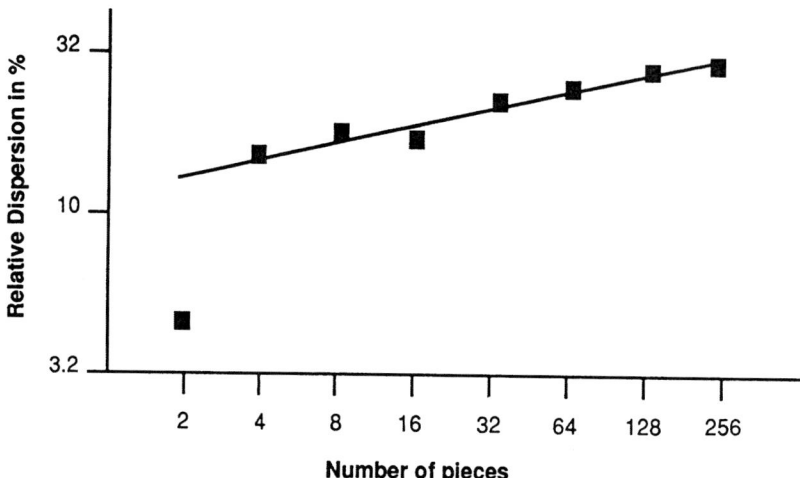

Fig. 7. Bassingthwaighte (1988) found that the relative
dispersion of the blood flow in the heart increases
as the blood flow is measured at finer resolution in
a heart divided into a larger number of pieces.

DIMENSION

The description of fractal objects and processes uses ideas
from the mathematical fields of topology and measure theory.
Since these fields are less familiar to most scientists than
algebra and differential èquations we will also provide
qualitative descriptions of the formal definitions.

Topology is the study of the most general properties of
objects and spaces. As noted by Kelley (1955, p. 88), "A
topologist is a man who doesn't know the difference between a
doughnut and a coffee cup." That is, both have one hole and
thus are topological equivalent so that each can be
continuously transformed into the other.

One way to study the properties of a space is to measure
distances between points in the space. A space where such
measurement can be done is called a metric space. In such a
space the distance $d(A,B)$ between points A and B has the
properties (Felschetz, 1954, p. 28)

$$d(A,B) = 0 \quad \Leftrightarrow \quad A = B, \quad \text{and} \tag{7}$$

$$d(B,C) \leq d(A,B) + d(A,C). \tag{8}$$

Since the distance between points is defined, we can define the
region that consists of all the points within a given distance
of one central point. Such a region is called a ball. Another
way to study the properties of a space is to study the
connectivity of points within the space.

Topological Dimension

There are different definitions of dimension which have different properties. What we normally refer to as the dimension, Mandelbrot (1983) calls the topological dimension. If we cover a metric space with balls, the balls must overlap to cover all the points in the space. Thus, each point in the space will lie within N balls. The topological dimension is equal to N-1 (Pontrygagin, 1952, pp. 16-17). For example, in a plane balls are called circles, a covering of the plane by overlapping circles requires that each point in the plane lie within three circles, and thus the topological dimension of the plane is 3 - 1 = 2, consistent with our expectation that the dimension of a plane is 2. The topological dimension can also be defined through the connectivity of the points that form the space. For example, consider a line. It may be so wiggly that it may nearly cover an area, but if we draw a ball of vanishingly small radius it will cover only points fore and aft on the same line, and not touch the next segment of the line that lies nearby.

Fractal Dimension - Hausdorff Measure

Another dimension that measures a different property of a space is called the Hausdorff measure, or Hausdorff-Besicovitch dimension, or fractal dimension (Mandelbrot, 1983). To help understand the fractal dimension, we will present it as a generalization of the simpler self-similarity dimension shown in Fig. 8. This dimension tell us as we look at finer details, how many new pieces of the object appear. Note that each new piece is self-similar to the whole original object, which justifies the name self-similar dimension. If we reduce the scale by factor S and N new pieces appear then the self-similar dimension D is given by $N = S^D$. For example, if we divide a line by 3, we find 3 pieces, $3=3^1$, thus the self-similar dimension of a line is 1. If we divide a square by 3, we find 9 pieces, $9 = 3^2$, thus the self-similar dimension of a square is 2. Note, that for these non-fractal objects the self-similar dimension is the same as our intuitive ideas of length, area, and volume. To generalize this definition to irregularly shaped objects, as shown in Fig. 9, we cover the object with balls and determine how many additional balls are needed to cover the object as the size of the balls decreases. More formally, the volume of a ball of radius r and dimension D is given by

$$V(r,D) = [\pi^{D/2}/\Gamma(1+D/2)]r^D \tag{9}$$

and if N(r) is the number of such balls required to cover the object then the fractal dimension is the value of D for which

$$\underset{r\to 0}{\text{Lim}} \quad N(r)V(r,D) \tag{10}$$

exists and is not equal to zero.

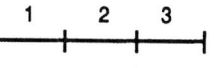

$$N = S^D$$

N = Number of congruent pieces
S = Division factor
D = Dimension

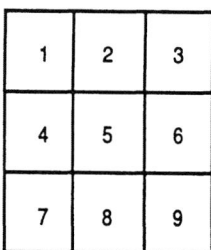

S=3, N=3, $3=3^1$ D=1 LENGTH

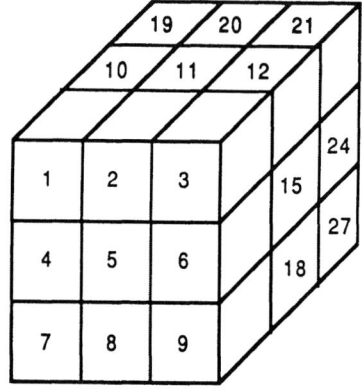

S=3, N=9, $9=3^2$ D=2 AREA

S=3, N=27, $27=3^3$ D=3 VOLUME

Fig. 8. The self-similarity dimension tell us as we look at finer detail how many new pieces we find.

Hausdorff-Besicovitch Dimension

N(x) = Number of balls of
radius x needed to cover the set

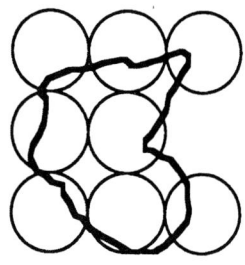

$$D = \operatorname*{Lim\,sup}_{x \to 0} \frac{\text{Log } N(x)}{\text{Log } (1/x)}$$

Fig. 9. The fractal dimension is a generalization of the self-similarity dimension to arbitrarily shaped objects. It tells us as we look at finer resolution how many new balls are needed to cover the object.

Definition of a Fractal

Using the definitions above we can now understand the definition of a fractal (Mandelbrot, 1983, p. 361):

Def. A set defined on a metric space is fractal if and only if its fractal dimension exceeds its topological dimension.

The topological dimension sort of defines the space where the object lives and the fractal dimension sort of defines how wiggly the object is in that space. That is, when the fractal dimension exceeds the topological dimension then many new pieces keep appearing as we look at finer detail. This is the central characteristic of fractal objects and leads to the properties of self-similarity, scaling, and monstrous moments described in the previous section. Thus, the fractal dimension can be evaluated using these properties. The exponent d of the power laws of Eqs. (2), (3), (4), and (5) is usually a simple function of the fractal dimension D and depends on the details of the problem and the topological dimension (Voss, 1988). For example, for ion channel kinetics $d = 1 - D$.

NOISE VS. CHAOS

A time series, such as the measurement of current through an open ion channel, that appears to be random may actually be either 1) "noise" produced by complex random processes or 2) "chaos" produced by a simple deterministic system. It is remarkable that simple deterministic systems can produce such a wealth of unexpected behavior. An excellent non-technical review of such systems was written by Gleick (1987) and more technical descriptions are given by Thompson and Stewart (1986) and Guckenheimer and Holmes (1983). It is now possible to analyze a time series to determine if it is actually random noise or deterministic chaos (Mayer-Kress, 1986).

For example the Lorenz equations (Thompson and Stewart, 1986; Guckenheimer and Holmes, 1983)

$$dx/dt = 10(y-x) \tag{11}$$

$$dy/dt = 28x - y - xz \tag{12}$$

$$dz/dt = -8z/3 + xy \tag{13}$$

produce an orbit that circles one of two centers and switches between them in a chaotic way. That is, these switches appear to be random but are actually produced by the deterministic Eqs. (11) - (13). The orbit of these equations in phase space (x,y,z) is limited to a small volume called a strange attractor that has fractal dimension 2.05. The fractal dimension of noise due to random events is infinite. The fact that the Lorenz equations have an attractor with low dimensionality demonstrates that this is a chaotic and not a random system.

Thus, the goal is to determine the fractal dimension of the orbit in phase space. If this orbit is of low fractal dimension then the signal is due to chaos produced by a deterministic system. If this orbit has high (infinite) fractal dimension then the signal is noise produced by truly random processes.

In analyzing experimental data the equations of the orbit are unknown. However, there is an absolutely remarkable theorem by Takens that proves the orbit in phase space can be reconstructed from the time series of only one of the variables! The time series of $x(t)$ alone can be used to determine the fractal dimension of all three Eqs. (11) - (13). This can be done because the equations are intimately, non-linearly coupled. More formally, the orbit obtained will be related to the actual orbit by an affine transformation and the topological properties, including the fractal dimension, are invariant under such an affine transformation.

To reconstruct the orbit in phase space the time series

$$x(t) = x(t_1), \ x(t_2), \ x(t_3), \ \ldots \tag{14}$$

is embedded in a space of topological dimension D_t. One way this can be done is by selecting D_t values of $x(t)$ equally spaced in time. These are then plotted as a point with those coordinates. For example, the points R_i to be plotted will be

$$R_i = \{x(t_i), \ x(t_i+\tau), \ x(t_i+2\tau), \ldots\}. \tag{15}$$

Then the fractal dimension D is determined for the set R_i. The embedding theorems require that if the system has n degrees of freedom that $D_t \geq 2n + 1$. Thus, the embedding is repeated for increasing values of D_t. If as D_t is increased, the fractal dimension D approaches a finite limit, then the experimental data is low dimensional chaos. However, if as D_t is increased, the fractal dimension also continues to increase, then the experimental data is random noise. The details of the procedures for determine τ and the fractal dimension are described by Mayer-Kress (1986), Farmer and Sidorowich (1988), and Albano et al. (1988).

To summarize the method. A time series, such as that recorded from an experiment, is turned into a geometric object so that its topological properties can be analyzed. If we find that this object is of low dimensionality then the system can be represented by a small set of deterministic equations. However, if we find that it is of high dimensionality then the system can only be represented by a complex set of truly random processes.

For example, this analysis has shown that some signals which appear to be random noise are actually low dimensional processes produced by deterministic systems. These include electroencephalographic signals recorded from monkeys and humans (Rapp et al., 1985; Skarda and Freeman, 1987), electrocardiograms (Glass and Mackey, 1988), and water dripping from a faucet (Packard and Shaw, 1986). On the other hand it has also been shown that noise from some sub-micron sized transistors (Webb and Gershenfeld, 1987) and the noise fluctuations of an ion channel within a state (Liebovitch, unpublished) are actually truly random and cannot be represented by a simple deterministic system.

ACKNOWLEDGEMENTS

This work was done during the tenure of an established investigatorship from the American Heart Association and was also supported in part by grants from the Whitaker Foundation and from the National Institutes of Health, EY6234. I thank Dr. Leo Levine for his help in some of the mathematical concepts.

REFERENCES

Albano, A. M., Muench, J., Schwartz, C., Mees, A. I., and Rapp, P. E., 1988, Singular-value decomposition and the Grassberger-Procaccia algorithm, *Phys. Rev.,* in press.

Avnir, D., Farin, D., and Pfeifer, P., 1984, Molecular fractal surfaces, *Nature,* 308:261-263.

Bassingthwaighte, J. B., 1988, Physiological heterogeneity: fractals link determinism and randomness in structures and functions, *News in Physiol. Sci.,* 3:5-10.

Brown, S. R. and Scholz, C. H., 1985, Broad bandwidth study of the topography of natural rock surfaces, *J. Geophys. Res.,* 90:12575-12582.

Crutchfield, J. P., Farmer, J. D., Packard, N. H., and Shaw, R. S., 1986, *Sci. Amer.,* Dec., pp. 46-57.

Cox, D. R., 1962, "Renewal Theory," Science Paperbacks, London.

Farmer, J. D. and Sidorowich, J. J., 1988, Exploting chaos to predict the future and reduce noise, Los Alamos Preprint LA-UR-88-901.

Glass, L. and Mackey, M. C., 1988, "From Clocks to Chaos," Princeton Univ., Princeton, NJ.

Gleick, J., 1987, "Chaos, Making a New Science," Viking, New York.

Guckenheimer, J. and Holmes, P., 1983, "Non-linear Oscillations, Dyanmical Systems, and Bifurcations of Vector Fields," Springer-Verlag, New York.

Hamill, O. P., Marty, A., Neher, E., Sakmann, B., and Sigworth, F. J., 1980, Improved patch-clamp techniques for high-resistance current recording from cells and cell-free membrane patches, *Pflügers Arch.,* 391:85-100.

Hille, B., 1984, "Ionic Channels of Excitable Membranes," Sinauer, Sunderland, MA.

Kagan, Y. Y. and Knopoff, L., 1981, Stochastic synthesis of earthquake catalogs, *J. Geophys. Res.,* 86:2853-2862.

Kelley, J. L., 1955, "General Topology," D. Van Nostrand, New York.

Kopelman, R., 1988, Fractal reaction kinetics, *Science,* 241:1620-1626.

Lavenda, B. H., 1985, Brownian motion, *Sci. Amer.,* Feb., pp. 70-84.

Lefschetz, S., 1954, "Introduction to Topology," Princeton Univ., Princeton, NJ.

Lewis, M. and Rees, D. C., 1985, Fractal surfaces of proteins, *Science,* 230:1163-1165.

Liebovitch, L. S., 1988a, The fractal random telegraph signal: signal analysis and applications, *Ann. Biomed. Engr.,* 16:483-494.

Liebovitch, L. S., 1988b, Analysis of fractal ion channel gating kinetics: kinetic rates, energy levels, and activation energies, *Math. Biosci.,* in press.

Liebovitch, L. S., Fischbarg, J., Koniarek, J. P., Todorova, I., and Wang, M, 1987a, Fractal model of ion-channel kinetics, *Biochim. Biophys. Acta,* 896:173-180.

Liebovitch, L. S., Fischbarg, J., and Koniarek, J. P., 1987b, Ion channel kinetics: a model based on fractal scaling rather than multistate Markov processes, *Math. Biosci.,* 84:37-68.

Liebovitch, L. S. and Sullivan, J. M., 1987, Fractal analysis of a voltage-dependent potassium channel from cultured mouse hippocampal neurons, *Biophys. J.,* 52:979-988.

Lovejoy, S., 1982, Area-perimeter relation for rain and cloud areas, *Science,* 216:185-187.

Mandelbrot, B. B., 1967, How long is the coast of Britain? Statistical self-similarity and fractional dimension, *Science,* 156:636-638.

Mandelbrot, B. B., 1983, "The Fractal Geometry of Nature," W. H. Freeman & Co., San Francisco.

Mayer-Kress, G., ed., 1986, "Dimensions and Entropies in Chaotic Systems," Springer-Verlag, New York.

Meakin, P., 1986, Computer simulation of growth and aggregation processes, *in* "On Growth and Form: Fractal and Non-Fractal Patterns in Physics," Martinus Nijhoff, Dordrecht, The Netherlands.

Morgan, J. D., McCammon, J. A., and Northrup, S. H., 1983, Molecular dynamics of ferrocytochrome c: time dependence of the atomic displacements, *Biopolymers,* 22:1579-1590.

Paumgartner, D., Losa, G., and Weibel, E. R., 1981, Resolution effect on the sterological estimation of surface and volume and its interpretation in terms of fractal dimensions, *J. Microsc.,* 121:51-63.

Pontryagin, 1952, (Trans. F. Bagemihl, H. Komm, and W. Seidel), "Foundations of Combinational Topology," Graylock Press, Rochester, NY.

Richardson, L. F., 1961, The problem of continguity: an appendix to *Statistics of Deadly Quarrels, Gen. Systems Yearbook,* 6:139-187.

Rapp, P. E., Zimmerman, I. D., Albano, A. M., deGuzman, G. C., Greenbaun, N. N., and Bashore, T. R., 1985, Experimental studies of chaotic neural behavior: cellular activity and electroencephalographic signals, *in* "Non-linear Oscillations in Biology and Chemistry," H. G. Othmer, ed., Springer-Verlag, New York.

Sakmann, B. and Neher, E., eds., 1983, "Single-Channel Recording," Plenum, New York.

Sander, L. M., 1986, Fractal growth processes, *Nature,* 322:789-793.

Sander, L. M., 1987, Fractal growth, *Sci. Amer.,* Jan., pp. 94-100.

Shlesinger, M., 1984, Williams-Watts dielectric relaxation: a fractal time stochastic process, *J. Stat. Phys.,* 36:639-648.

Skarda, C. A. and Freeman, W. J., 1987, How brains make chaos in order to make sense of the world, *Behav. Brain Sci.,* 10:161-195.

Thompson, J. M. T. and Stewart, H. B., 1986, "Non-linear Dynamics and Chaos," John Wiley and Sons, New York.

van Beek, J. H. G. M., Bassingthwaighte, J. B., and King, R. B., 1988, A fractal vascular network explains regional flow heterogeneity, *Biophys. J.*, 53:401a.

Voss, R. F., 1988, Fractals in nature: from characterization to simulation, *in:* "The Science of Fractal Images," Springer-Verlag, New York.

Webb, W. W. and Gershenfeld, N. A., 1988, The dimension of 1/f noise, *Bull. Amer. Phys. Soc.*, 32:482.

West, B. J. and Goldberger, A. L., 1987, Physiology in fractal dimensions, *Amer. Sci.*, 75:354-365.

Williams G. and Watts, D. C., 1970, Non-symmetrical dielectric relaxation behavior arising from a simple empirical decay function, *Trans. Faraday Soc.*, 66:80-85

TOWARDS A PHYSICS OF NEOCORTEX

Paul L. Nunez

Department of Biomedical Engineering
Tulane University
New Orleans, Louisiana 70118

INTRODUCTION

The cooperative efforts of physical and biological scientists over the past half century have resulted in a tentative, first order understanding of biological membranes and the behavior of single neurons. Nevertheless, recent work has further exposed the neuron as a highly complex system whose behavior is likely to be the subject of study for many years to come. In fact, the vast discrepancy between real neurons and the "neurons" of neural network theory has prompted some researchers to label the latter "morons." However, even complete understanding of the operation of single neurons or simple neural systems is, by itself, unlikely to add much to our understanding of higher brain function. Partly for this reason, there is an active interest in the field of neural networks, or much more generally, brain dynamics.

The distinction between neural network theory and brain dynamics is an important one because the field of neural networks has, thus far, largely avoided several issues which appear critical for both further theoretical developments and experimental verification in very complex neural systems like neocortex. On the experimental side is the relationship between "activity" in the network and data that can be recorded with a finite number of sensors of nonzero size. Recording of extracellular fields or other data involves space-averaging at spatial scales dependent on the size and location of the sensors. Temporal scales are also an issue. For example, mapping the magnetic field response to a stimulus with a single channel sensor can take several hours; the resulting field is due only to stationary sources, typically in primary sensory cortex. A multichannel EEG can, by contrast, distinguish changes in intracortical covariance over time scales in the 10 ms range. Other measures of brain function take place on spatial scales that span about five orders of magnitude and temporal scales that vary over about ten orders of magnitude (Churchland and Sejnowski 1988). From the theoretical perspective, it must also be recognized that neocortical dynamic function takes place at multiple spatial and temporal scales. There is good reason to believe that interactions take place across these scales so that formal methods to bridge the gaps between hierarchical levels are required.

This chapter is concerned with several issues related to "neocortical dynamics," that is, the dynamic interaction of the estimated 10^{10} neurons of human neocortex, experimental measures of the resulting electric and magnetic fields, and correlations between spatial-temporal properties of these fields and measured cognitive functions. The emphasis here is on "neocortical" rather than "brain" dynamics for several reasons:

1. Much, if not most, of the available data on humans - either electromagnetic fields, cognitive measures, PET scan data, etc. - are believed due to processes in neocortex.
2. Mammalian neocortex exhibits a macroscopic homogeneity and isotropy in directions tangent to its surface; thereby making it more accessible to theoretical description.

3. Human neocortex has a dense system of roughly 10^{10} subcortical interconnections, called corticocortical or association fibers, which form most of the white matter. The human neocortex is much more strongly connected to itself than to other brain structures, thereby providing some justification for its treatment as a separate system. In fact, this high density of corticocortical fibers in humans as opposed to lower mammals has been cited as a major factor in making the human brain "human" (Braitenberg 1978 ; reviewed by Katznelson 1981).

EEG AND COGNITIVE PROCESSING

It is suggested that neocortex be viewed as a distinct state of matter like solid, liquid, gas, or plasma with its properties at various spatial and temporal scales subject to both theoretical prediction and measurement. This optimistic view may, to some, appear inconsistent with observations of the complexity of even a single neuron. A system containing 10^{10} strongly interacting neurons might then appear hopelessly complex. One reason for optimism is that ample EEG data is available in humans at the level of the regional column of cortex (several cm^2), and this data is at least moderately correlated with cognitive function. A theoretical description of neocortical dynamics at large spatial scales which provides explanation of EEG observations is, by itself, a laudable scientific goal with important consequences for medicine. However, the implications of a theory of EEG at larger spatial scales goes much further than medical applications. It has long been appreciated that EEG is a measure, albeit a crude measure, of conscious experience. For example, the human alpha rhythm is a 10 Hz, near sinusoidal oscillation, which occurs in 95% of the population when sitting quietly and relaxed with eyes closed. In many subjects, mental calculations cause "blocking" of the alpha rhythm. That is, the 10 Hz rhythm with a typical bandwidth of 1 or 2 Hz is replaced by low amplitude potentials which may appear in the frequency domain as low pass noise. Another well-known example is the EEG recorded during the REM (rapid eye movement) stage of sleep, which is closely related to dreaming. In general, the human spontaneous EEG exhibits the following characteristics: high amplitude, spatially coherent, sinusoidal potentials distributed over the entire scalp are associated with lack of cognitive processing as in the states of deep sleep, anesthesia, and coma. (The awake alpha rhythm is somewhat of an intermediate case, being of moderate amplitude and easily recordable only over the back half of the head.) Low amplitude, broad spectrum EEG occurs in the awake subject with eyes open or during difficult cognitive task performance.

In addition to a history of 60 years of EEG data, researchers over the past 25 years have produced a large body of evoked (or event related) potential data. Both transient and steady state evoked potentials have been recorded (Regan 1988). The transient response to some stimulus (auditory, visual, or somatosensory) is averaged over many stimuli to remove at least some of the spontaneous EEG, biological noise, and system noise to produce a temporal waveform in the few hundred ms following the stimulus. It has been observed that transient responses are determined not just by stimulus properties or motor activity ("exogenous factors"), but also by cognitive processing ("endogenous factors"). For example, the "contingent negative variation" refers to a shift in scalp potential as the result of an expected stimulus. Another well-known transient response, called "P300" (positive potential over the top of the head with respect to ear reference at a latency of roughly 300 ms from the stimulus), is dependent on the perceived significance of the stimulus (review by Gevins and Cutillo 1986). For example, a series of auditory tones of constant frequency are presented, randomly interspersed with a few tones of a different frequency. The subject is asked to count the odd-ball tones. Whereas the odd-ball tones evoke a P300 response, regular tones do not.

With the use of multichannel recordings (typically 16 to 128), spatial as well as temporal properties of EEG and event related potentials have been studied in relation to cognitive processing (Livanov 1977). For example, visual imagery can change location of maximum alpha rms amplitude on the scalp (Lehmann 1988). Changes in interhemispheric and intrahemispheric

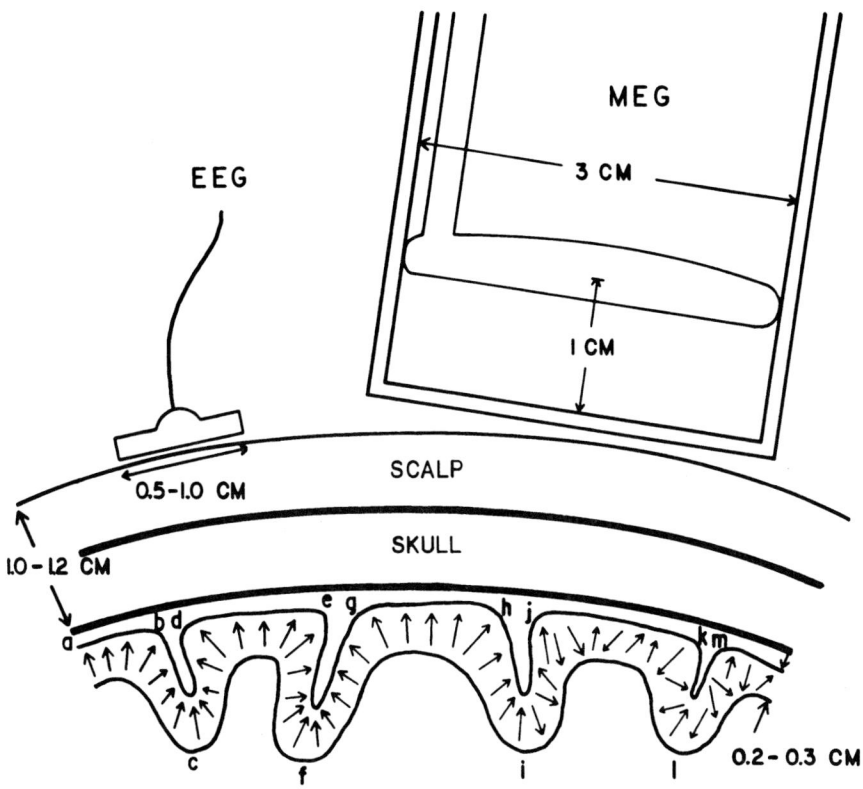

Fig. 1. Extracranial recording of electroencephalographic (EEG) and
magnetoencephalographic (MEG) data. EEG is most sensitive to correlated
dipole layer in gyri (ab, de, gh), less sensitive to correlated dipole layer in
sulcus (hi), and insensitive to opposing dipole layer in sulci (bcd, efg) and
random dipole layer (ijklm). MEG is most sensitive to correlated dipole layer in
sulcus (hi) and much less sensitive to all other sources shown because of
opposing dipoles, randomly oriented dipoles, or radial dipoles. (The external
magnetic field due to a radial dipole in a medium of spherically symmetric
conductive layers is zero.) MEG is measured with a SQUID (superconducting
quantum interference device), typically consisting of several coils used to cancel
spatially constant magnetic fields. Only the coil closest to the head is shown
here.

coherency patterns of alpha rhythm are associated with mental arithematic, mental reading, or
listening to music (Pockberger et al. 1988; Rappelsberger and Petsche 1988). Scalp covariance
patterns of event related potentials have been shown to change distinctively in preparation for a
cognitive task, and depend on whether the task was later accurately or inaccurately performed
(Gevins et al. 1980, 1981, 1983, 1985). These later data suggest the formation of specific
functional neural networks appropriate for specific types of cognitive processing on time scales
in the 10-100 ms range and spatial scales in the cm range. These networks typically involve
neurons in many widely spaced regions of both neocortical hemispheres.

MULTIPLE SPATIAL SCALES OF NEOCORTEX

Human scalp recordings of spontaneous EEG represent the space averaged current source activity from a minimum of several cm^2 of neocortex as depicted in Fig 1. The region of contributing cortex then typically contains 10^7 to 10^{10} neurons (Nunez 1981a). The attenuation of the spectral power between cortex and scalp may also be expressed in terms of wavenumber (spatial frequency). Nearly all scalp power occurs at wavenumbers $k < 0.5$ cm^{-1} (Nunez 1981a, 1988a). Evoked potentials in primary sensory cortex may result from somewhat smaller neural populations, but these potentials are typically only recordable at the scalp by averaging over many stimuli. Magnetic field recordings (MEG) avoid field spreading by the skull (which has a low electrical conductivity), but also suffer from very limited spatial resolution due to the large size and distance of the sensing coil from the sources (Wikswo and Roth 1988). Micro-EEG recordings, obtained in mammalian cortex and in a very limited number of human patients, are obtained with electrodes of tip diameter in the 10^{-3} to 10^{-1} cm range (Petsche et al. 1984) and typically record the summed activity of between 10^2 and 10^5 neurons (Abeles 1982). Thus, much of the available physiological data and nearly all the data directly reflecting human cognitive function is available only at spatial scales much larger than that of the single neuron.

Table 1. Spatial Scales of Human Neocortex

Name of Structure	Typical Diameter(mm)	Number of Neurons	Description
Minicolumn	2 to 5 x 10^{-2}	10^2	Spatial extent of inhibitory connections.
Dendritic column	10^{-1}	10^2 to 10^3	Spatial extent of apical column of dendritic bundles of pyramidal cells, i.e. scale for input to a single pyramidal cell.
Corticocortical column	3×10^{-1}	10^3 to 10^4	Width of aborization column of corticocortical afferents, i.e. input scale for specific long range connections.
Macrocolumn	3	10^5 to 10^6	Extent of axon collateral system of single pyramidal cell, i.e. longest spatial scale for intracortical, excitatory output.
Regional column	30	10^7 to 10^8	Average length of corticocortical fibers, i.e. one spatial scale for long range excitatory output. Typical scale of cytoarchitectonic area. EEG must be correlated over area of this size in order to be measurable on scalp without averaging.
Entire neocortex	400	10^{10}	Longest corticocortical fibers.

Even if abundant data relating to cognitive function were available at the level of the single neuron, this level of neocortical hierarchy is perhaps no more appropriate for study than any of several other levels as implied by Table 1. On the one hand, a scientist with a "microscopic mentality" may cite the complexity of single neurons and suggest study at lower hierarchical levels. But, an equally valid approach involves studies at higher levels - cortical columns at spatial scales ranging from the minicolumn (diameter ~ 20-50 μm and containing ~ 100 neurons) to the regional column (diameter ~ 1 cm and containing ~ 10^6 to 10^8 neurons). This view is partly supported by the fact that each cortical pyramidal cell can interact synaptically simultaneously with about 10^5 nearby cells. In addition, there is evidence for direct interaction by extracellular fields in the absence of active chemical synapses (Taylor and Dudek 1984a, b ; Adey 1988). Thus, a dominant feature of neural dynamics is one of collective interactions. (Physicists will note the similarity to collective interactions in a typical laboratory plasma having about 10^5 particles in a Debye sphere.) The viewing of cortical columns rather than neurons as basic functional units of neocortex is also supported by physiological data indicating redundancy of neural firings in tangential directions within columns of visual cortex (Hubel and Weisel 1962; Abeles 1982). Other relevant physiological data includes evidence that control of motor activity by primate motor cortex is related to population properties rather than individual neurons (Georgopoulos et al. 1986), and that olfactory information is coded in spatial patterns of potential recorded from the surface of the olfactory bulb in cats (Freeman 1987, 1988). Freeman has, in fact, long championed the concept of the neural mass (at various hierarchical levels), in which the density of interconnections is sufficiently large so that new, more macroscopic properties emerge which are not predictable at the level of the single neuron (Freeman 1975).

Different spatial scales in neocortex are distinguished by important anatomical as well as physiological characteristics as indicated by Table 1. It is of interest to note the apparent self-similarity of columnar structures at different scales, a property characteristic of the fractile geometry observed in many physical and biological systems. It is to be expected that different rules will govern interactions at different hierarchical levels. For example, interactions between minicolumns, for which inhibitory synaptic connections are numerous, will require different quantitative description than interactions between macrocolumns which are apparently dominated by excitatory synapses. A principal goal of neocortical physics is then to relate rules of interaction across hierarchical levels. An example would be to predict statistical properties of macroscopic potentials recorded in a particular experiment in humans from a combination of theory and physiological recordings obtained at smaller spatial scales in animal cortex. Rather than attempting to somehow model the cortex in terms of the interaction of 10^{10} neurons or 10^8 minicolumns, the idea is to move progressively up the spatial scale, with the rules at each hierarchical level derived from those at the next lower levels. In this manner, one might hope to create a manageable theory appropriate for the scale of a particular experiment. Another approach, discussed further in a later section, is to postulate plausible input/output relations at higher levels and consider the resulting predictions of EEG data.

Over the past several years, an ambitious attempt has been undertaken to formulate a comprehensive theory of neocortical interaction at multiple spatial-temporal scales (Ingber 1982, 1983, 1984, 1985a,b). The approach of this "statistical mechanics of neocortical interactions" is to follow the conditional probability of a neuron firing given earlier interactions with surrounding neurons in a manner similar to standard neural network theory. Activity at this microscopic scale is then aggregated up to a mesoscopic scale to establish a mesoscopic Lagrangian which defines a short-time probability distribution of firings in a minicolumn, given its just previous interactions with all other neurons in its macrocolumnar surround. This theory suggests a basis for short-term memory based on discrete minima in the Lagrangian, and the likelihood that EEG rhythms are generated by multiple mechanisms at multiple scales of interaction. The theory then allows local and global approaches to complement each other at a common level of formal analysis. More recently, effects of long range (corticocortical) interactions have been included (Ingber and Nunez 1988). A major challenge of this approach is to identify equation parameters more closely with physiological measurements. Possibly this can be accomplished indirectly by working back from EEG data. A major difficulty in evaluation of this formal theory is that the number of scientists in the world versed in modern statistical field theories, neocortical physiology/anatomy, and EEG is vanishingly small. But, this situation can be expected to change as the field of neocortical dynamics grows.

The important issue of linearity/nonlinearity of neocortical dynamics is intimately related to that of scale. A relatively simple example of this relationship which may provide useful insights into neocortical studies is that of a linear stretched string with attached springs (Ingber and Nunez 1988). The springs may be nonlinear; they may, in fact, undergo chaotic motion. The string displacement $u(x,t)$ is expanded in the usual series of eigenfunctions

$$u(x,t) = \sum_{m=1}^{\infty} G_m(t) \sin k_m x \tag{1}$$

Whereas the motion of the string at large wavenumbers may be highly nonlinear, any real sensor (like the scalp EEG electrode) will have only limited spatial resolution; that is, it will record only the motion

$$\bar{u}(x,t) = \sum_{m=1}^{M} G_m(t) \sin k_m x \tag{2}$$

One might expect to describe the macroscopic string displacement u(x,t) as a linear or quasilinear variable, but influenced by the local nonlinear behavior which crosses the hierarchical level from mesoscopic (analogous to local circuit effects in neocortical columns) to macroscopic (analogous to scalp EEG). Macroscopic string displacement was obtained by renormalization of the mesoscopic mechanical Lagrangian, i.e., by intergrating over large wavenumbers to smooth out their influence on macroscopic displacement (Ingber and Nunez 1988).

These issues of hierarchical level in neocortex are partly analogous to problems encountered in physical media having multiple length scales for interaction between functional units of the media (Ma 1976 ; Wilson 1979). The latter reference, an excellent popular account by a principal contributor to the field, provides a number of general ideas that appear to apply to neocortex. Consider, for example, the two-dimensional lattice shown in Fig 2 which may, in this account, represent either a region of ferromagnetic material or neocortical surface. In the magnetic case, each location in the lattice (a "microunit") shows the magnetic dipole moment associated with a single atom. A star represents a positive dipole moment, a blank space indicates a negative dipole moment. At high temperatures (Fig. 2b), the dipole moments are random so that a macroscopic region produces no net magnetization. When the temperature drops to the Curie point, the dipole moments of individual microunits become highly correlated to produce a net magnetizaion in the material (Fig. 2a). However, fluctuations at smaller scales persist below the Curie temperature and all scales of length must be included in any comprehensive theory of the ferromagnet. As Wilson puts it,

> "Small fluctuations are not suppressed; they merely become fine structure superimposed on the larger one. An ocean of positive dipole moments may contain an island of negatives, which, in turn, surrounds a lake of positives with an islet of negatives."

The dramatic increase in apparent correlation length as the temperature reaches the Curie point is due only to nearest-neighbor interactions. That is, a molecule with a positive dipole moment influences adjacent molecules so that they have a greater than even probability, dependent on temperature, of also having positive dipole moments. When this probability becomes sufficiently high, as happens at the Curie point when the influence of random molecular vibration is reduced, long range order emerges from short range interactions. Wilson's contribution, the renormalization group, is a method of constructing theories to cross hierarchical levels. That is, a theory of ferromagnetism is obtained by breaking down a large problem with multiple length scales into a sequence of smaller problems, each of which is confined to a single scale. For example, the lattice of microunits might be divided into M miniunits, each consisting of N contiguous microunits. The magnetic dipole moment of the m th miniunit is then just the average of the dipole moments of the N microunits, p_n, i.e.

$$p_m = \frac{1}{N} \sum_{n=1}^{N'} p_n \quad , \qquad m = 1, M \tag{3}$$

This procedure is illustrated in Fig. 2 at d and e with N=6300 microunits. The original lattice of microunits (c) is then replaced by a lattice of miniunits (b), and the problem has been moved up one level in the hierarchy. This procedure may be repeated so that the lattice of

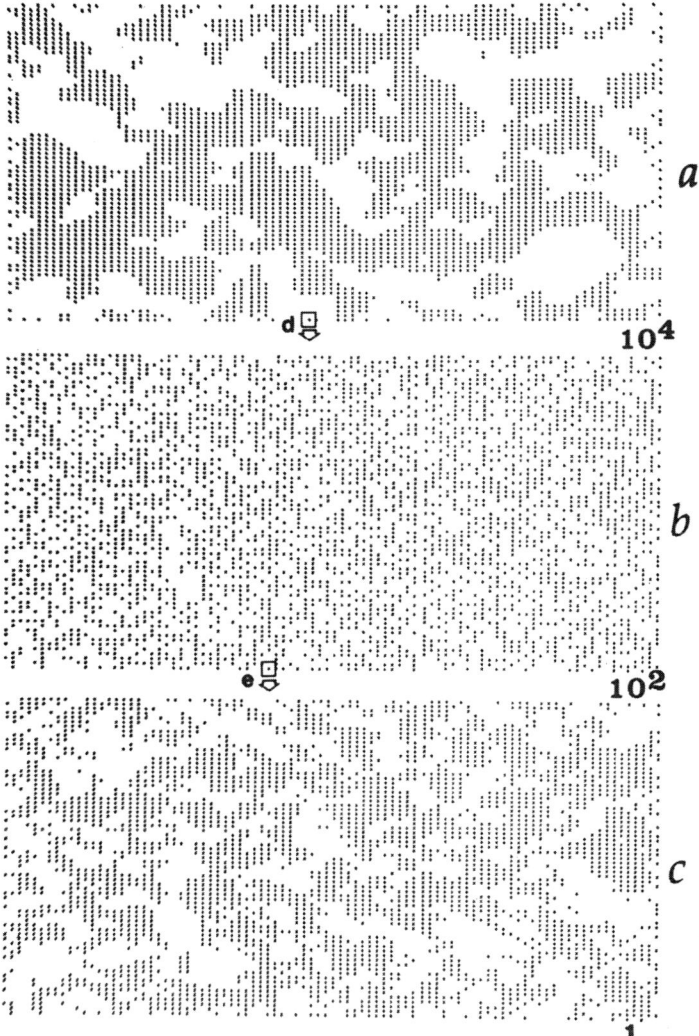

Fig. 2. Lattice of units representing either magnetic dipole moment vector in ferromagnetic materials or instantaneous polarization vector in neocortical columns at various scales. The appropriate vector can, in general, vary continuously, but for purposes of illustration, a filled space indicates a positive vector and a blank space a negative vector. In the magnetic example, (a) represents a material with temperature below the Curie point (magnetized) and (b) represents an unmagnetized material at high temperatures. In the neocortical example, three spatial scales are depicted, spanning roughly four orders of magnitude. Highly correlated dipole layer at say the regional column scale (a) is examined more closely by expanding small region (d) to reveal quasi-random dipole layer at intermediate scale (b), which is, in turn, examined more closely by expanding region (e) to obtain moderately correlated dipole layer at say sub-minicolumn scale. This figure is for illustrative purposes only; it is not based on any neurophysiological calculations.

miniunits is replaced by a lattice of mesounits (a), which in turn is replaced by a lattice of macrounits, and so on. The success of the renormalization group is in allowing calculation of the rules governing interactions at one level from the rules that apply at lower levels. Some mathematical features of the renormalization group have been applied to neocortex (Ingber and Nunez 1988). In the statistical mechanical treatment of neocortex (Ingber 1983, 1984), analogs of temperature (in magnetic materials) might be various neuromodulators of synaptic activity. The analogy between magnetic materials and neocortex cannot, of course, be taken too far. One issue is that the renormalization group treatment requires continuity across scales on each side of the Curie temperature. In neocortex, different scales may have quite different characteristics so that EEG theory, for example, is expected to require much more specific connection to the appropriate columnar scale as illustrated in a following section. Also, corticocortical fibers or long intracortical fibers can be expected to enhance long-ranged correlations which might be just barely supported by the shorter-ranged columnar interactions.

Figure 2 also illustrates dipole fields at three successive spatial scales. Note that the mesounits (a) shown here appear to exhibit a long correlation length, the miniunits (b) a short correlation length, and the microunits (c) a moderate correlation length, i.e. the dipole field can not be described by a single correlation length. Correlation functions estimated from experimental data will be very much dependent on the size and location of the sensor. This point appears to be poorly understood in electrophysiology where recordings span four orders of magnitude of spatial scales. The picture shown in Fig. 2 is only descriptive; it is not based on any neocortical theory. However, the complexity of neocortical dynamics, including the fact that synaptic interaction lengths span about seven orders of magnitude, makes this kind of speculation plausible.

STATE VARIABLES OF NEOCORTEX

Suppose that the lattice structures shown in Fig. 2 represent the surface of neocortex. We can characterize the state of each elemental column of neocortical tissue by its current dipole moment per unit volume $P(r,t)$, measured in amps/m^2. $P(r,t)$ is a quantity analogous to the electric polarization vector used to describe the state of a dielectric. The term "current polarization" is used here to distinguish it from the usual dielectric charge polarization. $P(r,t)$ is related to the current source function $s(r,t)$, measured in amps/m^3, due to individual synapses and the passive sources distributed along membrane surfaces required for current conservation by

$$P(r,t) = \frac{1}{\Delta V} \int_{\Delta V} r' \, s(r',t) \, d^3 r' \tag{4}$$

where ΔV is the volume of the column. To close approximation brain electric and magnetic fields at EEG frequencies are quasistatic (Plonsey 1969 ; Nunez 1981a). The electric potential $\Phi(r,t)$ at any location due to neocortical current sources is then given by Poisson's equation

$$\nabla \cdot \sigma \nabla \Phi = s(r,t) \tag{5}$$

which is just a statement of current conservation external to the elemental current source regions. Here $\sigma(r)$ is the local conductivity of tissue, defined at the same scale as the potential and source function. For the purposes of scalp recordings, which occur at distances large compared to cortical thickness, an appropriate approximation is

$$s(r,t) \cong S(x,y,t) \left\{ \delta \left[z - \frac{1}{2} z_1(x,y) \right] - \delta \left[z + \frac{1}{2} z_1(x,y) \right] \right\} \tag{6}$$

which identifies $s(r,t)$ as a current dipole layer in the cortex with effective pole separation $z_1(x,y)$ and $S(x,y,t)$ as an effective current density across the cortex. If the column area is sufficiently small so that $S(x,y,t)$ is approximately constant over its surface, Eqs. (4) and (6) combine to yield the current polarization vector for a cortical column

$$P(x,y,t) \cong \frac{z_1(x,y)}{d(x,y)} S(x,y,t)\, a_z(x,y) \tag{7}$$

where $d(x,y)$ is cortical thickness and $a_z(x,y)$ is a unit vector everywhere normal to the cortical surface. $P(x,y,t)$ can, for example, be defined at the minicolumn level by Eq. (7) and at higher levels by the coarse graining procedure indicated by Eq. (3). The potential difference across the cortex at any location is directly measurable in mammalian cortex and is given by

$$\Delta\Phi(x,y,t) \cong \frac{z_1(x,y)}{\sigma_c(x,y)} S(x,y,t) \tag{8}$$

where $\sigma_c(x,y)$ is cortical electrical conductivity, defined at the same scale as the polarization vector. For example, $\Delta\Phi$ is of the order of 100-200 μV in the case of spontaneous EEG in dog cortex (Lopes da Silva and Storm van Leeuwen 1978). The special case of dipole sources isolated at tangential locations of cortex (x_1, y_1) is

$$S(x,y,t) \cong f(t)\, \delta(x - x_1)\, \delta(y - y_1) \tag{9}$$

where $f(t)$ may be either deterministic or random. Equation (9) may apply approximately to evoked potential sources in primary sensory cortex and to a subset of epileptic spikes, but does not apply generally to observed EEG phenomena which are due to widely distributed sources (Nunez 1981a, 1988c ; Gevins et al. 1980, 1981, 1983, 1985). Recent publications of MEG data which may imply localization of sources can be explained by the procedure of averaging over nonstationary sources and/or the very selective sensitivity of MEG to dipoles in half sulci as indicated by Fig. 1 (Nunez 1986).

In addition to the polarization vector $P(r,t)$, another possible state variable for neocortex is a probability density function analogous to that used in the theory of hot plasma. In the case of neocortex, the probability density function $f(r,P,t)$, defined so as to be proportional to the probability that a mass of neural tissue located between r and $r+dr$ has polarization between P and $P+dP$ at time t. The function f can be defined at a number of spatial scales and bears a direct relationship to fields measured on the same scale. Other state variables like probability of firing action potentials or synaptic action density may be more convenient in theoretical developments, but their relationships to measurable quantities must be derived in separate, nontrivial analyses (Freeman 1975 ; Nunez 1981a ; Ingber and Nunez 1988).

LINEAR LIMITING CASES OF MACROSCOPIC EEG

Attempts to formulate quantitative descriptions of neocortex at any hierarchical level are likely to lead to very complicated, nonlinear equations. A number of theoretical effects have suggested possible mechanisms for various brain functions, for example visual psychophysics (Wilson and Cowan 1972, 1973 ; Cowan 1982), short-term memory (Ingber 1984), and short wavelength EEG (Lopes da Silva et al. 1974 ; Freeman 1975 ; van Rotterdam et al. 1982 ; Ingber 1985a). These and other approaches at several hierarchical levels span a spectrum of fully nonlinear, quasilinear, and linear theories, including integrodifferential equations and classical linear partial differential equations of the parabolic, hyperbolic, and elliptic types. From the experimental side, apparently linear EEG (Nunez 1981a), chaotic behavior of epileptic EEG (Babloyantz and Destexhe 1986), and nonlinear EEG in cat olfactory bulb (Freeman 1987) and other midbrain structures in cat (Roschke and Basar 1988) have been reported. Brain interactions are, in fact, sufficiently complicated that one may suggest that, in a mathematical sense, nearly anything that can happen outside the brain can happen inside the brain.

Given the complexity of neocortical interactions and the very limited available quantitative data needed to establish the neural parameters of nonlinear theories, it is important to search for linear limiting cases of neocortical function. With this goal in mind, consider the scalp EEG observed in patients under the influence of halothane anesthesia, with inspired concentration varied with time as shown in Fig 3. The EEG is of large amplitude (~100 μV) over the entire scalp. The peak frequency follows closely (after a short time delay) the concentration of

EEG POWER SPECTRUM [HALOTHANE]$_i$

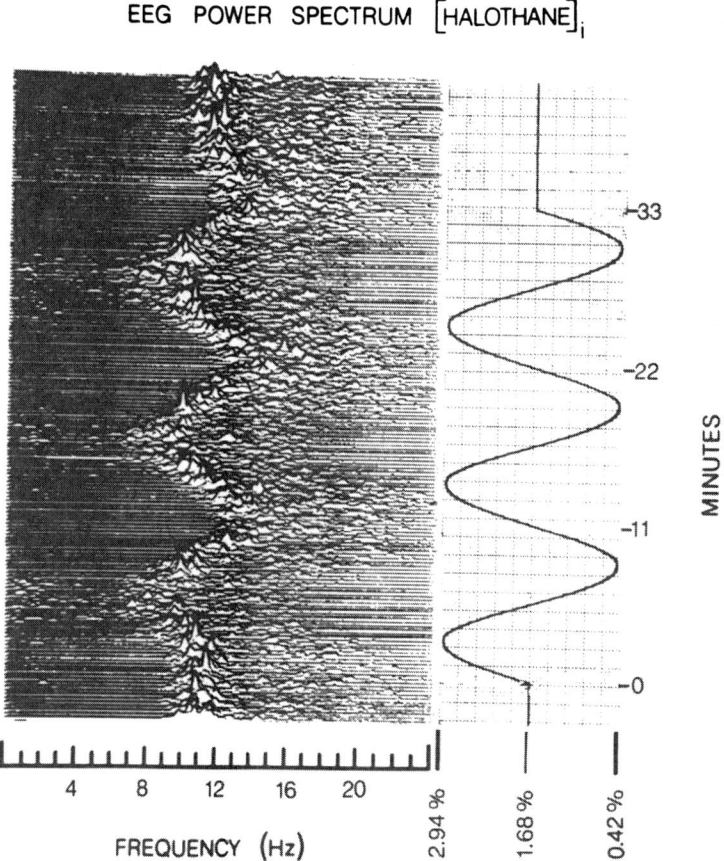

Fig. 3. Compressed spectral array of EEG recorded from the scalp of a patient under halothane anesthesia. Inspired concentration of halothane varies over a period of 33 minutes as indicated by the plot at right. One channel of eight channel recording is shown. Other channels reveal nearly identical spectra, i.e. EEG is of large amplitude and spatially coherent over the entire scalp. Data from one subject in a study of about 10 subjects is shown. All subjects showed the same qualitive behavior of EEG, but with quantitative differences in number and intensity of apparent modes, which move up and down in frequency as halothane concentration is varied. Data recorded in the operating room by J.J. Stockard, N.T. Smith, and P.L. Nunez. From (Stockard 1976).

halothane. In this manner the neocortex can be "tuned" to any frequency between about 4 Hz (high concentrations) and 16 Hz (low concentrations). Furthermore, in this and other experiments (Nunez 1981a), the EEG contains a number of "modes", which move together up and down in the frequency domain. The word "mode" is used purposely here to suggest a connection to standing wave phenomena, or at least to a system of many coupled oscillators.

What general characteristics must a medium possess to support standing waves? First, there must be some mechanism for propagation of signals at finite velocity. In neocortex, this is provided by the 10^{10} corticocortical fibers which carry action potentials with distributed velocities, peaked at about 6-9 m/s (Nunez 1981a). Second, since there is no apparent physiological mechanism for reflection of waves from boundaries; periodic boundary conditions are required. That is, the surface must be at least partially closed so that waves traveling in one direction can coalesce with waves in other directions to form standing waves. Neocortex provides such a surface. Of course, there are inhibitory physiological mechanisms to cause wave damping so that such waves may or may not be observed in EEG data. To see if this idea makes any sense let us first consider some rough semi-quantitative ideas. Consider the case of nondispersive waves on the surface of a hollow spherical shell of radius R. (It doesn't matter whether these are sound waves, electromagnetic waves, or any other nondispersive waves provided interaction with surrounding media can be ignored.) The normal mode frequencies are

$$\omega_{\ell} = \frac{v}{R} \sqrt{\ell(\ell + 1)} \quad , \qquad = 1, 2, 3, \ldots \qquad (10)$$

where v is the medium propagation velocity. An effective radius of the cortex, R~11 to 18 cm, accounting for stretching due to fissures and sulci, may be obtained from reports of surface area of 1600 to 4000 cm^2. Using the 6-9 m/sec range for corticocortical fiber propagation, the lowest mode is

$$\omega_1 \sim 47 \text{ to } 113 \text{ sec}^{-1}$$

$$\text{or} \qquad\qquad\qquad\qquad (11)$$

$$f_1 \sim 8 \text{ to } 18 \text{ Hz}$$

that is, a frequency within a factor of two of the alpha or halothane frequency. Of course, we have no apriori reason to assume nondispersive waves; actual dispersion relations must be derived from the physiology/anatomy.

Consider another experiment - the steady state response of the human neocortex to sinusoidally modulated light (Regan 1988) as shown in Fig. 4. In Fig. 4b, the lowest normal mode frequencies for nondispersive standing waves on the surface of a prolate spheroidal shell are shown on the same scale (Morse and Feshback 1953 ; Nunez 1988c). Only a little fudging is required to make the lower modes line up with the experimental results. That is, a velocity v=6 m/sec (at the low end of propagation velocity for corticocortical fibers) was chosen. The prolate spheroid, which here represents a single brain hemisphere, has an area and eccentricity which reasonably matches anatomical data. While the physiological and anatomical data used to construct the theoretical/ experimental comparison of Fig. 4 are known only approximately, it should be emphasized that <u>no free parameters</u> were used to make this comparison.

No claim is made that Fig. 4 represents an explanation for EEG analogous to say the Balmer series for the hydrogen atom. For one thing, the agreement is no better than a factor of about two. More importantly, oscillatory EEG phenomena in neocortex can evidently result from local delays (mostly rise times of postsynaptic potentials), global delays due to finite propagation velocity of action potentials, or a combination of both. Furthermore, oscillatory phenomena may perhaps originate at a number of spatial scales of neocortex. Neocortical dynamic theory at sub-macrocolumn scales, based on local delays and intracortical (short range) connections, has predicted traveling cortical waves with wavemenbers k~100's of cm^{-1} and propagation velocities v~several cm's/sec (Wilson and Cowan 1973 ; van Rotterdam et al. 1982 ; Ingber 1985a). Predictions of EEG frequencies (~ 40 Hz) for cat olfactory bulb and cortex (Freeman 1974, 1987) and dog or human cortex (~12 and 28 Hz, Lopes da Silva et al 1974 ; van Rotterdam et al.

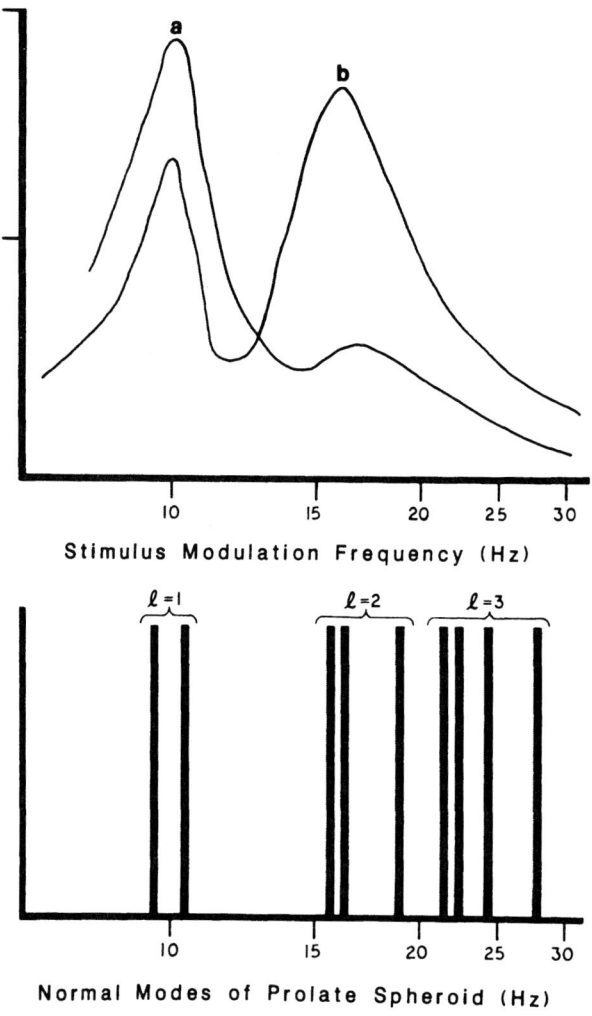

Fig. 4. Top. RMS amplitude of steady state scalp evoked potential. Response of human brain to sine wave modulated light, red on blue (a) and red (b). Centrally fixated stimulus field of 60 degrees subtense; mean retinal illumination 3900 trolands; luminance modulation depth 33%. From (Regan 1988). Other experiments on humans show that the response of the brain is sensitive to stimulus parameters. Also, both linear and nonlinear responses are apparently possible depending on parameters and subject (van der Tweel and Verduyn Lunel 1965).

Bottom. Lowest normal mode frequencies $\omega_{\ell m}$ ($\ell = 1, 2, 3$) for nondispersive waves on the surface of a prolate spheroidal shell constructed of a medium with propagation velocity = 6 m/sec, semi-major axis before stretching to account for fissures and sulci = 8.4 cm, stretching factor = 2.2, and eccentricity = 0.8. These data are based on estimates of propagation velocity in corticocortical fibers and actual shape of one brain hemisphere (Nunez 1981a). Uncertainty in location of lowest mode (assuming nondispersive waves) is perhaps a factor of two. Relative uncertainty of higher modes is probably less. When eccentricity is zero, normal mode frequencies are those of a spherical shell, Eq(10), independent of m index. Departures from prolate spheriodal geometry or inhomogeneity/anisotropy of neocortical connections can be expected to cause further splitting of degenerate modes. Modes calculated from Helmholtz equation with classical methods (Morse and Feshbach 1953 ; Nunez 1988c).

1982 ; Zhadin 1984) based on local delays have been published. Predictions of higher frequencies appear quite plausible, but it is not clear that local delays are, by themselves, sufficiently long to account for the typical range of human EEG frequencies of 1-15 Hz (scalp) or 1-30 Hz (cortex) (Pfurtscheller and Cooper 1975).

Over the past 15 years, a number of issues relating to possible standing neocortical waves, or more generally, global resonances have been considered (Nunez 1974a, b, 1981a, b, c, 1988a, b, c ; Nunez et al. 1977 ; Katznelson 1981). Some of the results are summarized as follows: Since EEG reflects the space-averaged activity of at least several cm^2 of cortical surface, the number density of synaptic and action potential firings at the scales of the macrocolumn/ regional column were followed with a macroscopic theory that avoids much of the complexity of neural network models. Cortical columns are connected by both short (intracortical) and long range (corticocortical) fibers. Delays caused by both finite propagation velocity and postsynaptic potential rise times are included in a more general version of the theory (Nunez 1988b); solutions that ignore local delays have been obtained for waves on the surface of a homogeneous sphere (Katznelson 1981). For purposes of both brevity and clarity, the simplest one-dimensional version with negligible local delays is outlined here. The number, length distribution, and associated velocity distribution of corticocortical fibers is only partly known. It is assumed that white matter contains N overlapping fiber systems, where the number of connections between macrocolumns separated by distance $|x|$ falls off as $\exp(-\lambda_n|x|)$, n=1, N. For large N, this assumption can be made to fit nearly all distributions.

The global theory is based on plausible input/output relations at the macrocolumn scale. That is, the number of action potentials g, fired at any time from within a column depends on the number of excitatory h_E and inhibitory h_I inputs to the column, i.e., if local delays are negligible

$$g = g[h_E(x,t), h_I(x,t)] \tag{12}$$

The procedure is to follow fluctuations δg about some local equilibrium g_0, as might be expected in a fixed physiological state. The modulation of action potential activity is then

$$\delta g = \left(\frac{\partial g}{\partial h_E}\right)_0 \delta h_E + \left(\frac{\partial g}{\partial h_I}\right)_0 \delta h_I \tag{13}$$

which leads to two coupled, linear integral equations for the synaptic action modulations, δh_E and δh_I. The dispersion relation relating complex frequency $p = j\omega + \gamma$ to wavenumber k is obtained as a solution to the coupled integral equations, that is

$$1 - \sum_{n=1}^{N} A_n \int_0^\infty \frac{\lambda_n^2 v^2 + p\lambda_n v}{(\lambda_n v + p)^2 + k^2 v^2} f_n(v) \, dv = 0 \tag{14}$$

The sum is over N excitatory fiber systems and $f_n(v)$ is the velocity distribution function for the Nth system. To examine the general character of the dispersion relation, let each fiber system carry action potentials with fixed velocity, ie $f_n(v) = \delta(v-v_n)$. Also, note that $|p/v_n| \sim (60 \text{ s}^{-1})/(6\text{-}9 \text{ m/s}) \sim 0.1 \text{ cm}^{-1}$ for scalp EEG. Thus, all intracortical and short corticocortical systems have $\lambda_n \gg |p/v_n|$. If all systems but one (n=1) satisfy this condition, the dispersion relation reduces to

$$1 - B \frac{\lambda_1^2 v_1^2 + p\lambda_1 v_1}{(\lambda_1 v_1 + p)^2 + k^2 v_1^2} = 0 \tag{15}$$

where the positive nondimensional parameter

$$B \sim -\left(\frac{\rho_1}{\rho_I}\right) \frac{\left(\frac{\partial g}{\partial h_E}\right)_0}{\left(\frac{\partial g}{\partial h_I}\right)_0} \tag{16}$$

Here ρ_1 and ρ_I are the number density of long range (excitatory) fibers and the number density of inhibitory (short range) fibers, respectively.

Physiological data suggest that B is probably of order one. Input from the brainstem reticular formation is known to regulate sleep/waking; we cannot stay awake without this diffuse and apparently inhibitory influence on the cortex. It is then suggested that the excitability parameter B is a macroscopic descriptor of physiologic state, with increasing B making it more easy for the "average" cortical neuron to fire an action potential. Equation (15) has the following solution for the frequency ω and temporal damping of "brainwaves"

$$\omega_m^2 = v_1^2 \left(k_m^2 - \lambda_1^2 \frac{B^2}{4}\right) \ , \qquad m = 1, \infty \tag{17}$$

$$\gamma = -\lambda_1 v_1 \left(1 - \frac{B}{2}\right) \tag{18}$$

where the subscript m has been added to ω and k to indicate that only discrete frequencies are expected in a finite medium. Thus, as B approaches zero, the waves are nondispersive, $\omega_1 = v_1 k_1$ and the higher modes tend to have less relative damping; that is, $|\gamma/\omega| \sim \lambda_1/k_1$. In an anisotropic medium, waves will tend to propagate along the directions of the longest fibers, which largely determine the dispersion relation. If the longest fibers cut across fissures, their effective lengths and propagation velocities are doubled; they then yield the same delay between cortical regions separated by x, the surface distance measured in and out of fissures and sulci.

Multiple long-range fiber systems, corresponding to the inclusion of terms n>1 in Eq. (14), result in polynominal equations of higher order for the complex frequency p. Multiple branches of the dispersion relation then occur; that is, more than one type of brain wave can exist simultaneously in the cortex in a manner analogous to the occurrence of optical and acoustical waves in crystals or plasmas. Corticocortical fibers carry action potentials with velocities distributed according to the functions $f_n(v)$ in Eq. (14) rather than the single velocity used to obtain Eqs. (17) and (18). Diameter histograms of white matter fibers suggest that the dominant $f_n(v)$ may be approximated by a Gaussian with peak velocity v_0 in the 6-9 m/s range. The half-width of the distribution $\Delta v/v_0$ is of the order of 0.2-0.4 (Katznelson 1981). The main effect of a spread in the velocity distribution is to increase the damping of higher modes (Nunez 1988b,c) In the two-dimensional version of the theory (waves on a spherical surface), multiple branches of the dispersion relation are predicted even when only one (N=1) long range fiber system occurs (Katznelson 1981), but the general semiquantitative character of weakly damped modes is preserved (Nunez 1988b).

EXPERIMENTAL SUPPORT FOR THE LINEAR MACROSCOPIC THEORY

For small B, the frequency of each mode m is approximately independent of B. However, as B approaches $2k_m/\lambda$, there is a sharp drop in the frequency of the k_m mode in a manner suggestive of the transition from the awake to the sleeping state. Furthermore, higher modes are reduced to a lesser extent by increasing B. This observation, together with arguments presented in a following section, suggest the possibility that non-delta activity observed during sleep is composed of a number of these higher modes. If the threshold parameter B is small, the frequency of the lowest expected mode may be estimated from Eq. (17). The longest "circumference" (in and out of fissures and sulci) of one hemisphere is $L \sim 100$ cm. With propagation velocities in the 6-9 m/s range (or doubled for fibers cutting across fissures), standing waves in a closed strip have discrete wavenumbers determined by continuity of both the potential and its derivatives

$$k_m = \frac{2m\pi}{L} \quad , \qquad m = 1, \infty \qquad (19)$$

$$k_1 \approx 0.06 \text{ cm}^{-1}$$

$$\text{or} \qquad\qquad\qquad (20)$$

$$f_1 \approx \frac{\omega_1}{2\pi} \approx 6\text{--}18 \text{ Hz}$$

Thus, either of the two lowest modes appear to be possible candidates for the alpha rhythm, because of uncertainty in the parameter estimates. With the assumption of an effective length (twice the actual length) of 10-50 cm for the longest corticocortical fibers, Eqs. (17) and (18) yield damping estimates for the case where B approaches zero; that is

$$\left|\frac{\gamma}{\omega_1}\right| \approx \frac{\lambda_1}{k_1} \approx 0.3\text{--}1.7$$

$$\qquad\qquad\qquad (21)$$

$$\left|\frac{\gamma}{\omega_2}\right| \approx \frac{\lambda_1}{k_2} \approx 0.16\text{--}0.8$$

with damping somewhat reduced as B is increased. Because of uncertainty of homogeneity/isotropy of the long-range connections, detailed investigation of boundary conditions does not appear warranted at this stage, but some qualitative connections to EEG are evident. For example, lack of symmetry in a brain-like shape can be expected to cause splitting of the modes for each m index. The splitting of the $m=1$ or $m=2$ modes may account for the double-peaked alpha rhythm observed in some subjects (Nunez 1981a). Even in subjects with a single sharp alpha peak of width ~1 Hz, the spatial distribution of frequency components just below the peak significantly differs from that just above the peak, suggesting the occurrence of multiple modes near 10 Hz in all subjects.

An interesting similarity between physical wave phenomena and the pattern visual evoked potential concerns the latter's resonance-like response at particular spatial frequencies in the pattern (Teyler et al. 1978). That is, in primates, a topological projection of the visual field into visual cortex exists so that spatial frequencies in the visual field are correlated with spatial frequencies in the visual cortex. In the macaque, the spatial frequency selectivity to the visual stimulus is observed in visual cortex but not in the lateral geniculate body, suggesting a cortical rather than thalamic resonance (De Valois et al. 1977).

The existence of a dispersion relation for EEG has been suggested by means of frequency-wavenumber spectral analysis of the alpha rhythm (Nunez 1974b, 1981a) using linear electrode arrays along the midline. While these experiments were limited by poor spatial resolution, all subjects in the studies showed an expected increase in wavenumber Δk across the alpha band $\Delta\omega$. A very rough estimate of the group velocity $(d\omega/dk)$ indicated an approximate range of 4-20 m/s, in qualitative agreement with the dispersion relation Eq. (17). The relative contributions of free and forced oscillations to this frequency-wavenumber shift are unknown. The recorded wavenumber spectra appeared roughly consistent with the lower modes $m=1$ or 2. In a study of phase and coherence of scalp EEG in 189 subjects, the average anterior/posterior phase shift of alpha rhythm translates to a phase velocity in the 7-16 m/s range (Thatcher et al. 1986). Long-range coherence estimates indicate that corticocortical connections are anisotropic; they are consistent with anterior/posterior standing wave patterns.

Whereas the frequency of each mode is determined by the parameters (λ, v, B), for which only rough estimates can be made, it is generally true that if other factors are held constant, larger systems should produce lower frequencies. This prediction was verified in a study of alpha frequency versus head size in a adult sample of 123 having peaked alpha rhythm (Nunez et al. 1977); that is, there is a weak (significance = 0.02), negative correlation between head size and alpha frequency. This study excluded children if only because myelination of axons during maturation can be expected to change velocity distributions.

Since mammalian cortices are quite similar, one might inquire whether the size-frequency relationship can be applied to nonhumans. This idea is complicated by the fact that the ratio of corticocortical to afferent fibers is much higher in man than in other mammals, as well as by other anatomical differences. Nevertheless, rough estimates of expected lower modes can be obtained from linear scale factors, which are the cube roots of the ratios of human to nonhuman brain volumes. For the dog, this ratio is $(1500/70)^{1/3} \sim 2.8$; for the cat, the ratio is $(1500/30)^{1/3} \sim 3.6$. The dog does indeed exhibit a cortical EEG rhythm which is peaked near 28 Hz (Lopes da Silva et al. 1970a). Also, driving the visual system of the dog with sine wave-modulated light produces a resonance-like peak in the same frequency range (Lopes da Silva et al 1970b), which appears analogous to a similar effect in humans near 10 Hz (van der Tweel and Verduyn Lunel 1965 ; Regan 1988). Under halothane anesthesia, the dog may produce rhythms which are also roughly twice the frequency of the human halothane rhythm (Nunez 1981a). But, these theoretical/experimental comparisons are clouded by the existence of a 12-Hz rhythm in the dog, which is recorded in parts of the cortex and thalamus, is attenuated by opening eyes, and is considered by some to be the analog of the human alpha rhythm (Lopes da Silva et al. 1973). Possible explanations of the 12-Hz dog rhythm are that it is analogous to a lower mode than the human alpha, is more dominated by local delays, involves nonlinear phenomena, or any combination of effects.

The cat produces a cortical rhythm near 40 Hz which is believed to be generated by local delays in the olfactory bulb (Freeman 1975). It is interesting to speculate that this apparent matching of local and global delays may not be fortuitous; perhaps the local delays are "learned" as the result of the influence of global modes. That is, this could be an example of a top-down influence across spatial scales, as has been proposed for human EEG (Ingber 1985a). Other animal experiments are also of interest to this theory. Lesions of midbrain structures in cats are shown to change the damping, but not the dominant frequencies of the EEG, thereby supporting the idea that the cortex is the medium responsible for resonance (Wright and Kydd 1984). Also, lesions in the cortex of cats indicate that EEG is dramatically altered only when the white matter is also lesioned (Gloor et al. 1977), a result consistent with the theoretical idea that the long-range connections may dominate normal modes.

ACKNOWLEDGEMENTS

This work was supported by NSF grant ECE 852 1101 and NIH grant 1RO1 NS243314. The author expresses his appreciation to Lester Ingber for very helpful comments on statistical field theory.

REFERENCES

Abeles, M., 1982, "Local Cortical Circuits," Springer-Verlag, New York.

Adey, W. R., 1988, Electromagnetic field interactions in the brain, in: "Dynamics of Sensory and Cognitive Processing by the Brain," E. Basar, ed., Springer-Verlag, Berlin.

Babloyantz, A., Destexhe, A., 1986, Low-dimensional chaos in an instance of epilepsy, Proc Natl Acad Sci., 83: 3513.

Braitenberg, V., 1978, Cortical architectonics: general and areal, in: "Architechtonics of the Cerebral Cortex," M. A. B. Brazier and H. Petsche, eds., Raven, New York.

Churchland, P. S., Sejnowski, T. J., 1988, Perspectives on cognitive neuroscience, Science, 242: 741.

Cowan, J. D., 1982, Spontaneous symmetry breaking in large scale nervous activity, Int J Quant Chem., 22: 1059.

DeValois, R. L., Albrecht, D. G., Thorell, L. G., 1977, Spatial tuning of LGN and cortical cells in monkey visual system, in: "Spatial Contrast," H. Spekreijse and L. H. van der Tweel, eds., North-Holland, Amsterdam.

Freeman, W. J., 1975, "Mass Action in the Nervous System," Academic, New York.

Freeman, W. J., 1987, Analytic techniques used in the search for the physiological basis of the EEG, in: "Handbook of Electroencephalography and Clinical Neurophysiology," A. S. Gevins and A. Remond, eds.,(revised series, Vol. 1), Elsevier, New York.

Freeman, W. J., 1988, Nonlinear neural dynamics in olfaction as a model for cognition, in: "Dynamics of Sensory and Cognitive Processing by the Brain," E. Basar, ed., Springler-Verlag, Berlin.

Georgopoulos, A. P., Schwartz, A. B., Kettner, R. E., 1986, Neuronal population coding of movement direction, Science, 233: 1416.

Gevins, A. S., Doyle, J. S., Schaffer, R. E., Callaway, E., Yeager, C., 1980, Laterized cognitive processes and the EEG, Science, 207: 1005.

Gevins, A. S., Doyle, J., Cutillo, B., Schaffer, R., Tannehill, R., Ghannam, J., Gilcrease, V., Yeager, C., 1981, Electrical potentials in human brain during cognition: new method reveals dynamic patterns of correlation of human electrical potentials during cognition, Science, 213: 918.

Gevins, A. S., Schaffer, R. E., Doyle, J. C., Cutillo, B. A., Tannehill, R. L., Bressler, S. L., 1983, Shadows of thought: rapidly changing, asymmetric brain-potential patterns of a brief visuomotor task, Science, 220: 97.

Gevins, A. S., Doyle, J. C., Cutillo, B. A., Schaffer, R. E., Tannehill, R. S., Bressler, S. L., 1985, Neurocognitive pattern analysis of a visuospatial task: rapidly-shifting foci of evoked correlations between electrodes, Psycholphysiol, 22: 32.

Gevins, A. S., Cutillo, B. A., 1986, Signals of cognition, in: "Handbook of Electroencephalography and Clinical Neurophysiology," F. H. Lopes da Silva, W. Storm van Leeuwen, and A. Remond, eds., (revised series, Vol. 2), Elsevier, New York.

Gloor, P., Ball, G., Shaul, N., 1977, Brain lesions that produce delta waves in the EEG, Neurology, (Minneap) 27: 326.

Hubel, D. H., Wiesel, T. N., 1962, Receptive fields, binocular interaction, and functional architecture in the cat's visual cortex, J Physiol (Lond)., 160: 106.

Ingber, L., 1982, Statistical mechanics of neocortical interactions. I. Basic formulation, Physica 5D, 83.

Ingber, L., 1983, Statistical mechanics of neocortical interactions. Dynamics of synaptic modification. Phys Rev A., 28: 395.

Ingber, L., 1984, Statistical mechanics of neocortical interactions. Derivation of short term memory capacity, Physiol Rev A., 29: 3346.

Ingber, L., 1985a, Statistical mechanics of neocortical interactions. EEG dispersion relations, IEEE Trans Biomed Eng., 32: 91.

Ingber, L., 1985b, Statistical mechanics of neocortical interactions: Stability and duration of the 7±2 rule of short-term-memory capacity, Phys Rev A, 31: 1183.

Ingber, L., Nunez, P. L., 1988, Multiple scales of statistical physics of neocortex: application to electroencephalography, Mathl Comput Modelling., in press.

Katznelson, R. D., 1981, Normal modes of the brain: neuroanatomical basis and a physiological theoretical model, in: "Electric Fields of the Brain: The Neurophysics of EEG," P. L. Nunez, Oxford University Press, New York.

Lehmann, D., 1988, EEG maps and map analysis, in: "Statistics and Topography in Quantitative EEG," D. Samson-Dollfus, ed., Elsevier, Paris.

Livanov, M. N., 1977, "Spatial Organization of Cerebral Processes," Wiley, New York.

Lopes da Silva, F. H., van Rotterdam, A., Strom van Leeuwen, W., Tielew, A. M., 1970a, Dynamic characteristics of visual evoked potentials in the dog. I. Cortical and subcortical potentials evoked by sine wave modulated light, Electroencephal Clin Neurophysiol., 29: 246.

Lopes da Silva, F. H., van Rotterdam, A., Storm van Leeuwen, W., Tielew, A. M., 1970b, Dynamic characteristics of visual evoked potentials in the dog. II. Beta frequency selectivity in evoked potentials and background activity, Electroencephal Clin Neurophysiol., 29: 260.

Lopes da Silva, F. H., Hoeks, H., Smits, H., Zetterberg, L. H., 1974, Model of brain rhythmic activity, Kybernetic, 15: 27.

Lopes da Silva, F. H., Storm van Leevwen, W., 1978, The cortical alpha rhythm in dog: the depth and surface profile of phase, in: "Architectonics of the Cerebral Cortex," M. A. B. Brazier and H. Petsche, eds., Raven Press, New York.

Ma, S. K., 1976, "Modern Theory of Critical Phenomena," Benjamin/Cummings, Reading, MA.

Morse, P. M., Feshbach, H., 1953, "Methods of Theoretical Physics," McGraw-Hill, New York, 2: 1502.

Nunez, P. L., 1974a, The brain wave equation: a model for the EEG, Math Biosci., 21: 279.

Nunez, P. L., 1974b, Wave-like properties of the alpha rhythm, IEEE Trans Biomed Eng., 21: 473.

Nunez, P. L., 1981a, "Electric Fields of the Brain: The Neurophysics of EEG," Oxford University Press, New York.

Nunez, P. L., 1981b, A study of origins of the time dependencies of scalp EEG. I. Theoretical basis, IEEE Trans Biomed Eng., 28: 271.

Nunez, P. L., 1981c, A study of the origins of the time dependencies of scalp EEG. II. Experimental support of theory, IEEE Trans Biomed Eng., 28: 281.

Nunez, P. L., 1986, The brain's magnetic field: some effects of multiple sources on localization methods, Electroencephal Clin Neurophysiol., 63: 75.

Nunez, P. L., 1988a, Spatial filtering and experimental strategies in EEG, in: "Statistics and Topography in Quantitative EEG," D. Samson-Dollfus, ed., Elsevier, Paris.

Nunez, P. L., 1988b, Generation of human electroencephalographic rhythms by a combination of long and short-range neocortical interactions, Brain Topography, submitted.

Nunez, P. L., 1988c, Global contributions to cortical dynamics: theoretical and experimental evidence for standing wave phenomena, in: "Dynamics of Sensory and Cognitive Processing by the Brain," E. Basar, ed., Springer-Verlag, New York.

Nunez, P. L., Reid, L., Bickford, R. G., 1977, The relationship of head size to alpha frequency with implications to brain wave model, Electroencephal Clin Neurophysiol., 44: 344.

Petsche, H., Pockberger, H., Rappelsberber, P., 1974, On the search for the sources of the electroencephalogram, Neuroscience, 11: 1027.

Pfurtscheller, G., Cooper, R,. 1975, Frequency dependence of the transmission of the EEG from cortex to scalp, Electroencephal Clin Neurophysiol., 38: 93.

Plonsey, R., 1969, "Bioelectric Phenomena," McGraw Hill, New York.

Pockberger, H., Rappelsberger, P., Petsche, H., 1988, Cognitive processing in the EEG. in: "Dynamics of Sensory and Cognitive Processing by the Brain," E. Basar, ed., Springler-Verlag, Berlin.

Rappelsberger, P., Petsche, H., 1988, Probability mapping: power and coherence analysis of cognitive processes, Brain Topography, 1: 46.

Regan, D., 1988, Human visual evoked potentials. in: "Handbook of Electroencephalography and Clinical Neurophysiology," T. W. Picton, ed., (revised series, Vol. 3), Elsevier, New York.

Roschke, J., Basar, E., 1988, The EEG is not simple noise: strange attractors in intracranial structure, in: "Dynamics of Sensory and Cognitive Processing by the Brain," E. Basar, ed., Springer-Verlag, Berlin.

Stockard, J. J., 1976, "Epileptogenicity of General Anestheics. Basic Mechanisms and Clinical Significance," Ph.D. Dissertation, University of California at San Diego.

Szentragothai, J., 1978a, The neural network of the cerebral cortex: a functional interpretation, Proc R Soc Lond., 201: 219.

Szentagothai, J., 1978b, Specificity versus (quasi) randomness in cortical connectivity, in: "Architectonics of the Cerebral Cortex," M. A. B. Brazier and H. Petsche, eds., Raven Press, New York.

Taylor, C. P., Dudek, F. E., 1984a, Excitation of hippocampal pyramidal cells by an electrical field effect, J Neurophysiol., 52: 126.

Taylor, C. P., Dudek, F. E., 1984b, Synchronization without active chemical synapses during hippocampal afterdischarges, J Neurophysiol., 52: 143.

Teyler, C. W., Apkarian, P., Nakayama, K., 1978, Multiple spatial frequency tuning of electrical responses from human visual cortex, Exp Brain Res., 33: 535.

Thatcher, R. W., Krause, P. J., Hrybyk, M., 1986, Corticocortical associations and EEG coherence: a two compartmental model, Electroencephal Clin Neurophysiol., 64: 123.

van der Tweel, L. H., Verduyn Lunel, H. F. E., 1965, Human visual responses to sinusoidally modulated light, Electroencephal Clin Neurophysiol., 18: 587.

van Rotterdam, A., Lopes da Silva, F. H., van der Ende, J., Viergever, M. A., Hermans, A. J., 1982, A model of the spatial-temporal characteristics of the alpha rhythm, <u>Bull Math Biol</u>., 44: 283.

Wikswo, J. P., Roth, B. J., 1988, Magnetic determination of the spatial extent of a single cortical current source: a theoretical analysis, <u>Electroenceph Clin Neurophysiol</u>., 69: 266.

Wilson, H. R., Cowan, J. D., 1972, Excitatory and inhibitory interactions in localized populations of model neurons, <u>Biophysics</u>, 12: 1.

Wilson, H. R., Cowan, J. D., 1973, A mathematical theory of the functional dynamics of cortical and thalmic nervous tissue, <u>Kybernetic</u>, 13: 55.

Wilson, K. G., 1979, Problems in physics with many scales of length, <u>Sci Amer</u>., 241: 158.

Wright, J. J., Kydd, R. R., 1984, A test for constant natural frequencies in electorcortical activity under lateral hypothalamic control, <u>Biol Cybern</u>., 50: 83.

Zhadin, M. N., 1984, Rhythmic processes in the cerebral cortex, <u>J Theor Biol</u>., 108: 565.

THE USE OF WIENER-VOLTERRA METHODS IN THE ANALYSIS

AND SYNTHESIS OF SIMULATED NEURAL NETWORKS

Hugh King

Private Medical Practice, 1909 Ivanhoe St.
Denver, Colorado

INTRODUCTION

This paper applies the Wiener-Volterra method of system analysis to some simple simulated neural networks. It then shows that a broad class of more complicated systems can be decomposed into these simple networks and that the overall system Wiener-Volterra representation is the algebraic composition of the representations for the simple networks. A simulated neural network as used here consists of a collection of elements each of which has a relatively simple transfer function. These elements accept input from outside the system and from other elements, process it, and pass the resulting output to other elements and to the system output. The word neural in the name comes from the parallel, distributed nature of the system: the transfer functions of the elements are neither claimed nor constrained to match the transfer functions of nerve cells. However, by choosing the transfer function of the elements properly, these networks can be used as models for physiologic neural networks. Such simulated networks are useful in that by proper configuration they can mimic a wide range of systems and in some cases, by allowing a correction term to be fed back to modify the interconnections or the elements, the network can 'learn' the proper configuration itself.

The overall objective of this project is to determine relations between the Wiener-Volterra representation of a system and the architecture of neural networks that approximate the system. Of course, if the original system is a neural network then the goal is to determine the architecture of the network itself. The ideal would be an algorithm that if given a Wiener-Volterra representation of a system would produce a realization of the system as a neural network and, conversely, if given a neural network the algorithm would produce the Wiener-Volterra representation. Consideration of the simple networks presented here gives some insight into the general goal.

MOTÍVATION

Consider a causal system that has as input a time series that changes only at integral multiples of a time unit. For example, a pseudorandom sequence as discussed in Marmarelis and Marmarelis (1978) or the sample and hold of any input process. Suppose also that the output is observed only at integral multiples of the time unit and take this time unit to be 1. The Volterra series can be formally written:

$$y(n) = h_0 + \int\limits_{s=0}^{\infty} h_1(s)x(n-s)ds + \iint\limits_{s_1,s_2=0}^{\infty\infty} h_2(s_1,s_2)x(n-s_1)x(n-s_2)ds_1 ds_2 + \dots$$

Splitting the domain of integration into areas of constant x gives:

$$y(n) = h_0 + \sum_{i=1}^{\infty} \int\limits_{i-1}^{i} h_1(s)x(n-s)ds + \sum_{i=1}^{\infty}\sum_{j=1}^{\infty} \int\limits_{i-1}^{i}\int\limits_{j-1}^{j} h_2(s_1,s_2)x(n-s_1)x(n-s_2)ds_1 ds_2 + \dots$$

Then remove the x from under the integral:

$$y(n) = h_0 + \sum_{i=1}^{\infty} x(n-i) \int\limits_{i-1}^{i} h_1(s)ds + \sum_{i=1}^{\infty}\sum_{j=1}^{\infty} x(n-i)x(n-j) \int\limits_{i-1}^{i}\int\limits_{j-1}^{j} h_2(s_1,s_2)ds_1 ds_2 + \dots$$

And so:

$$y(n) = k_0 + \sum_{i=1}^{\infty} k_1(i)x(n-i) + \sum_{i=1}^{\infty}\sum_{j=1}^{\infty} k_2(i,j)x(n-i)x(n-j) + \dots \ .$$

The $k_1(i)$'s thus represent the effect of the input i time units in the past on the current output, the $k_2(i,j)$'s represent the effect of the inputs i time units ago and j time units ago, etc. In the case of a system with memory of two time steps this becomes:

$$\begin{aligned} y(n) = \ &k_0 + k_1(1)x(n-1) + k_1(2)x(n-2) + \\ &k_2(1,1)x^2(n-1) + k_2(1,2)x(n-1)x(n-2) + \\ &k_2(2,1)x(n-2)x(n-1) + k_2(2,2)x^2(n-2) + \text{higher order terms.} \end{aligned} \qquad (1)$$

Now consider a simple three element simulated neural network as shown in Fig. 1.

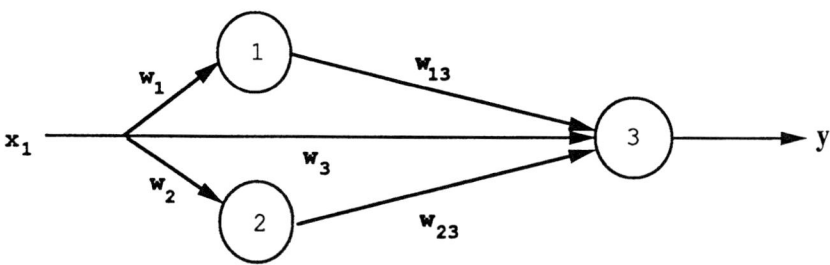

Fig. 1. Three element system with one input.

Let the transfer function of each element be to sum its inputs, perform a generally non-linear operation on this number and output the result on the next time cycle. (This type of transfer function in widely used. See, for example, Rumelhart and McClelland (1987)). These outputs are then the inputs to the next element after attenuation by multiplication by the "weight" connecting the two elements (the w's in the diagram).

Now suppose that the non-linear operation of each element can be approximated by a polynomial series:

$$H_i(x) = a_{0i} + a_{1i}x + a_{2i}x^2 + \ldots + a_{Mi}x^M.$$

(For physical systems one can argue that this is reasonable because the transfer function is continuous and the Stone-Weierstrass theorem is applicable.)

Now use the above element transfer function to calculate the output of the network. The input to element number one is $w_1 x(k)$ on time cycle k and the output from element number 1 is:

$$O_1(k+1) = a_{01} + a_{11}w_1 x(k) + a_{21}w_1^2 x^2(k) + \ldots + a_{M1}w_1^M x^M(k)$$

on time cycle k+1. Continuing, the input to element number three for time cycle k+1 is thus:

$$I_3(k+1) = w_{13}O_1(k+1) + w_{23}O_2(k+1) + w_3 x(k+1).$$

Passing this through element three gives the system output:

$$O_3(k+1) = a_{03} + a_{13}I_3(k+1) + a_{23}I_3^2(k+1) + \ldots + a_{M3}I_3^M(k+1).$$

Substitute for the I's, collect terms, shift the time axis and rename the coefficients to get that the output of this network is:

$$y(n) = g_0 + g_1(1)x(n-1) + g_1(2)x(n-2) + \qquad (2)$$
$$g_2(1,1)x^2(n-1) + g_2(1,2)x(n-1)x(n-2) +$$
$$g_2(2,1)x(n-1)x(n-2) + g_2(2,2)x^2(n-2) + \text{higher order terms.}$$

Comparing this (Eq. 2) with the general Wiener-Volterra (WV) representation derived above (Eq. 1) we see that they have exactly the same form. This shows explicitly that the neural network is calculating its WV series as it processes its inputs and suggests that further consideration of the relation between the network and its WV representation may be of interest.

Also, notice that each g_i is a function of the inter-element connecting weights and the coefficients of the transfer functions:

$$g_i = G_i(w_1, w_2, \ldots, w_{11}, w_{12}, \ldots, a_{01}, a_{02}, \ldots).$$

A simultaneous inversion of this system of equations would, thus, determine a network that matches the y output above. This means that given a WV representation, the g_i's above are determined by the kernel functions and then inverting the g_i's gives a network that approximates the original system.

As the network becomes more complicated with longer paths through it the output expression gains more terms corresponding to more memory in the system and finally approaches:

$$y(n) = g_0 + \sum_{i=1}^{\infty} g_1(i)x(n-i) + \sum_{i=1}^{\infty} \sum_{j=1}^{\infty} g_2(i,j)x(n-i)x(n-j) + \ldots$$

An interesting sidelight of this is that the value of a non-linearity in a neural network may be to allow the input variables to interact. To see this consider a simple three element system as above, but with no time delay at the elements and two inputs as shown in Fig. 2.

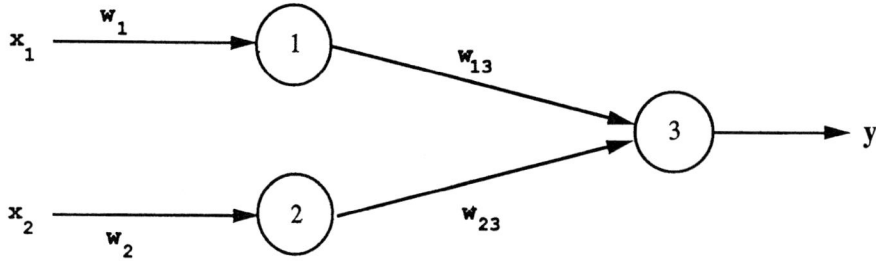

Fig. 2.   Three element system with two inputs.

Assume that the transfer function of each element is linear (i.e., z in implies $a_i z$ out). An analysis like the one above shows the system output to be:

$$y = a_1 a_3 w_{13} w_1 x_1 + a_2 a_3 w_{23} w_2 x_2 .$$

This output function introduces degrees of freedom to adjust the relative contribution of x_1 and x_2 to the output but does not allow x_1 to modulate the effect of x_2 or vice versa.

Now suppose that the transfer function of the elements is a non-linear function, say for simplicity that $H(x) = 1 + x + x^2$. Then the system output becomes (after an analysis similar to the above):

$$
\begin{aligned}
y = &(1+w_{13}+w_{23}+w_{13}{}^2+w_{23}{}^2+2w_{13})+(w_1 w_{13}+2w_1 w_{13}{}^2+2w_1 w_{13} w_{23})x_1+ \\
&(w_2 w_{23}+2w_2 w_{23}{}^2+2w_2 w_{13} w_{23})x_2+(w_{13} w_1{}^2+2w_{23} w_{13} w_1{}^2+w_1{}^2 w_{13}{}^2)x_1{}^2+ \\
&(w_{23} w_2{}^2+2w_{13} w_{23} w_2{}^2+2w_2{}^2 w_{23}{}^2+w_2{}^2 w_{23}{}^2)x_2{}^2+2w_1 w_2 w_{13} w_{23} x_1 x_2+ \\
&w_1{}^3 w_{13}{}^2 x_1{}^3+w_2{}^3 w_{23}{}^2 x_2{}^3+2w_1 w_{13} w_{23} w_2{}^2 x_1 x_2{}^2+2w_2 w_{23} w_{13} w_1{}^2 x_1{}^2 x_2{}^2+ \\
&w_{13}{}^2 w_1{}^4 x_1{}^4+w_{23}{}^2 w_2{}^4 x_2{}^4 .
\end{aligned}
$$

The point here being not the complexity of the coefficients but the emergence of cross terms- $x_1 x_2$, $x_1 x_2{}^2$, etc. that allow the variables to interact.

LOGICAL NETWORKS AS SYSTEM BUILDING BLOCKS

Consider now some simple systems that evaluate logical functions. For example, let the inputs x_1 and x_2 take the values 1 (for logical true) and 0 (for logical false). For example, if we define the system output to be 0 when each input is 0, 1 when $x_1=1$ and $x_2=0$ or $x_1=0$ and $x_2=1$, and 0 when $x_1=1$ and $x_2=1$ we get the system that calculates the exclusive or (XOR) function as shown in Fig. 3.

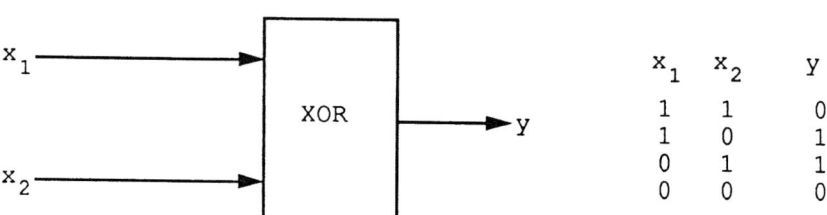

x_1	x_2	y
1	1	0
1	0	1
0	1	1
0	0	0

Fig. 3.   System that calculates the XOR.

There are corresponding systems that calculate AND, OR, NOT, NAND and other logical functions. These systems are interesting in their own right but are even more important because they provide a step in the decomposition of a large class of systems to neural networks. In fact, any sampled input, sampled output, digital system with finite memory can be represented as compositions of them.

To see this, first note that for the set of all sequences of true and false of length N there is a logical function composed of AND, OR and NOT that maps any given sequence to true or false in a specified manner. The general proof is constructive but for here consider a simple example. Let N = 3 and suppose we require the truth table shown in Table 1.

Table 1.   Sample Truth Table.

x_1	x_2	x_3	y
0	0	0	0
0	0	1	0
0	1	0	0
1	0	0	0
0	1	1	1
1	0	1	0
1	1	0	0
1	1	1	1

The function defined by Table 1. is true if and only if (x_1 = 0 AND x_2 = 1 AND x_3 = 1) OR (x_1 = 1 AND x_2 = 1 AND x_3 = 1). Thus, the logical function is y = [(NOT x_1) AND x_2 AND x_3)] OR [x_1 AND x_2 AND x_3]. It is clear that for any given truth table one can construct such a function from AND, OR, NOT. (The y values in the above table were chosen at random and so the argument applies to any function.)

Now consider a system whose input is sampled and digitized into M bit words and whose output consists of K bit words. Let the system be causal with finite memory = n. This system can be modeled as a system that at each time cycle maps n*M bits into an output of K bits. From the analysis above it is possible to construct a system from ANDs, ORs and NOTs that properly maps the n*M input bits to any of the K output bits. Let S_i, i=1,K represent the AND, OR, NOT network for the ith output bit (O_i) and the system can be represented as in Fig. 4.

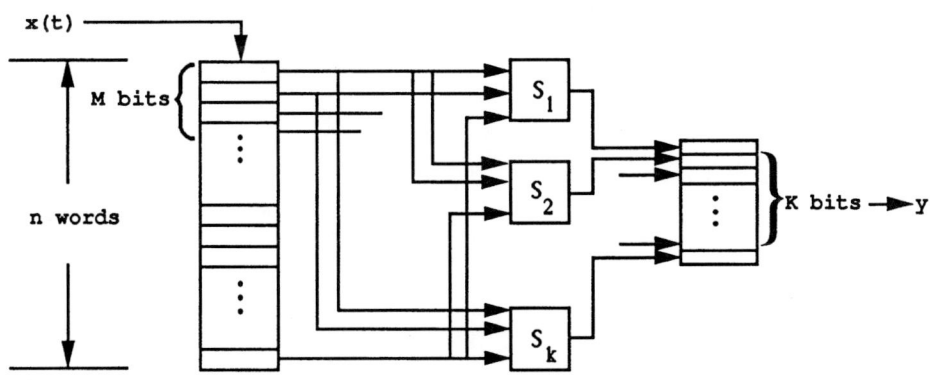

Fig. 4.   Example of Sampled, Digitized System.

Let $R(x(t),x(t-1),\ldots,x(t-n+1))$ be the function that maps the n dimensional real input space into the n*M dimensional binary space that represents the contents of the shift register. The system output can be determined by tracing the steps in Fig. 4.

$$y(t) = 2^{K-1}O_K + 2^{K-2}O_{K-1} + \ldots + 2^0O_1 =$$

$$2^{K-1}S_K\{R(x(t),\ldots,x(t-n+1))\} + 2^{K-2}S_{K-1}\{R(x(t),\ldots,x(t-n+1))\} + \ldots + 2^0S_1\{R(x(t),\ldots,x(t-n+1))\}.$$

First note that each component of the R function maps an n dimensional real space into 0 or 1. Once we know the memory of the system and the word length of the encoded input R is defined and a WV series can be obtained either by the GWN approach or analytically. Thus, if we had a Volterra type representation for the S_i's we could get a Volterra type series representation for the whole system by algebraic composition of the S_i's with R. Further, if we had a Volterra representation of the simple logical functions AND, OR, NOT we could determine a Volterra representation for the composite system S_i by algebraic composition of the simple representations.

One further fact to note here is that AND, OR, NOT have realizations in neural networks. For example, the network in Fig 5. calculates AND as is easily seen by considering the cases. $()^2$ means that the element sums the inputs and squares, Sigma means that the element simply sums the input. Other examples are easy to construct as will be shown later.

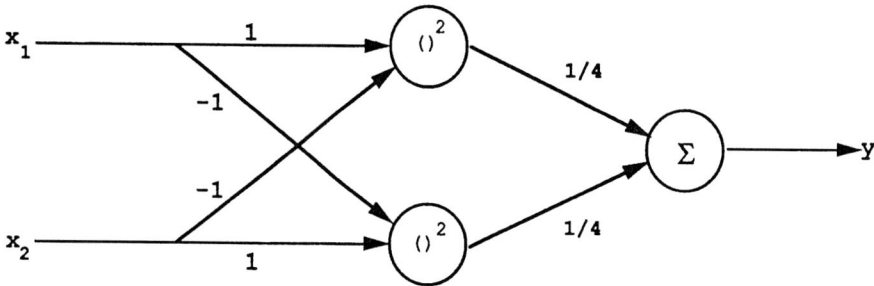

Fig. 5. Network That Calculates AND.

WV REPRESENTATION OF LOGICAL NETWORKS

In an attempt to determine the WV representation of the simple logical functions the author first applied zero mean Gaussian white noise to specific simulated networks that model the function and applied the Lee-Schetzen method. The first function studied was XOR. (Truth table shown in Table 2.)

Table 2. XOR

x_1	x_2	$y=XOR(x_1,x_2)$
0	0	0
1	0	1
0	1	1
1	1	0.

The first network that was studied is shown in Fig. 6 (i.e., Rumelhart and McClelland (1987)).

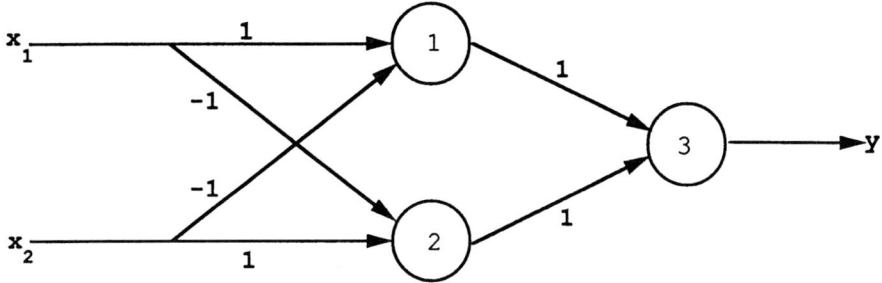

Fig. 6. First XOR network studied.

The transfer function of the elements is:

$$H_i(I) = \begin{cases} I, & I>a \\ 0, & I<a \end{cases} \qquad 0<a<1.$$

It is also assumed that there is no time delay.

With this transfer function the network clearly calculates XOR- ie. $x_1 = x_2 = 0$ means that the input to element 1 is $1(0)+(-1)(0) = 0$. Similarly, the input to element 2 is $(-1)(0)+(1)(0) = 0$ and, thus, the input to element 3 is 0 and so is the total system output, as desired. Calculations like this for the other cases yield appropriate results.

However, this network is defined for all real values of x_1 and x_2 and so it is possible to test the network with GWN and determine the WV representation using the Lee-Schetzen technique as diagrammed in Fig. 7. (Schetzen 1980)).

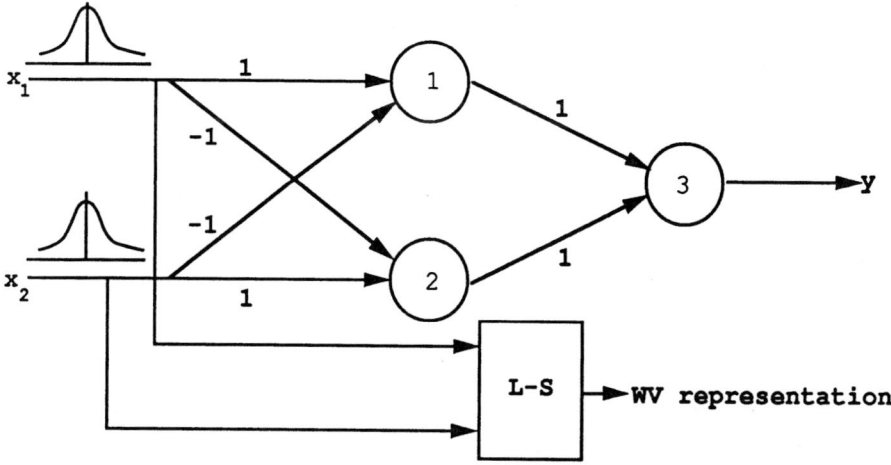

Fig. 7. Technique to determine the wv series.

This system yielded the WV representation of:

$h_{11}(0) = .4$, $h_{22}(0) = .4$, $2h_{12} = .8$.

All other kernel values were zero- no memory and only these quadratic terms. Thus, the WV representation is:

$y(t) = .4[x_1(t)]^2 + .4[x^2(t)]^2 - .8x_1(t)x_2(t)$.

This suggests that, when consideration is taken for the power lost because H maps any input less than a to zero, the system is "trying" to have a WV representation of $x_1{}^2 + x_2{}^2 - 2x_1x_2$, which clearly represents XOR.

The next question is whether any network that represents XOR would approach this function. This is not the case as seen by considering the network in Fig. 8.

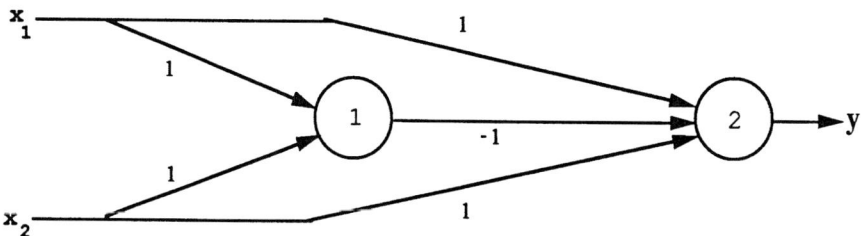

Fig. 8. Another Network That Calculates XOR.

The transfer function of element 2 is to sum its inputs and output that sum. The transfer function of element 1 is to sum its inputs and then:

$$H(I) = \begin{cases} 1, & I > 1.5 \\ 0, & I < 1.5. \end{cases}$$

Consideration of the cases shows that this network also calculates XOR. However, in this case the WV series determined as above is $y = .8x_1 + .8x_2$, with all other terms through the third degree nearly zero.

This function clearly does not calculate the XOR. One problem is that the zero mean Gaussian input is testing the system in the wrong part of its range. Since the sample inputs are heavily peaked around zero the probability that $x_1 + x_2$ reaches 1.5 (the point of the non-linearity) is small and so the representation is nearly linear.

This leads to an important point when testing any system: Use input values of interest and only the values of interest whenever possible. This seems obvious in retrospect but, as here, may be hidden in the problem. In the first example, all the input values less than zero (which are not really defined for XOR) lead to zero output and, thus, reduce the magnitude of the coefficients in the series. In the second example, the test samples generally fall below an interesting value $x_1 = x_2 = 1$ and miss the non-linearity altogether.

In both of the above cases, since the XOR is only defined for inputs of 0 or 1, we are testing with many input values we don't really care about. This leads to the idea of testing with a binary input sequence. Since we actually define only two input values this would match the input range exactly. The binary method does not work as we will see shortly, but its attempted application is interesting and presented here. One way to do this would be to use the pseudorandom sequences discussed in Marmarelis and Marmarelis (1978). However, because the system has no memory it is simpler to calculate directly as follows: Let the input x_i be either -1 or 1, each with probability .5 and such that x_1 and x_2 are independent. Pass the x_i first through a system with output $(1+x_i)/2$ and then through XOR. We can remove the effect of the $(1+x_i)/2$ later. This system is shown in Fig. 9.

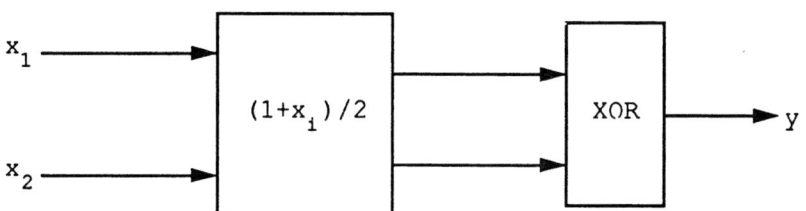

Fig. 9. XOR System With +-1 as Input.

First notice that the appropriate $\dot{\text{W}}\text{V}$ series is:

$$y = h_0 + h_1 x_1 + h_2 x_2 + h_{11} x_1^2 + h_{22} x_2^2 + 2h_{12} x_1 x_2 - (h_{11} + h_{22}) + \ldots,$$

where the $-(h_{11} + h_{22})$ is necessary for orthogonality. It also leads to a problem general to the binary input case because the Lee-Schetzen technique can not be applied to give the "diagonal coefficients"- h_{ii}. To see this, try to apply the algorithm to determine h_{11}. The procedure as outlined in Marmarelis and Marmarelis (1978) calls for sequentially determining $y_1 = y_0 - h_0$ and $y_2 = y_1 - h_1 x_1 - h_2 x_2$ and then multiplying both sides of the equation:

$$y_2 = h_{11} x_1^2 + h_{22} x_2^2 + 2h_{12} x_1 x_2 + \ldots$$

by x_1^2 and taking expectations. Doing this gives:

$$Ey_2 x_1^2 = h_{11} Ex_1^4 + h_{22} Ex_1^2 x_2^2 + 2h_{12} Ex_1^3 x_2 - (h_{11} + h_{22}) Ex_1^2.$$

However, $Ex_1^2 = 1^2(1/2) + (-1)^2(1/2) = 1$, $Ex_1^2 x_2^2 = 1^2*1^2(1/4) + 1^2(-1)^2(1/4) + (-1)^2(1)^2(1/4) + (-1)^2(-1)^2(1/4) = 1$ and $Ex_1^3 x_2 = 0$.

Thus, $Ey_2 x_1^2 = h_{11}*1 + h_{22}*1 - (h_{11} + h_{22})*1 = 0$ and the method breaks down.

The next idea to try is ternary sequences. In this case we must choose another input value, say zero and so allow one input that is not explicitly defined by XOR, but this should still be better than the continuum of extra values allowed in the GWN case. Easy calculation shows that the new orthogonal WV series is:

$$y = h_0 + h_1 x_1 + h_2 x_2 + h_{11} x_1{}^2 + h_{22} x_2{}^2 + 2h_{12} x_1 x_2 - (2/3)(h_{11} + h_{22})$$
$$+ \ldots.$$

where $x_i = \begin{cases} 1, & \text{prob}(1/3) \\ -1, & \text{prob}(1/3) \\ 0, & \text{prob}(1/3). \end{cases}$

For now assign the "extra" values to variables as in Table 3.

Table 3.

x_1	x_2	y	
1	1	0	
1	-1	1	XOR
-1	1	1	
-1	-1	0	
1	0	A	
0	1	B	
-1	0	C	New values
0	-1	D	
0	0	E	

Now apply the Lee-Schetzen technique. After much algebra the result is:

$h_0 = (1/9)(2+A+B+C+D+E)$,
$h_1 = (A-C)/6$,
$h_2 = (B-D)/6$,
$h_{11} = 1 - (2+A+B+C+D+E)/3 + (A+C)/2$
$h_{22} = 1 - (2+A+B+C+D+E)/3 + (B+D)/2$
$h_{12} = -1/4$
.
.
.

The WV series, thus, becomes:

$$y = (1/9)(2+A+B+C+D+E) + [(A-C)/6]x_1 + [(B-D)/6]x_2 +$$
$$[1-((2+A+B+C+D+E)/3)+(A+C)/2]x_1{}^2 + [1-((2+A+B+C+D+E)/3)+(B+D)/2]x_2{}^2 -$$
$$(1/2)x_1 x_2 - (2/3)[2-(2/3)(2+A+B+C+D+E)+(1/2)(A+B+C+D)] +$$
higher order terms.

This is a complicated expression, but if one assumes A=C, B=D and tries to minimize the power passed to terms higher than quadratic one finds that there is a solution that passes no power to terms of third degree and it is A=B=C=D= 1/4, E=0. When this is substituted and the $(1+x_i)/2$ transformation undone the representation again becomes:

$$y = x_1{}^2 + x_2{}^2 - 2x_1 x_2.$$

OTHER LOGICAL FUNCTIONS

By arguments similar to the above WV representations of other logical functions can be determined. In general the coefficients involve the choices made for A,B,C,D,E. These choices are arbitrary and arise from the necessity of using ternary sequences to calculate the diagonal elements. This introduces artifact because, as discussed above, they require definition of the system at points in the input space where the system is not "really" defined. However, starting with arbitrary A,B,C,D,E and then making constraints among them that simplify the result (as done for XOR above) one can show that the following are WV representations of some simple logical functions:

270

$$XOR(x_1,x_2) = y = x_1{}^2 + x_2{}^2 - 2x_1x_2,$$
$$OR(x_1,x_2) = y = x_1{}^2 + x_2{}^2 - x_1x_2,$$
$$AND(x_1,x_2) = y = x_1x_2,$$
$$NOT(x_1) = y = 1 - x_1. \quad (\text{NOT can even be done with a binary sequence.})$$

SUMMARY

We have, thus, constructed a WV representation for some simple logical functions. Note first that this means that from a network composed of these logical functions one can construct a WV representation of the network by algebraic composition. Take, for example, the system described in Section III. The system function derived there was:

$$y = [(NOT~x_1)~AND~x_2~AND~x_3]~OR~[x_1~AND~x_2~AND~x_3].$$

From the above work we can write immediately:

$$y = [(1-x_1)x_2x_3]^2 + [x_1x_2x_3]^2 - [(1-x_1)x_2x_3][x_1x_2x_3]$$
$$= x_2{}^2x_3{}^2 - 3x_1x_2{}^2x_3{}^2 + 3x_1{}^2x_2{}^2x_3{}^2.$$

This y calculates the system as can be checked by substitution. This method can be applied to any network composed of the logical functions.

We also have developed a way to construct a neural network from the WV representation for the class of finite memory digital systems. This can be done by using the WV representation to construct the equivalent network of simple logical functions as done in part IV and then realizing this network as a neural network. We have, thus, realized for a class of systems the objectives set in the introduction of constructing an algorithm that produces a WV representation from a system and a system from a WV representation.

ONGOING WORK

The above technique for going from the WV representation to the neural network representation is not satisfying- construction of the logical network requires determination of the output for all combinations of possible input bits. As the system becomes more complex the combinatoric explosion makes this a very long process. The above may be viewed as an existence proof and the search for a practical factorization continues. For example, one would like an algorithm that would reverse the steps in the summary above to determine that:

$$x_2{}^2x_3{}^2 - 3x_1x_2{}^2x_3{}^2 + 3x_1{}^2x_2{}^2x_3{}^2 \qquad \text{(the given WV representation)}$$

$$= \qquad \text{(via a factorization algorithm)}$$

$$[(1-x_1)x_2x_3]^2 + [x_1x_2x_3]^2 - [(1-x_1)x_2x_3][x_1x_2x_3]. \qquad \text{(a representation from which the logical network can be constructed)}$$

Such an algorithm would allow immediate realization of a logical network as its WV representation and, thus, as a neural network.

Another, related, question for further study is how, in general, to invert the mapping from the WV kernels to the weight, coefficient space as discussed in section II. An example of this is the XOR function itself. One can factor it as:

$$y = x_1{}^2 + x_2{}^2 - 2x_1x_2 = (x_1 - x_2)^2.$$

In this later form it is seen that the XOR can be simply realized if we allow the only non-linearity to be squaring, for example, as in the network in Fig. .10.

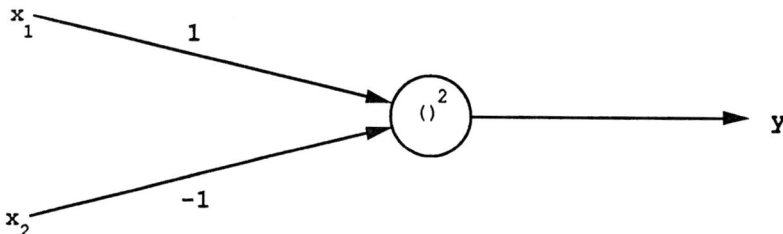

Fig. 10. XOR using only squaring non-linearity.

The author also used this technique to determine the AND example in Fig. 5. by noting $AND(x_1, x_2) = x_1{}^*x_2 = (1/4)[(x_1 + x_2)^2 - (x_1 - x_2)^2]$ from which the network for AND using only $()^2$ quickly follows. General factorization algorithms like these would allow immediate construction of both logical and neural network representations of a system from its WV representation.

A third question, of physiological interest, is whether the factorization above implies structure in real neural networks. If we know the WV representation of a real neural network and require the element transfer functions to model real neuron transfer functions, the factorization algorithm yields a network that simulates the real neural network. It is of interest whether this simulated network in any way approximates the structure of the real neural network.

Acknowledgement

The author would like to thank Dr. George Swanson, who introduced him to the WV method, Dr. Harvey Greenberg, whose course in neural networks provided a starting point for some of the above work and Mr. Mike Epstein, whose consultation regarding neural networks was very helpful.

REFERENCES

King,H., 1987, The Wiener Method of Non-Linear System Analysis Applied to Neural Networks and the Use of Neural Networks as Time Series Processors, in: "Report of Math Clinic on Neural Networks," H. Greenberg, ed., University of Colorado at Denver,Denver,Co.

Marmarelis,P.Z.,and Marmarelis,V.Z., 1978, "Analysis of Physiological Systems: The White-Noise Approach," Plenum Press, New York.

Rumelhart,D., and McClelland,J., 1987, "Parallel Distributed Processing, Vol 1 and 2," The MIT Press, Cambridge, Mass.

Schetzen,M., 1980, "The Volterra and Wiener Theories of Nonlinear Systems," Wiley, New York.

WIENER ANALYSIS OF THE HODGKIN–HUXLEY EQUATIONS

Spiridon H. Courellis and Vasilis Z. Marmarelis

Departments of Biomedical and Electrical Engineering
University of Southern California
Los Angeles, CA 90089-1451

INTRODUCTION

Scientific efforts to understand the basic mechanisms of information processing and coding in the nervous system have produced a lot of remarkable work on the generation mechanism of action potentials (spikes), the properties of excitable membranes and the dynamic transformations of spike trains as they travel through the nervous system. Since the pivotal work of Hodgkin and Huxley (1952), who first proposed a quantitative dynamic model for the generation mechanism of action potentials in the squid axon, many competent investigators have explored mathematical aspects (for review see Cronin, 1987) and modifications of the original Hodgkin–Huxley (H–H) model, motivated by a variety of reasons. Attempts to simplify the H–H model (i.e., reduce its complexity while maintaining its essential functional characteristics) have been motivated by the need to develop workable models of neuronal aggregates (networks) that incorporate the essential features of the spike generation mechanism while, at the same time, retain manageable computational complexity.

A number of such simplified models have been proposed, ranging from the simple integrate–and–fire model to models containing some type of refractory feedback and adaptive features (for partial review see Deutsch & Micheli–Tzanakou, 1987). Most of these models employ a simple threshold–trigger mechanism that amounts to a binary quantization of continuous signals. Experimental neurophysiologists, on the other hand, interested in the question of how neuronal spike trains are dynamically transformed by neuronal units, have been recording and analyzing the (analog) spike data as binary signals (point processes) whose information content is strictly found in the timing of these binary (spike) events and not in the detailed (analog) waveform of the physiological signal.

The purpose of this paper is three–fold: (1) to apply Wiener analysis to the simulated H–H model and explore the ability of this approach to extract the essential functional characteristics of the H–H model using only up to second–order Wiener kernels; (2) to explore the use of Wiener kernel measurements as a tool in developing simplified "block–structured" models that mimic the functional characteristics of the

H–H model; (3) to test the adequacy of the binary representation of spike trains (point processes) in studies of neural systems, using the obtained Wiener kernel measurements in each case as the basis of comparison.

THE HODGKIN–HUXLEY MODEL

The Hodgkin–Huxley (H–H) model describes the transmembrane potential (output signal) of an excitable nerve membrane in response to an applied transmembrane current (input signal) as a result of the flow of ionic currents through the membrane (Hodgkin & Huxley, 1952). It is given by the following set of nonlinear differential equations:

$$I(t) = C\frac{dV}{dt} + \bar{g}_K n^4(V - V_K) + \bar{g}_{Na} m^3 h(V - V_{Na}) + \bar{g}_L(V - V_L) \tag{1}$$

$$\frac{dn}{dt} = \alpha_n(V)[1 - n] - \beta_n(V)n \tag{2}$$

$$\frac{dm}{dt} = \alpha_m(V)[1 - m] - \beta_m(V)m \tag{3}$$

$$\frac{dh}{dt} = \alpha_h(V)[1 - h] - \beta_h(V)h \tag{4}$$

where $V(t)$ is the voltage output in $[mV]$ (measured with reference to the resting potential), $I(t)$ is the current stimulus in $[\mu A]$, and the voltage–dependent functions $\alpha_n, \beta_n, \alpha_m, \beta_m, \alpha_h, \beta_h$ are given by the expressions:

$$\alpha_n(V) = 0.01\frac{(10 - V)}{1 + e^{1-0.1V}} \tag{5}$$

$$\beta_n(V) = 0.125e^{-\frac{V}{80}} \tag{6}$$

$$\alpha_m(V) = 0.1\frac{25 - V}{1 + e^{2.5-0.1V}} \tag{7}$$

$$\beta_m(V) = 4\,e^{-\frac{V}{18}} \tag{8}$$

$$\alpha_h(V) = 0.07e^{-\frac{V}{20}} \tag{9}$$

$$\beta_h(V) = \frac{1}{1 + e^{3 - 0.1V}} \tag{10}$$

The quantities n, m, h are concentrations of ions in the ionic currents due to potassium (n) and sodium (m, h). The constants $\bar{g}_K, \bar{g}_{Na}$ represent conductivities for the ionic components due to potassium (K^+) and sodium (Na^+) respectively, while $\bar{g}_L$ is the conductivity corresponding to a "leakage current". All conductivities are measured in $[\frac{mmho}{cm^2}]$ and the capacitance C of the membrane in $[\mu F]$. Finally, V_K, V_{Na}, V_L are the equilibrium potentials of the components denoted by the corresponding subscript. Note that time is measured in $[msec]$, and the constants have the values: $\bar{g}_K = 36 \frac{mmho}{cm^2}$, $\bar{g}_{Na} = 120 \frac{mmho}{cm^2}$, $\bar{g}_L = 0.3 \frac{mmho}{cm^2}$, $V_k = -12mV$, $V_{Na} = 415mV$ and $V_L = 10.5989mV$, at the nominal temperature of 6.3° C. These physical units will be implied throughout the article.

If we adopt the notation:

$$\tau_r(V) = \frac{1}{\alpha_r(V) + \beta_r(V)} \tag{11}$$

$$\gamma_r(V) = \frac{\alpha_r(V)}{\alpha_r(V) + \beta_r(V)} = \alpha_r(V)\tau_r(V) \tag{12}$$

where r may represent n, m, or h, then Eqs. (2) through (4) assume the form:

$$\tau_r(V)\frac{dr}{dt} + r = \gamma_r(V) \tag{13}$$

Plots of the different $\tau(V)$'s and $\gamma(V)$'s as functions of the voltage V are given in Fig. 1 and Fig. 2, respectively. The advantage of the latter representation lies in the fact that the functions $\tau(V)$ and $\gamma(V)$ are bounded compared to the functions $\alpha(V)$ and $\beta(V)$. Furthermore, this representation is more system–oriented, because the $\tau(V)$'s represent voltage–dependent time constants. Note also that positive voltage values correspond to depolarization while negative voltage values correspond to hyperpolarization.

The first important feature of the H–H model is its "threshold" behavior in response to DC current inputs, i.e., it generates sustained spike trains only if the input DC value is above a certain threshold. Higher DC input values lead to a decrease in the interspike interval (increase of firing frequency) until a finite limit is reached. Further increase of the DC input value eventually leads to diminution of the output spike size. When the DC input is below this threshold value, the model generates only a short–time transient response whose shape depends on the DC value.

It is instructive to examine the behavior of the model for extreme values of the output signal $V(t)$, because of the asymptotic values of the voltage–dependent parameters shown in Figs. 1 and 2, and their implications for system stability. When

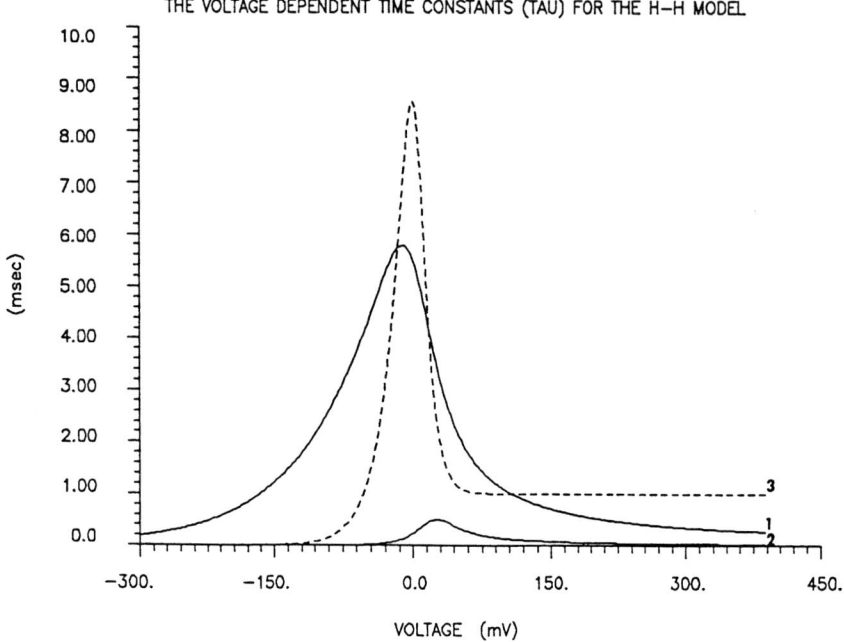

Fig. 1. Plots of the voltage-dependent time constants $\tau_n(V)$ (trace 1), $\tau_m(V)$ (trace 2), $\tau_h(V)$ (trace 3), of Eq.(13). Observe boundedness and asymptotic behavior for large absolute values of V.

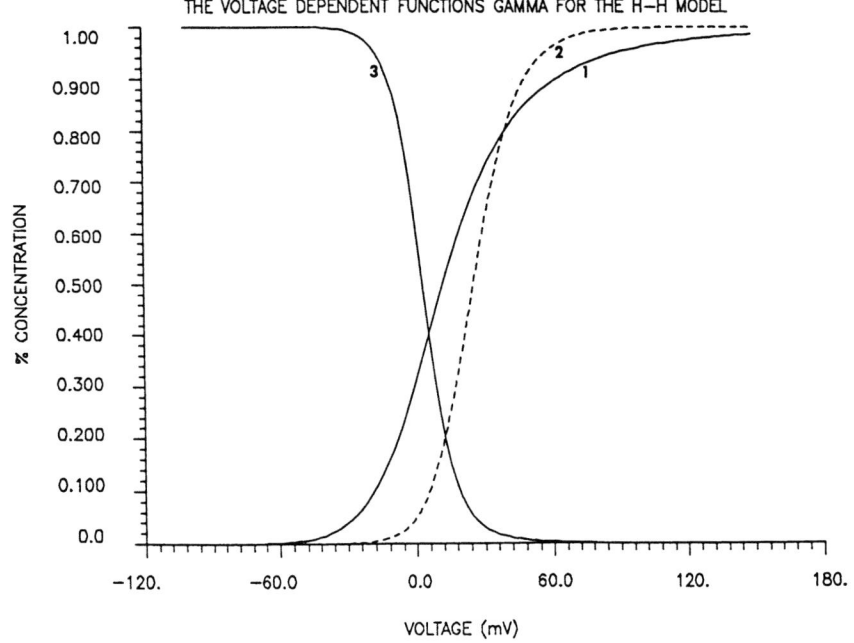

Fig. 2. The concentration functions in Eq.(13) $\gamma_n(V)$ (trace 1), $\gamma_m(V)$ (trace 2), $\gamma_h(V)$ (trace 3) are bounded functions of V and range between 0 and 1 in a sigmoidal fashion.

$V(t)$ is highly negative then the model parameters shown in Figs. 1 and 2, attain constant asymptotic values which lead to the simplified linear H–H model:

$$I(t) = C\frac{dV}{dt} + \bar{g}_L(V - V_L) \tag{14}$$

Likewise, when $V(t)$ is highly positive then the H–H model attains the linear form:

$$I(t) = C\frac{dV}{dt} + \bar{g}_K(V - V\kappa) + \bar{g}_L(V - V_L) \tag{15}$$

In both cases, the equivalent transfer function of the system has a single stable pole and, therefore, the response $V(t)$ is (uniformly) bounded for (uniformly) bounded input $I(t)$. This property implies the existence of a Volterra series expansion, as discussed in the following section.

We explore next the output of the model for sinusoidal input signals at different amplitudes and frequencies. The model produces spikes for certain combinations of frequencies and amplitudes, i.e., there is a specific domain of frequency–amplitude combinations that lead to the generation of spikes. This is illustrated in Fig. 3, where the model output is shown to exhibit the transition from "subthreshold" to "suprathreshold" behavior for the same input frequency ($5\ Hz$) and increasing amplitude (from 10 to 20 $\mu A/cm^2$), and in Fig. 4 where the same transition is observed for fixed input amplitude (10 $\mu A/cm^2$) and increasing frequency (from 5 Hz to 10 Hz). Frequency–amplitude combinations that do not lead to spike generation can be found at both the low and high end of the range of values of these sinusoidal input parameters. This "band–pass" type of behavior in terms of frequency and amplitude is due to the nonlinear dynamic nature of the model (FitzHugh, 1969), and suggests the suitability of Gaussian white noise (GWN) as a test input that covers, in principle, the entire frequency and amplitude range in a single experiment. In practical terms, the entire bandwidth of this system can be covered by use of band–limited GWN inputs, while the dynamic range of input amplitudes can be adjusted by varying the GWN power level. This is explored in the following section.

WIENER ANALYSIS OF THE H–H MODEL

Since the bounded–input/bounded–output stability of the H–H model was established in the previous section, the input–output relation can be expressed as a Volterra series expansion of the form:

$$V(t) = \sum_{n=o}^{\infty} \int \cdots \int_0^{\infty} k_n(\tau_1, \ldots, \tau_n) \prod_{i=1}^{n} I(t - \tau_i) d\tau_i \tag{16}$$

and the equivalent Wiener series expansion, when the input $I(t)$ is Gaussian white noise, of the form:

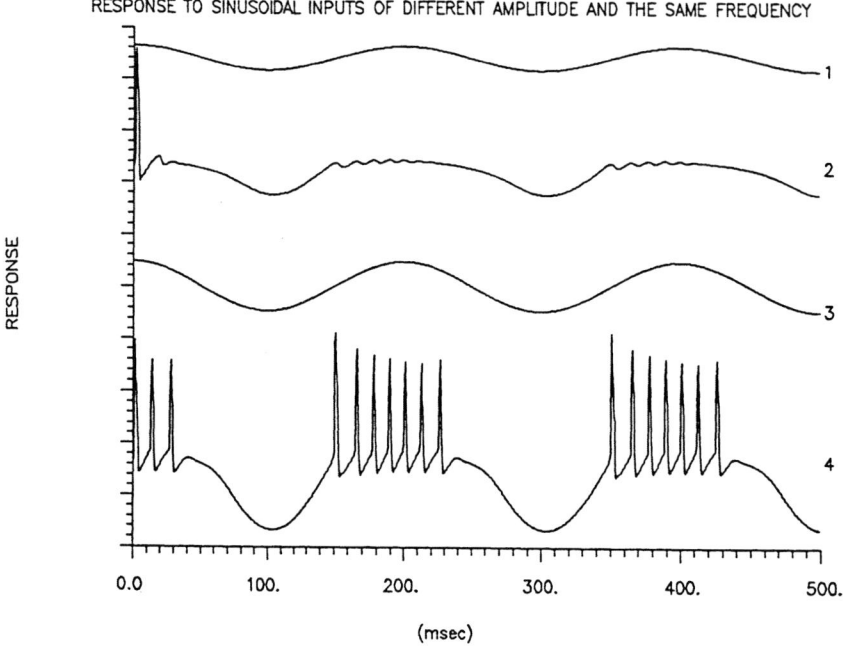

Fig. 3. Transition from subthreshold to suprathreshold region due to input amplitude. Traces 1 and 3 are the sinusoidal inputs with amplitudes 10 and 20 respectively; traces 2 and 4 are the corresponding responses.

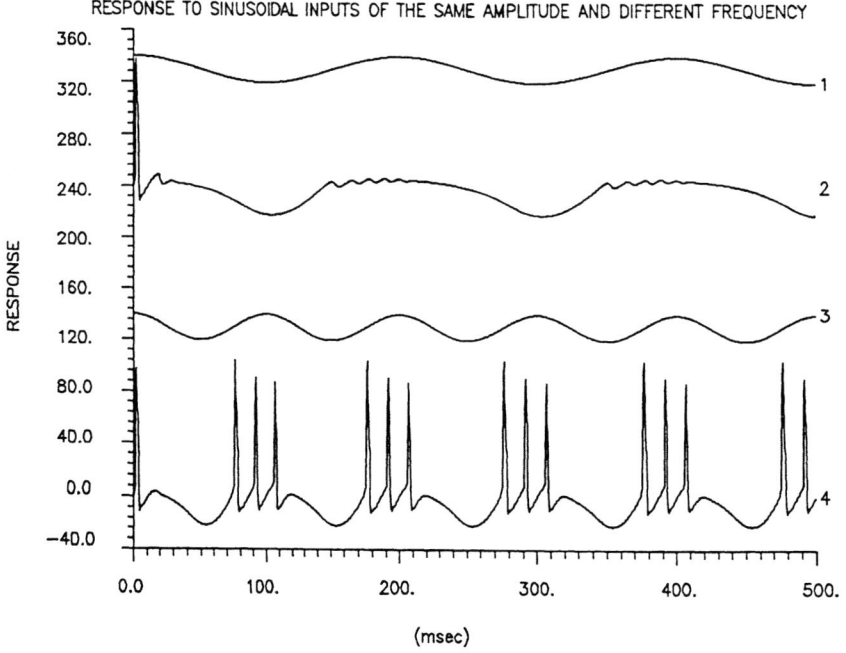

Fig. 4. Transition from subthreshold to suprathreshold region due to input frequency. Traces 1 and 3 are the sinusoidal inputs with frequencies 5Hz and 10Hz respectively ; traces 2 and 4 are the corresponding responses.

278

$$V(t) = \sum_{n=o}^{\infty} G_n[h_n; I(t')] \ , \ t' \le t] \tag{17}$$

where $G_n[\cdot]$ are the orthogonal Wiener functionals:

$$G_n[h_n; I(t'), t' \le t] = \sum_{m=0}^{\lfloor \frac{n}{2} \rfloor} \frac{(-1)^m n! P^m}{(n-2m)! m! 2^m} \int\int \cdots \int_0^{\infty} h_n(\tau_1, \ldots, \tau_{n-2m}, \lambda_1, \lambda_1, \cdots, \lambda_m, \lambda_m)$$
$$d\lambda_1 \ldots d\lambda_m \prod_{i=1}^{n-2m} I(t - \tau_i) d\tau_i \tag{18}$$

and k_n, h_n are the corresponding Volterra and Wiener kernels, while P is the power level of the GWN input. The main concern in this section is the estimation of Wiener kernels (of first and second order) from simulations of the H–H model. Theoretical background and applications of Volterra–Wiener analysis can be found in Marmarelis & Marmarelis, (1978) and Marmarelis (1987). First and second order Wiener kernels were estimated by use of the LYSIS package (Marmarelis & Herman, 1988), for different GWN input power and mean levels. Calculations of the kernels are limited to the second order kernel because of the computational burden associated with the estimation of higher order kernels.

The GWN input power levels $P = 0.5, 2.0, 8.0$ were used, representing respectively the regions of subthreshold, (near–) threshold and suprathreshold activities. There is no exact definition of the (dynamic) threshold region except that it is the neighborhood of P values for which sparse spikes will appear at the output. In addition to varying the input power level (P), the input mean level (μ) was varied to explore the regions of depolarization ($+4\mu A/cm^2$) and hyperpolarization ($-4\mu A/cm^2$).

Table 1 below summarizes the nine different combinations of GWN input mean and power levels used in the simulations, and gives the obtained mean values of the system output (response bias) for each case. These values represent the estimates of the zero–order Wiener kernel, h_0, in each case. According to the theory (Marmarelis, 1988), this constant kernel value is related to the Volterra kernels of the system (for given values of P and μ) by the expression:

$$h_0 = \sum_{m=0}^{\infty} \sum_{\ell=0}^{\infty} \frac{(2m+\ell)!}{m! \ell!} \left(\frac{P}{2}\right)^m \mu^{\ell}$$
$$\int \cdots \int_0^{\infty} k_{2m+\ell}(\lambda_1, \lambda_1, \ldots, \lambda_m, \lambda_m, \sigma_1, \ldots, \sigma_{\ell}) d\lambda_1 \ldots d\lambda_m d\sigma_1 \ldots d\sigma_{\ell} \tag{19}$$

where the Volterra kernels refer to zero input mean level. The values in Table 1 suggest a rising response bias as P increases for given value of μ. This rise becomes accelerated for higher values of μ. These observations are suggestive of a decompressive nonlinearity within the explored input range.

Table 1. Estimates of h_0 (response bias) for different P and μ values.

	$P = 0.5$	$P = 2$	$P = 8$
$\mu = -4$	-4.9224	-4.9172	-4.3898
$\mu = 0$	0.0143	0.1716	2.0534
$\mu = 4$	2.7676	3.5842	5.7312

Prior to computing the first and second order Wiener kernels, we must establish two parameters that are critical for this purpose: the effective memory and the effective bandwidth of the system. The former determines the maximum number of lags that ought to be estimated, and the latter indicates the appropriate bandwidth of (band–limited) GWN input that ought to be used. The effective memory of the system is determined by inspecting the autocorrelation function of the system output for all aforementioned combinations of GWN input power and mean levels. It appears that an effective memory of 30 $msec$ is adequate in all cases. The effective bandwidth of the system is examined by using the rather broad GWN input bandwidth of 5 kHz and inspecting the output spectrum for all aforementioned cases. It appears that a bandwidth of 1 kHz is adequate for all cases. According to the theoretical error analysis for Wiener kernel estimates (Marmarelis & Marmarelis, 1978) the use of broader GWN input bandwidth than necessary leads to considerable increase of estimation variance in the second–order kernels.

The first–order Wiener kernel estimates obtained for $\mu = 0$ and $P = 0.5$, 2, and 8 are shown in Fig. 5. We observe a gradual increase of the undershoot and broadening of the main (positive) lobe in the kernel waveform, as well as an increase in size as P increases. Likewise, the kernel estimates for $\mu = 4$ and $P = 0.5$, 2 and 8 are shown in Fig. 6. We observe again a gradual increase of the undershoot and size with increasing P, as well as decreased damping relative to the $\mu = 0$ case. However, the main (positive) lobe first broadens and then contracts as P increases. The kernel estimates for $\mu = -4$ and $P = 0.5$, 2 and 8 are shown in Fig. 7, exhibiting insignificant changes between $P = 0.5$ and 2, but considerable increase of the undershoot and size, as well as broadening of the main lobe for $P = 8$.

The FFT magnitudes of the kernels shown in Fig. 5 are given in Fig. 8. We observe increasing gain and band–pass characteristic (corresponding to the increased undershoot) as P increases, however the resonance frequency shifts gradually to lower values. This is indicative of decreasing stiffness or, equivalently, positive decompressive feedback (see Marmarelis, this volume).

The second–order Wiener kernel estimates for $\mu = 0$ and $P = 0.5$, 2 and 8 are shown in Figs. 9, 10 and 11 respectively. We observe relatively minor changes in the shape of these kernel estimates, however changes in size are discernible and they show a monotonic increase as P increases although the rate of this increase is reduced for the highest value of P. The second–order Wiener kernel estimates for $\mu = 4$ are shown in Fig. 12 ($P = 0.5$), Fig. 13 ($P = 2$) and Fig. 14 ($P = 8$). There are small changes in shape relative to the $\mu = 0$ case or among them as P increase, however the changes in size are intriguing since they show a substantial increase from $P = 0.5$ to $P = 2$ and then a slight decrease for $P = 8$. In the case $\mu = -4$, the second–order Wiener kernels are negligible for $P = 0.5$ and 2, but of size comparable to the previous cases for $P = 8$. The latter kernel is shown in Fig. 15 and exhibits a shape similar to the one shown in Fig. 10 ($\mu = 0$, $P = 2$).

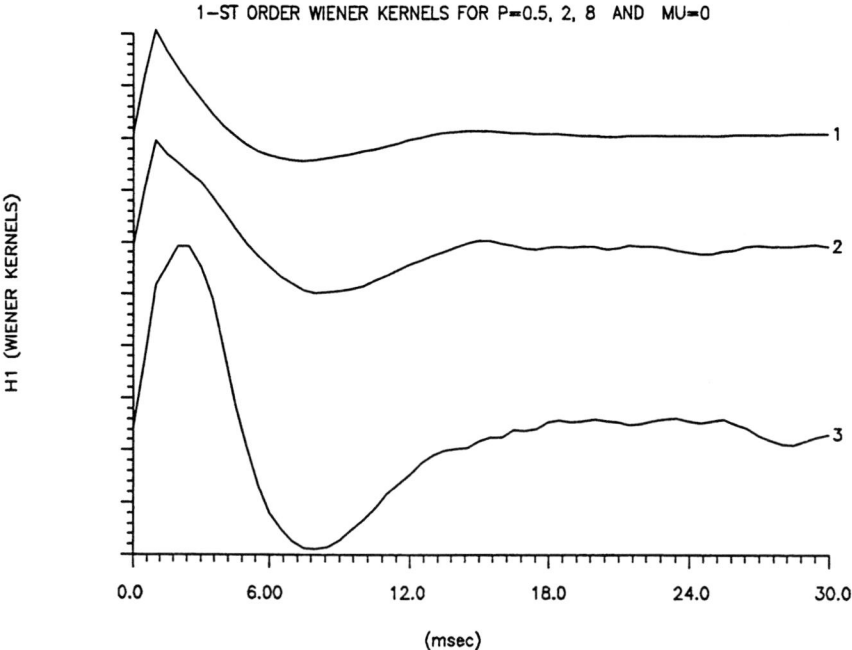

Fig. 5. First-order Wiener kernels for $\mu=0$ and P=.5 (trace 1), P=2 (trace 2)and P=8 (trace 3). Observe gradual increase of the undershoot and broadening of the main (positive) lobe as P increases.

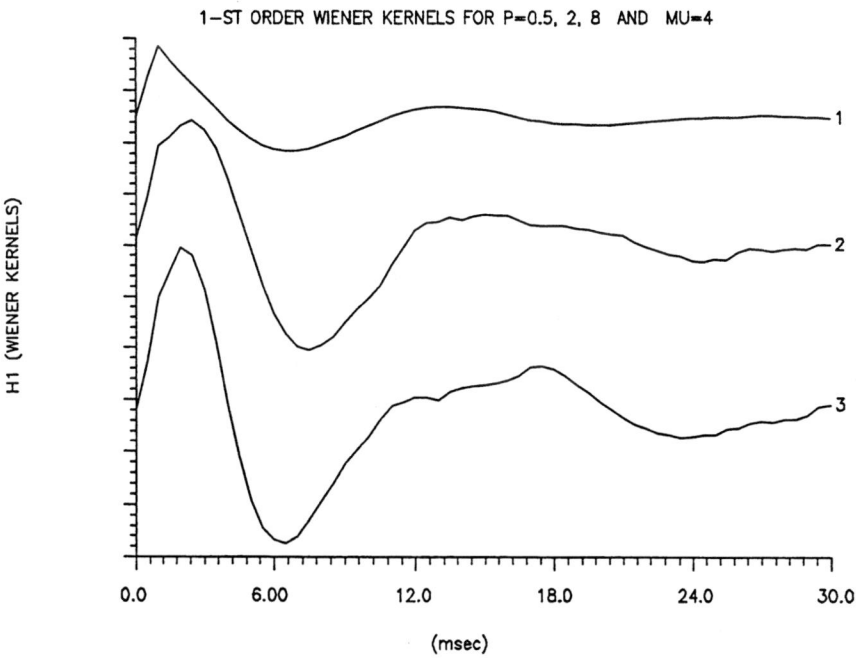

Fig. 6. First-order Wiener kernels for $\mu=4$ and P=0.5 (trace 1), P=2 (trace 2)and P=8 (trace 3). Increase in size and undershoot with P is observed here too. But the main lobe increases and then contracts as P increases.

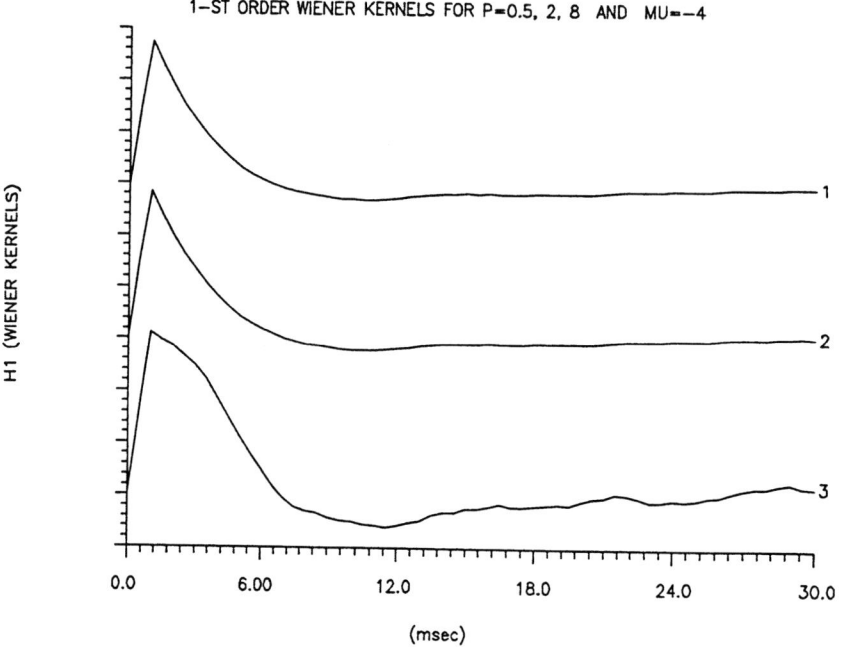

Fig. 7. First-order Wiener kernels for $\mu=-4$ and P=0.5 (trace 1), P=2 (trace 2), P=8 (trace 3). The considerable increase in size and undershoot for P=8 is clearly noticeable.

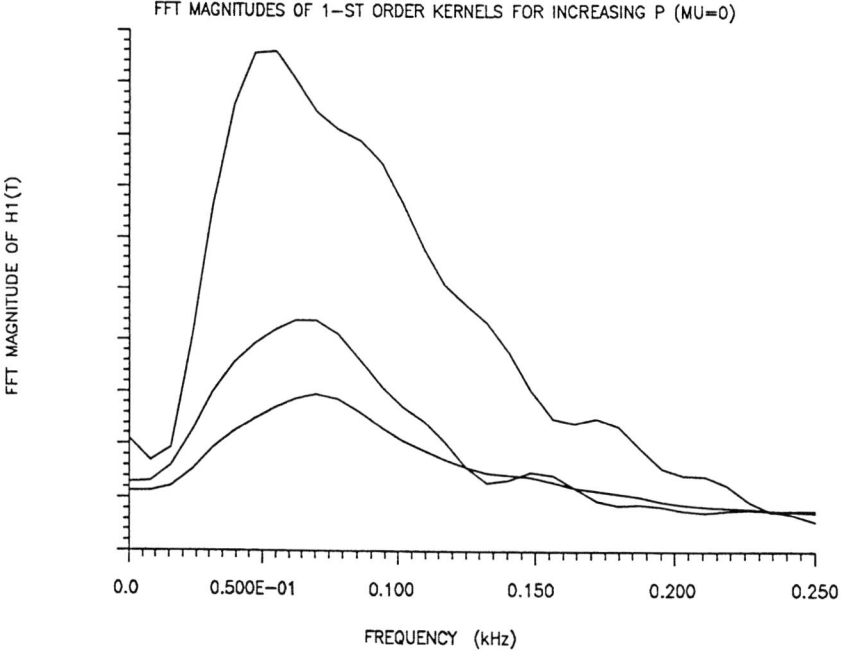

Fig. 8. FFT magnitudes of kernels shown in Fig.5. Observe increase in gain, decrease in damping and downward shift of the resonance frequency as P increases.

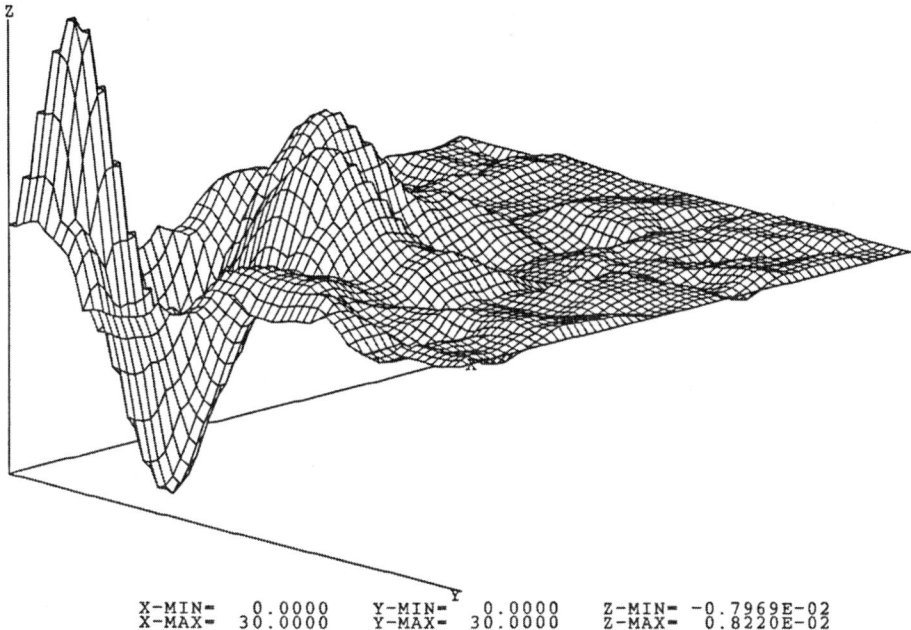

Fig. 9. Second-order kernel in the subthreshold region. The small values of the kernel suggest a (near-)linear behaviour of the system in this region.

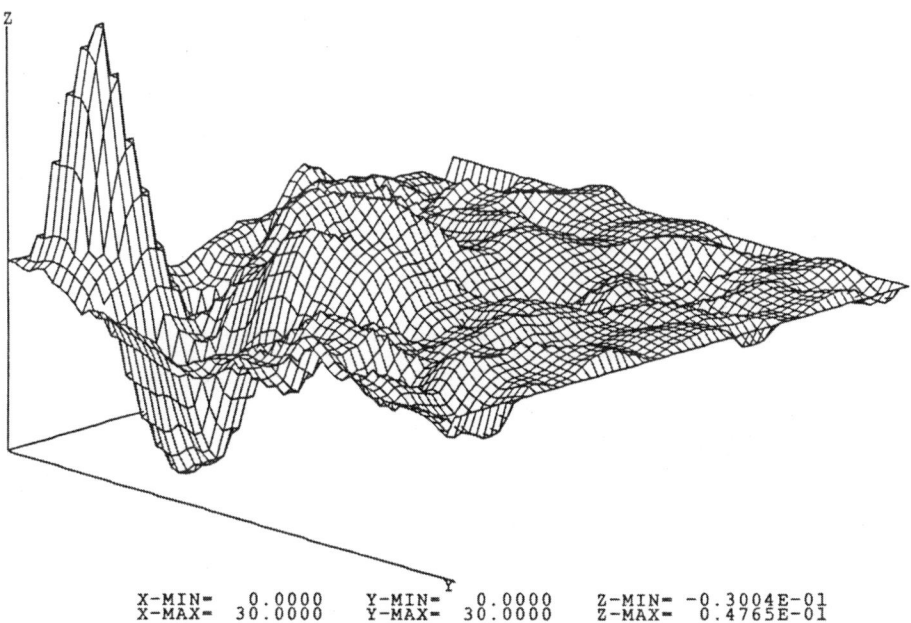

Fig. 10. Second-order kernel in the (near-)threshold region. The appearance of non-linear behaviour in the near-threshold region is reflected by the considerable increase in size relative to the subthreshold case.

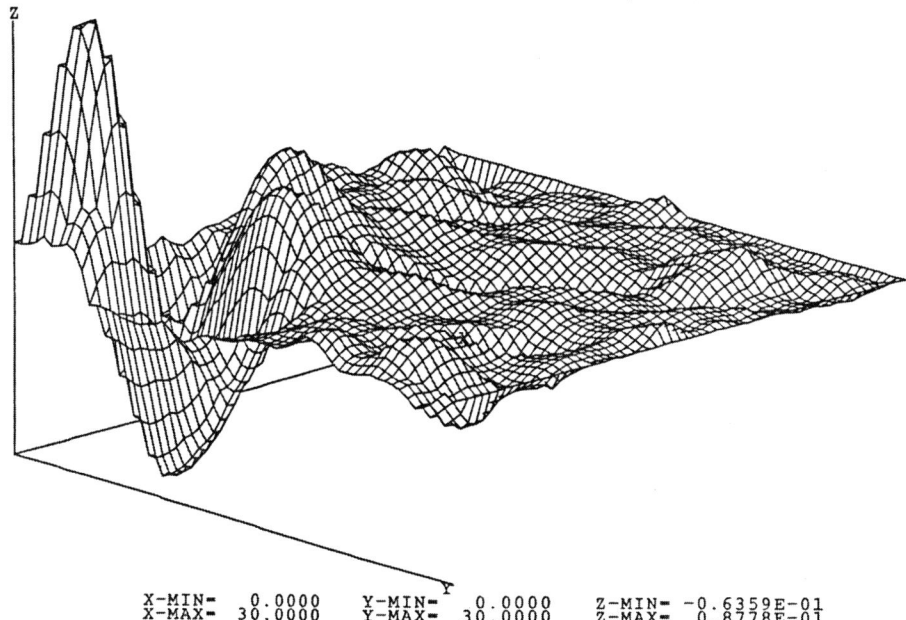

2-ND ORDER WIENER KERNEL FOR P=8, MU=0

```
X-MIN=   0.0000    Y-MIN=   0.0000    Z-MIN= -0.6359E-01
X-MAX=  30.0000    Y-MAX=  30.0000    Z-MAX=  0.8778E-01
```

Fig. 11. Second-order kernel in the suprathreshold region. Although the shape of the kernel has not significantly changed, increase in the size clearly indicates the the non-linear behaviour in this region.

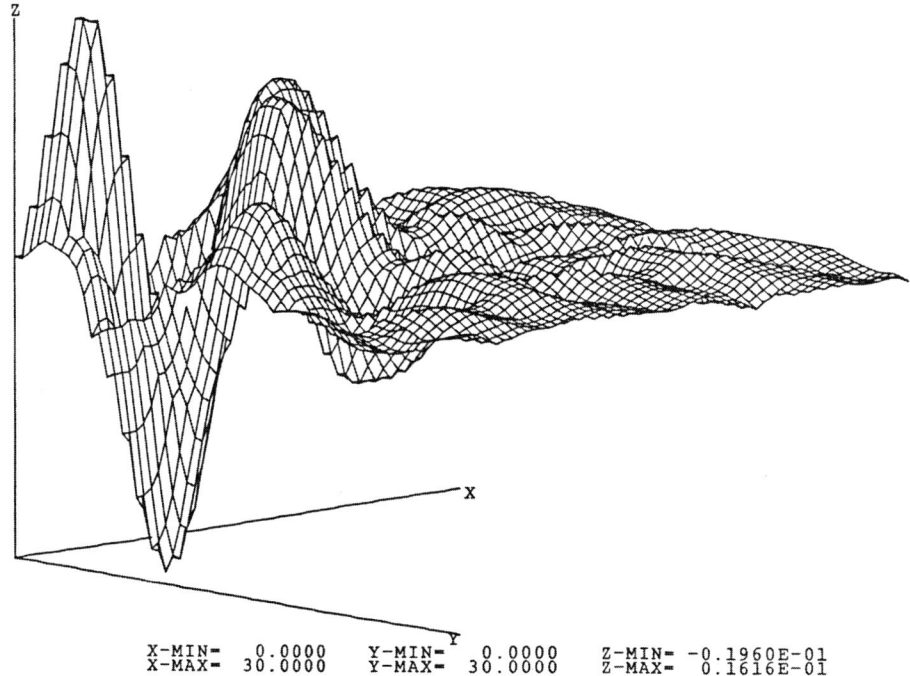

```
X-MIN=   0.0000    Y-MIN=   0.0000    Z-MIN= -0.1960E-01
X-MAX=  30.0000    Y-MAX=  30.0000    Z-MAX=  0.1616E-01
```

Fig. 12. Second-order kernel in the(near-)threshold region. Compared to the one at P=0.5, μ=0, this kernel depicts increased size rather than altered shape.

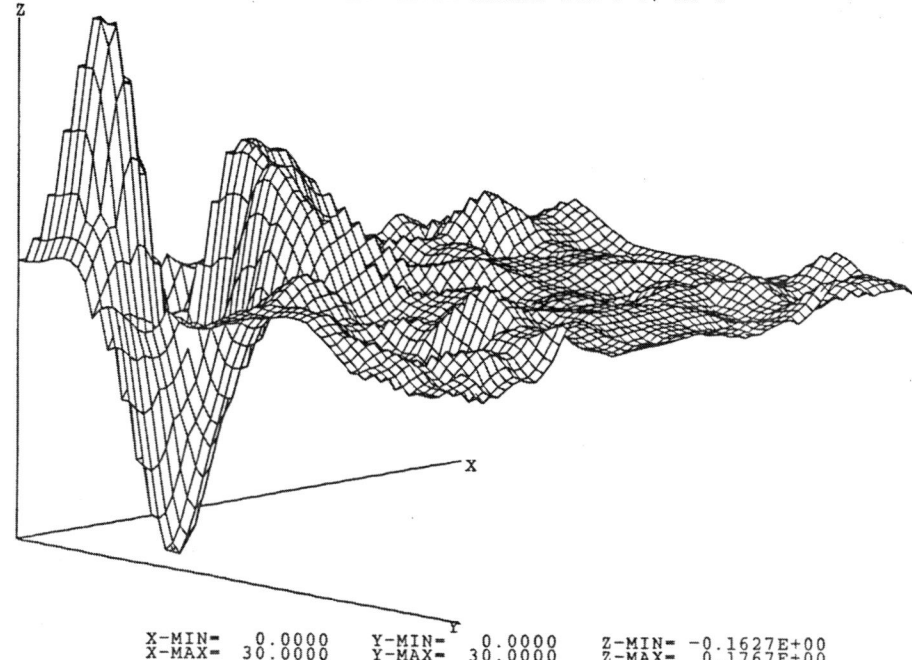

X-MIN= 0.0000	Y-MIN= 0.0000	Z-MIN= -0.1627E+00
X-MAX= 30.0000	Y-MAX= 30.0000	Z-MAX= 0.1767E+00

Fig. 13. Second-order kernel in the suprathreshold region. The minor shape changes in this case are followed by considerable increase in size.

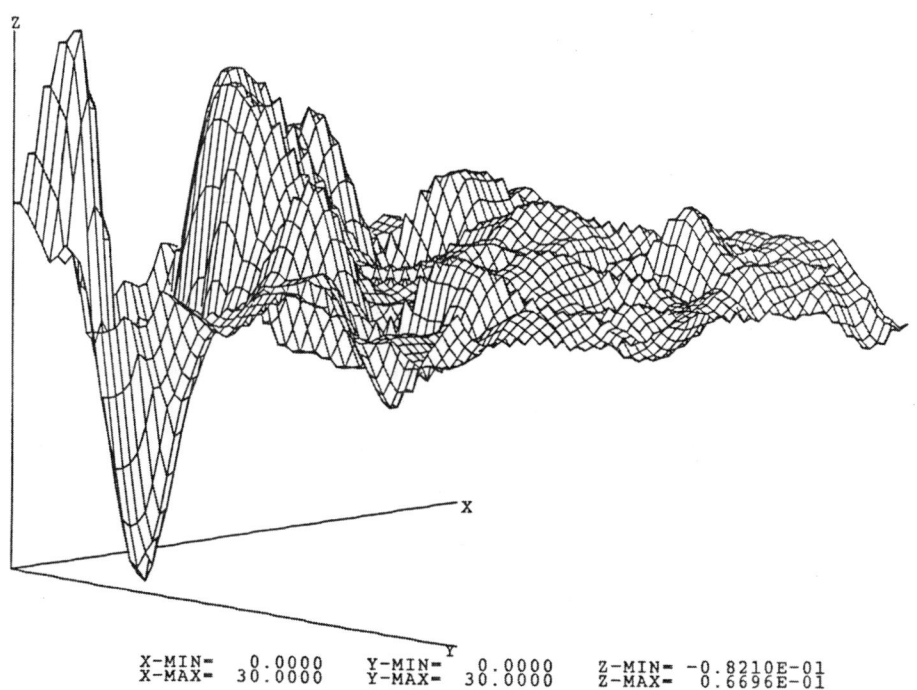

X-MIN= 0.0000	Y-MIN= 0.0000	Z-MIN= -0.8210E-01
X-MAX= 30.0000	Y-MAX= 30.0000	Z-MAX= 0.6696E-01

Fig. 14. Second-order kernel in the suprathreshold region. Reduction in size despite the increase in power constitutes the main characteristic of this kernel.

We must note that the observed changes in the waveform of the first–order Wiener kernel estimates are similar to the ones observed by Guttman et al. (1974) in experimental kernel measurements (cross–correlograms) from the squid axon membrane under analogous conditions of GWN stimulation. This fact corroborates the suitability of the H–H model in predicting the membrane response to broadband random stimulation as well.

In order to examine the relative contributions of these kernels to the system output, we show corresponding segments of the actual output, the first–order kernel contribution, the second–order kernel contribution and the combined first and second order contributions for $\mu = 0$, and $P = 0.5$ (subthreshold case) in Fig. 16, $P = 2$ (near–threshold case) in Fig. 17, and $P = 8$ (suprathreshold case) in Fig. 18. We observe increasing second–order kernel contribution as P increases. This contribution is most important in the suprathreshold case (as expected) and helps accentuate the presence of output spikes within the background activity. In the subthreshold and (near–) threshold cases, the second–order contribution improves the model prediction but only marginally. This behavior is expected since higher power levels introduce greater nonlinearities needed for the formation of spikes, which are partially reflected on the second–order Wiener kernel. It is also clear that the second–order Wiener kernel is not enough for the accurate prediction of spikes in the output and higher order kernels are needed. The normalized (by the output mean–square value) mean–square errors of the model predictions are summarized in Table 2 for these cases.

Table 2. Normalized mean–square errors of first and second order Wiener model predictions for $\mu = 0$.

	$P = 0.5$	$P = 2$	$P = 8$
1st–order model:	0.0527	0.7694	0.7997
2nd–order model:	0.0514	0.7112	0.6792

The obtained Wiener kernels $\{h_n\}$ for different power and mean levels of the GWN input have a specific relation with the Volterra kernels of the system $\{k_n\}$ which are independent of stimulus conditions (Marmarelis, 1988):

$$
h_n(\tau_1, \ldots, \tau_n) = \sum_{m=0}^{\infty} \sum_{\ell=0}^{\infty} \frac{(n + 2m + \ell)!}{n!\, m!\, \ell!} \left(\frac{P}{2}\right)^m \mu^\ell
$$
$$
\int \cdots \int_0^{\infty} k_{n+2m+\ell}(\tau_1, \ldots, \tau_n, \lambda_1, \lambda_1, \ldots, \lambda_m, \lambda_m, \sigma_1, \ldots, \sigma_\ell) d\lambda_1 \cdots d\lambda_m\, d\sigma_1 \cdots d\sigma_\ell \quad (20)
$$

and in the frequency domain:

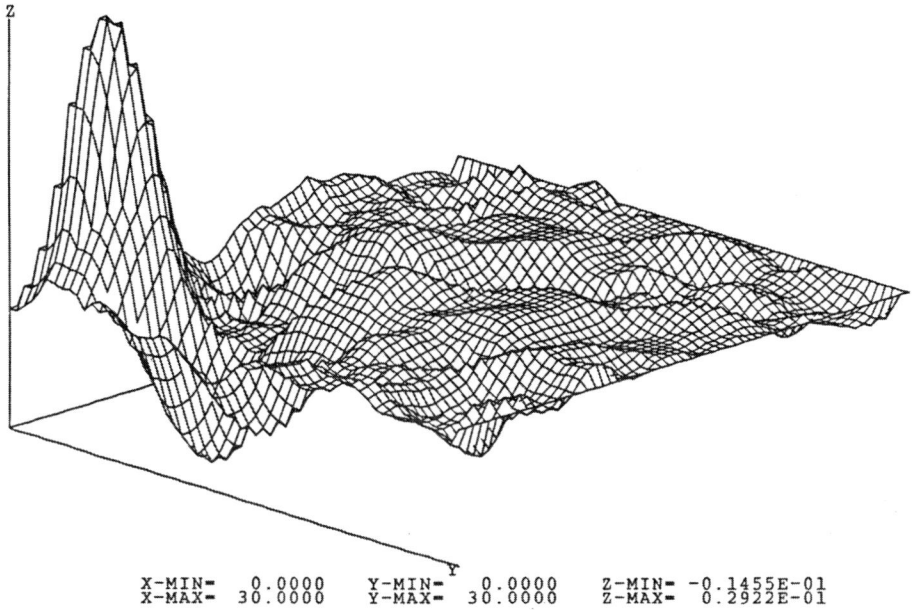

```
X-MIN=   0.0000    Y-MIN=   0.0000    Z-MIN= -0.1455E-01
X-MAX=  30.0000    Y-MAX=  30.0000    Z-MAX=  0.2922E-01
```

Fig. 15. Second-order kernel in the subthreshold region. Since μ=-4, it takes P=8 for the second-order kernel to obtain significant values.

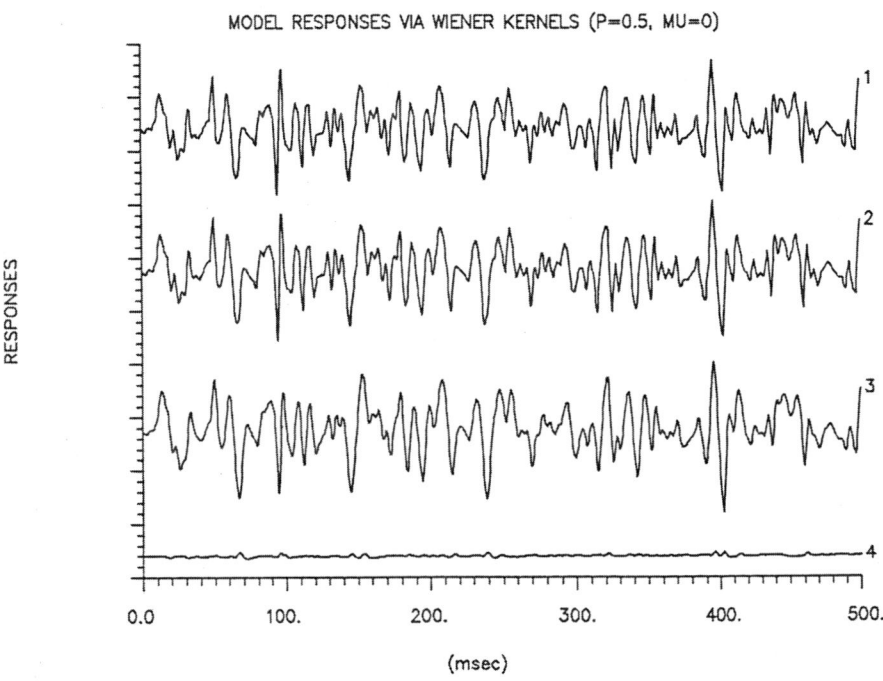

Fig. 16. Subthreshold region : (1) system response, (2)Wiener model response due to 1-st and 2-nd order kernels, (3) Wiener model response due to 1-st order kernel, (4) Wiener model response due to 2-nd order kernel.

287

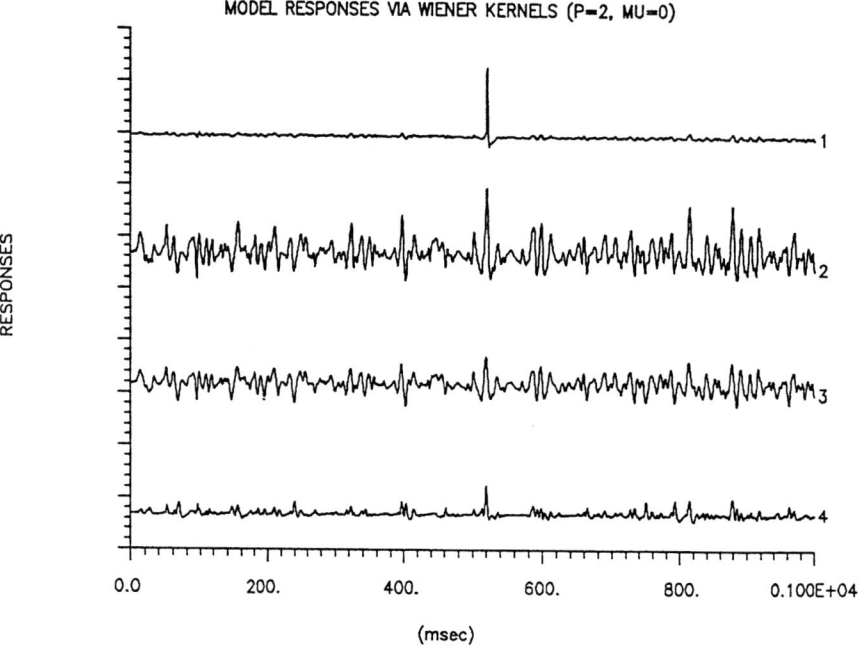

Fig. 17. Near-threshold region : (1) system response, (2) Wiener model response due to 1-st and 2-nd order kernels, (3) Wiener model response due to 1-st order kernel, (4) Wiener model response due to 2-nd order kernel.

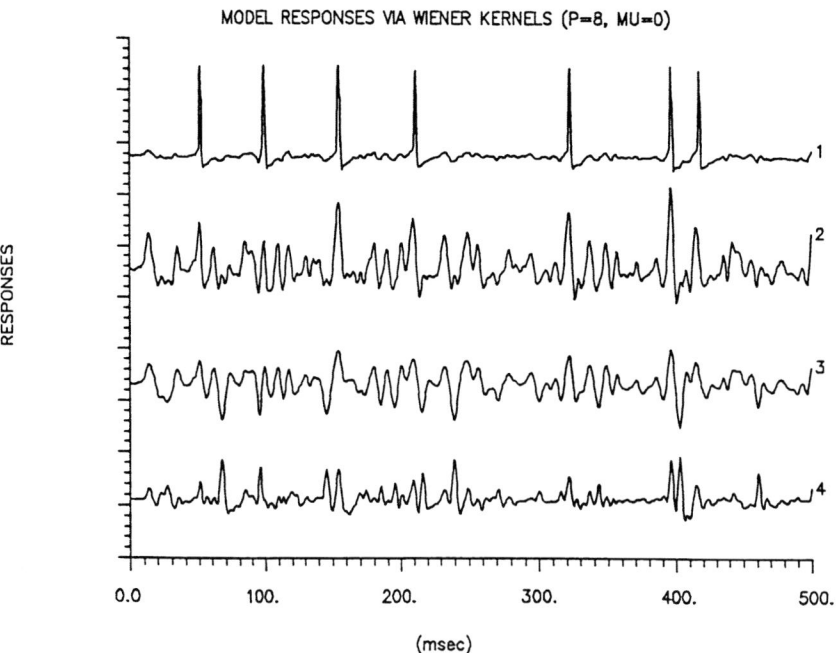

Fig. 18. Suprathreshold region : (1) system response, (2) Wiener model response due to 1-st and 2-nd order kernels, (3) Wiener model response due to 1-st order kernel, (4) Wiener model response due to 2-nd order kernel.

$$H_n(\omega_1, \ldots, \omega_n) = \sum_{m=0}^{\infty} \sum_{\ell=0}^{\infty} \frac{(n + 2m + \ell)!}{n!\, m!\, \ell!} \left(\frac{P}{4\pi}\right)^m \mu^\ell$$

$$\int_{-\infty}^{\infty} \cdots \int K_{n+2m+\ell}(\omega_1, \ldots, \omega_n, u_1, -u_1, \ldots, u_m, -u_m, 0, \ldots, 0)\, du_1 \cdots du_m \quad (21)$$

The above equations describe quantitatively the effect of varying GWN input power and mean levels on the Wiener kernel measurements, since the Volterra kernels remain invariant. The observed changes in the obtained first and second order Wiener kernel measurements, for different values of P and μ, indicate the presence of higher order kernels and can be used to infer functional properties of the system nonlinearities, as discussed in the following section.

INTERPRETATION OF THE WIENER KERNELS

Although the first three Wiener kernels (h_0, h_1, h_2) are not enough to replicate the behavior of the H–H model in the suprathreshold region, they may contain sufficient information to allow inferences about the functional organization of the system and suggest simplified models of the original H–H equations.

The values of the zero–order Wiener kernel are given in Table 1 for the different values of P and μ. For $\mu = 0$ an increase of P by a factor of 4 causes an increase of h_0 by a factor of 12. This increasing output bias for increasing P is introduced by the system nonlinearities which skew the output at accelerating rate towards positive values. When a positive GWN input mean is introduced $(4\mu A/cm^2)$ a larger positive output bias results but the rate of increase of h_0 with increasing P is lower than before. On the other hand, a negative $(-4\mu A/cm^2)$ input mean yields a significant negative output bias that rises only slightly with increasing P. In the context of Eq. (19) these observations suggest the presence of a large number of high–order Volterra kernels.

The first–order Wiener kernels for $\mu = 0$ exhibit a (band–pass) differentiation characteristic whose emphasis on differentiation (i.e., relative size of undershoot) increases with increasing values of P, while the resonance frequency shifts downward and the gain increases (see Fig. 8). When the positive input mean is introduced, these effects of increasing P on the kernel waveform are further accentuated, however the increase in gain appears to reach gradual saturation. When a negative input mean is introduced, these effects of increasing P are milder and the resonance frequency is considerably lower relative to the previous cases. A significant decrease of damping is also observed as μ increases. These observations indicate the property of the system to selectively accommodate a certain range of frequencies depending on the input mean level.

The second–order Wiener kernels assume significant values when the system produces spikes, as expected. Comparison to the first–order Wiener kernels shows that the corresponding second–order kernels do not constitute a scaled "Cartesian" product of the first–order kernel with itself, which would have indicated the plausibility of a cascade model comprised of a linear subsystem followed by a static nonlinearity (LN cascade). Likewise, a NL cascade (i.e., static nonlinearity followed by a linear subsystem) is ruled out since it would introduce a scaled version of the first–order Wiener kernel to the diagonal of the second–order one, while zero (or insignificant) values would be found in the rest of its (τ_1, τ_2) domain. A LNL cascade is also excluded since it would imply only scaling changes of the first–order kernel for different values of P and μ (a property shared by the LN and NL cascades as well). Therefore, simple cascade models cannot explain or reproduce the Wiener kernel measurements obtained from the H–H model. Can nonlinear feedback models account for the observed kernel measurements?

It has been shown (Marmarelis, this volume) that a class of nonlinear feedback systems with linear forward exhibit qualitatively this kind of behavior, viz., the waveform of their Wiener kernels changes as the GWN input power or mean level vary. Specifically, it was shown that the changes in the first–order Wiener kernels of this class of systems (that involve output and output–derivative feedback) are described in first approximation by the relation:

$$h_1(\tau) = k_1(\tau) + A_1(P)C_1(\tau) + B_1(P)C_1'(\tau) \tag{22}$$

where $k_1(\tau)$ is the first–order Volterra kernel of the system (representing its linear part), $C_1(\tau)$ is the convolution of k_1 with itself, and $C_1'(\tau)$ its derivative. The parameters A_1 and B_1 are functions of the quantities $(P\kappa)$ and $(P\kappa')$ where κ and κ' are the Euclidean norms of k_1 and its derivative respectively. The parametric functions $A_1(\cdot)$ and $B_1(\cdot)$ depend on the form of the nonlinear feedback and, specifically, its odd part. The above relation is for fixed input mean level (nominally $\mu = 0$). When μ changes, the parameters A_1 and B_1 change in a way determined by the form of the nonlinear feedback (see Marmarelis, this volume).

Let us examine the first–order Wiener kernel measurements of the H–H model, obtained for $\mu = 0$ and $P = 0.5, 2$ and 8, in the context of Eq. (22). To this purpose we need estimates of the functions $C_1(\tau)$ and $C_1'(\tau)$, which depend on $k_1(\tau)$. The latter can be estimated through input–output cross–correlation for very small values of P (since $h_1 \sim k_1$ for $P \to 0$). Estimates obtained for $P = 0.01$ were very similar to the h_1 estimate for $P = 0.5$, indicating the insignificance of nonlinearities in the low end of the subthreshold region. Consequently, the difference between the first–order Wiener kernel estimates obtained for higher P values (i.e., 2 and 8) and the kernel for $P = 0.5(\sim k_1)$ were computed and compared with the estimates of the functions $C_1(\tau)$ and $C_1'(\tau)$ using linear regression. The actual kernel differences and the linear regression fits using the functions $C_1(\tau)$ and $C_1'(\tau)$ are shown in Figs. 19 and 20 for the cases of $P = 2$ and $P = 8$, respectively. For $P = 2$, we observe a good fit except for large τ values (> 15). The regression coefficients in this case are: $A_1 = 0.294$ and $B_1 = -0.025$, suggesting primarily positive output feedback and only weak negative output–derivative feedback for this (near–) threshold region of input power. For $P = 8$, the fit is not as good, especially for τ values greater than 12. The regression coefficients in this case are: $A_1 = 1.010$ and $B_1 = 0.769$, suggesting significant output and output–derivative feedback in this suprathreshold region. Clearly, many more kernel measurements (for different values of P) are required for testing adequately the nonlinear feedback model. This example only serves the purpose of outlining the approach to searching for nonlinear feedback models of this specific class that may be equivalent to systems described by their Wiener kernels obtained for different GWN input power and mean levels.

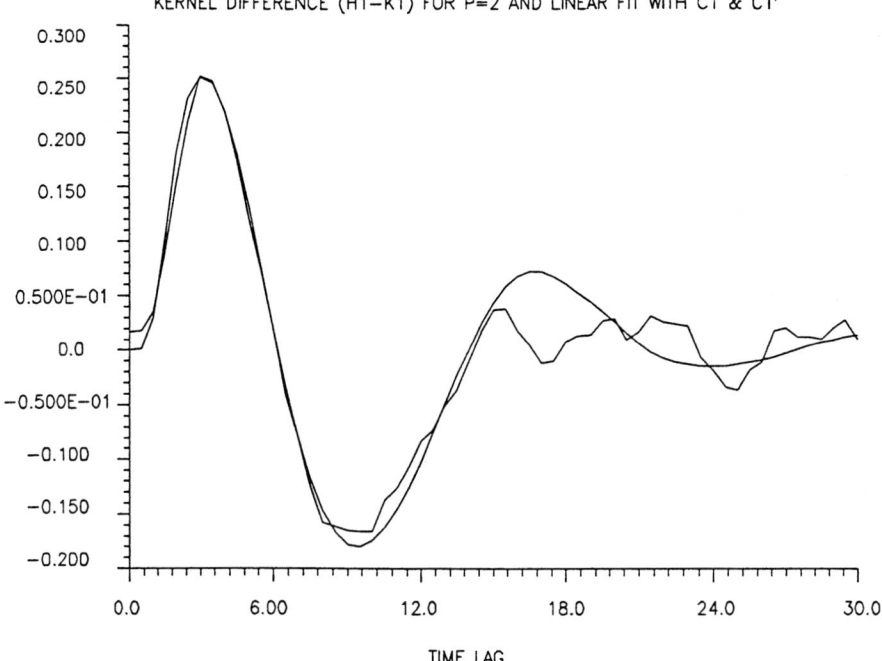

Fig. 19. Linear regression fit of the first order kernel difference for P=2 using the functions $C_1(\tau)$ and $C_1'(\tau)$ (cf. Eq.22).

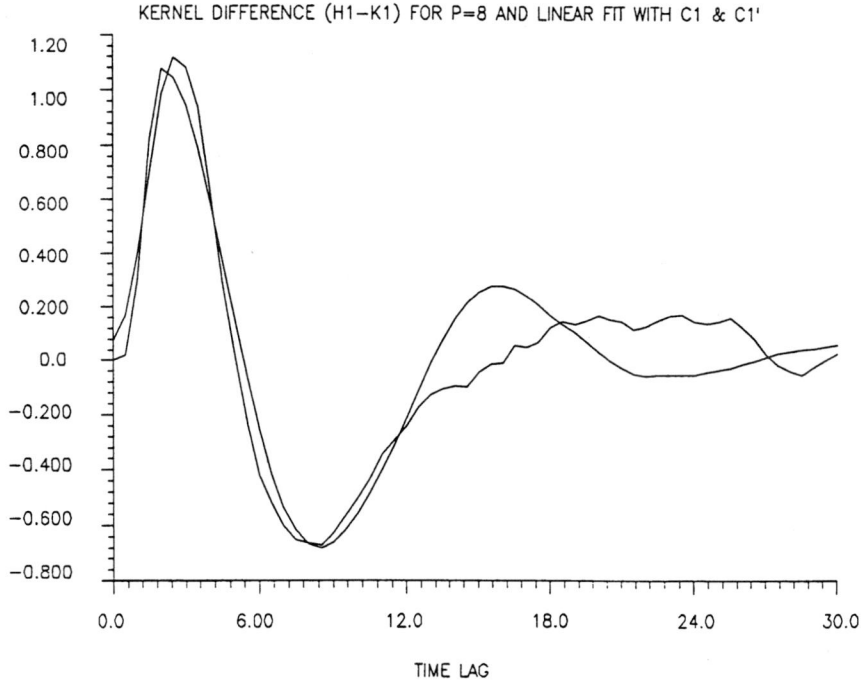

Fig. 20. Linear regression fit of the first order kernel difference for P=8 using the functions $C_1(\tau)$ and $C_1'(\tau)$ (cf. Eq.22).

The second–order Wiener kernels of this class of nonlinear feedback systems (for $\mu = 0$) have the approximate form (Marmarelis, this volume):

$$h_2(\tau_1, \tau_2) = A_2(P)C_2(\tau_1, \tau_2) + B_2(P)C_2'(\tau_1, \tau_2) + \Gamma_2(P)C_2''(\tau_1, \tau_2) \qquad (23)$$

where,

$$C_2(\tau_1, \tau_2) = \int_0^{min(\tau_1, \tau_2)} k_1(\lambda)k_1(\tau_1 - \lambda)k_1(\tau_2 - \lambda)d\lambda \qquad (24)$$

$$C_2'(\tau_1, \tau_2) = \left[\frac{\partial C_2(\tau_1, \tau_2)}{\partial \tau_1} + \frac{\partial C_2(\tau_1, \tau_2)}{\partial \tau_2} \right] \qquad (25)$$

$$C_2''(\tau_1, \tau_2) = \frac{\partial^2 C_2(\tau_1, \tau_2)}{\partial \tau_1 \partial \tau_2} \qquad (26)$$

and the parametric functions A_2, B_2 and Γ_2 are defined similarly to the previous functions A_1 and B_1. Note that (A_2, B_2, Γ_2) depend on the even part of the feedback nonlinearity, while (A_1, B_1) depend on the odd part. Evaluation of the functions C_2, C_2' and C_2'', using the previously estimated k_1, and comparison with the obtained second–order Wiener kernels, indicates that Γ_2 is insignificant (implying lack of even powers of the output–derivative in the nonlinear feedback) and that B_2 is much smaller than A_2 (implying preponderance of even powers of output feedback, as opposed to "mixed" terms involving both the output and its derivative). The resulting best fit for the case ($\mu = 0$, $P = 2$) is shown in Fig. 21. Comparison with the actual second–order kernel for this case (see Fig. 10) shows that the fit is lacking in several respects, especially in the secondary lobes, although it exhibits a basic similarity in shape with the actual kernel. Similar results were obtained in the other cases, indicating that the postulated nonlinear feedback model must be modified before it can reproduce the observed second–order Wiener kernels more accurately. This will be the subject of future studies.

EFFECT OF BINARY REPRESENTATION OF SPIKE DATA

As discussed in the introduction, investigators interested in modeling the dynamic transformations of spike sequences by neuronal units or the encoding of slow (continuous) potentials into spike sequences, have usually simplified the representation of spikes (action potentials) by converting them into binary form using a numerical threshold. This simplification in spike representation reduces immensely the computational burden. In this section, we explore the effects of this binary representation on the obtained Wiener kernel measurements.

To this purpose, the first and second order Wiener kernels were computed in the case ($\mu = 0$, $P = 8$) when the output signal is converted to a binary process by

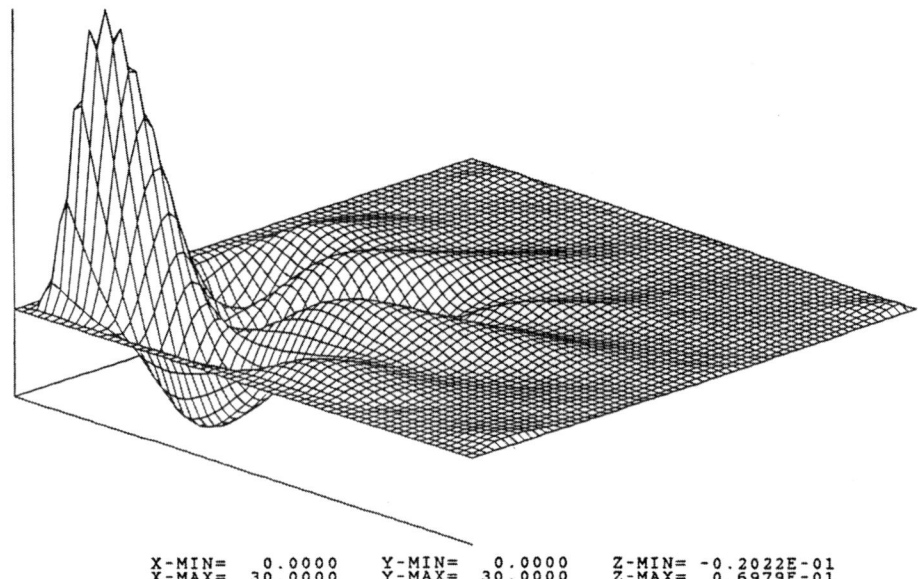

X-MIN= 0.0000   Y-MIN= 0.0000   Z-MIN= -0.2022E-01
X-MAX= 30.0000   Y-MAX= 30.0000   Z-MAX= 0.6979E-01

Fig. 21. Linear regression fit of the second order kernel difference for P=2 using the functions $C_2(\tau)$ and $C_2'(\tau)$ (cf. Eq.23).

1ST-ORDER WIENER KERNELS (P=8,MU=0) FOR: (1)BINARY OUTPUT, (2)CONTINUOUS OUTPUT

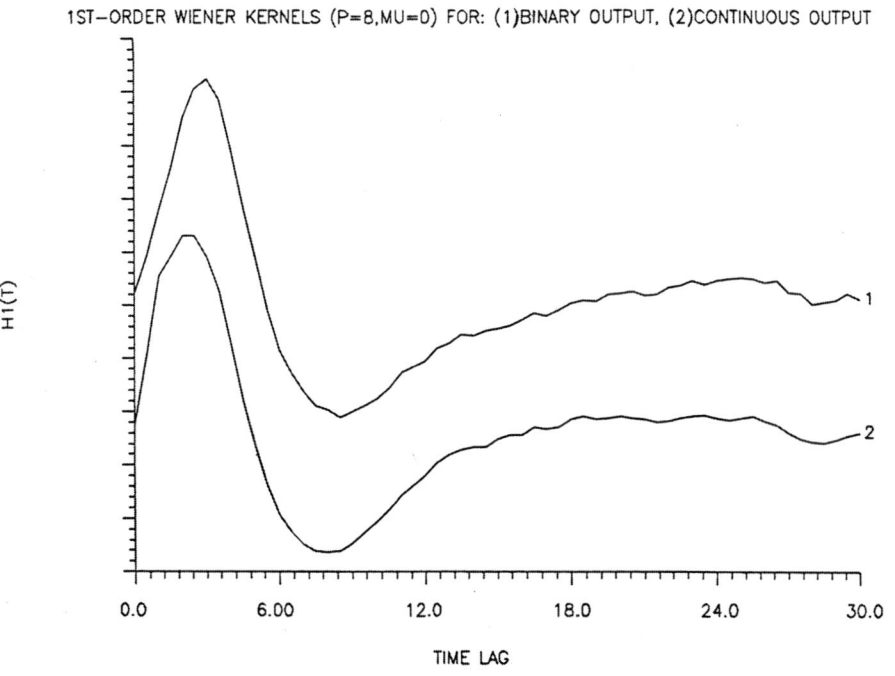

Fig. 22. First order kernel estimates for the case (P=8, μ=0) using continuous representation of the output (trace 2) and binary representation of the output (trace 1).

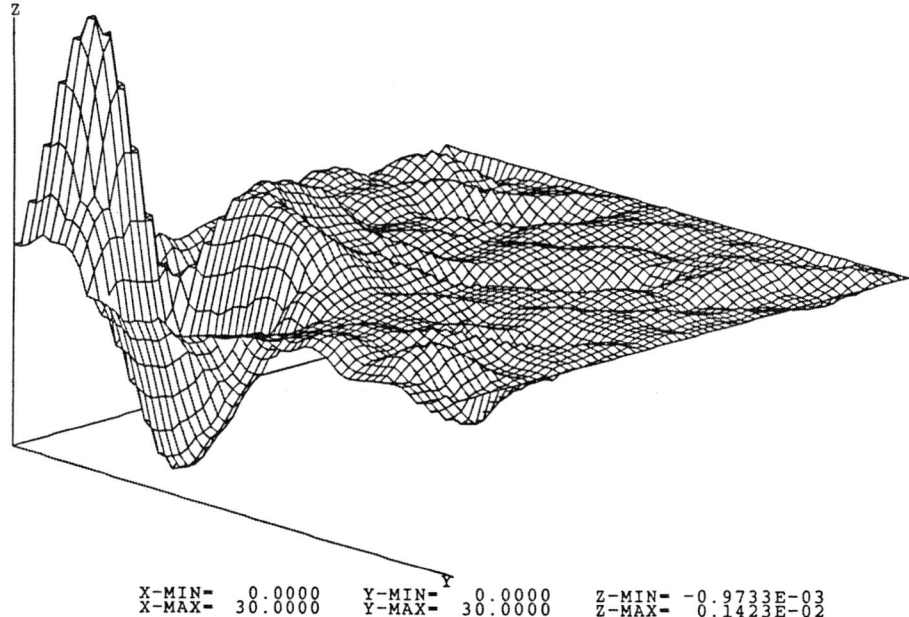

```
X-MIN=    0.0000    Y-MIN=    0.0000    Z-MIN=  -0.9733E-03
X-MAX=   30.0000    Y-MAX=   30.0000    Z-MAX=   0.1423E-02
```

Fig. 23. Second order kernel estimate for the case (P=8, μ=0) using binary representation of the output.

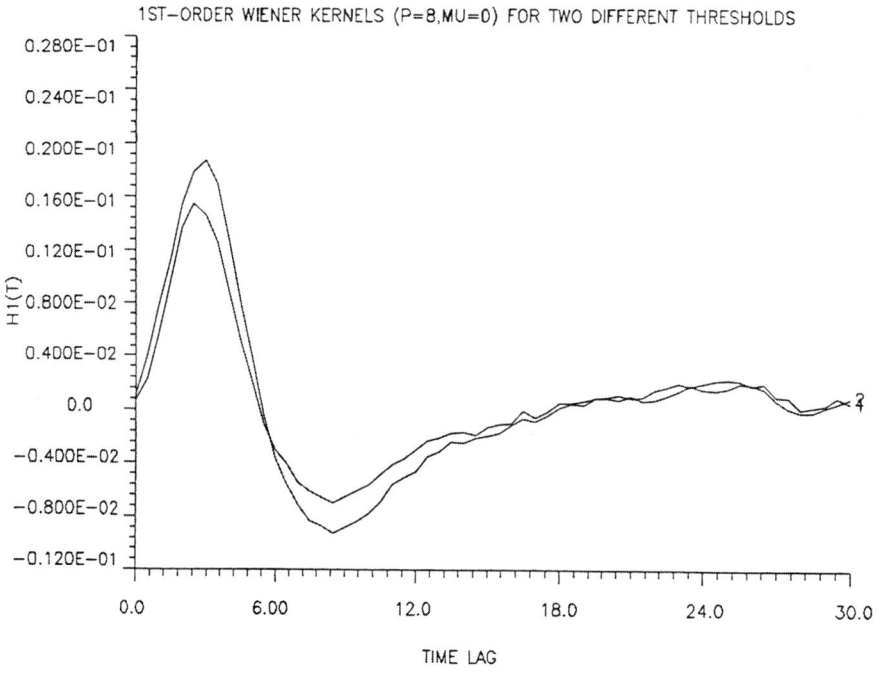

Fig. 24. First order kernel estimates using binary representation of the output by thresholding at two different values (10 and 50). Observe similarity of wave forms but decreased size for the higher threshold.

use of an appropriate threshold ($10\ mV$). The resulting first–order kernel is shown in Fig. 22, along with its counterpart computed previously from the exact (continuous) output signal. We observe some differences in waveform, especially in the rising phase of the principal lobe, due to the artificially imposed static nonlinearity (numerical threshold) at the actual output of the system. Scaling differences are also evident, as expected (Marmarelis et al., 1986). The second–order kernel obtained from the thresholded output is shown in Fig. 23, and comparison with its continuous–output counterpart (shown in Fig. 11) indicates some differences in waveform, especially in the secondary positive lobes and the diagonal, due to the thresholding operation. Note that, when a different numerical threshold is used, the resulting kernels exhibit only scaling differences. This is demonstrated in Fig. 24, where the first–order kernels for thresholds 10 and $50\ mV$ are shown superimposed. The smaller kernel corresponds to the higher threshold, as expected. Similar results were obtained for the second–order kernels.

The obtained results demonstrate the effects on estimated Wiener kernels when the binary representation of output spikes is used (see also Marmarelis et al., 1986; Marmarelis, 1989). These effects are not major and they are further mitigated in practice when additional dynamic transformations of narrower bandwidth (i.e., longer time constants) precede the generation of action potentials at the axon hillock. The latter being almost always the case in actual neuronal systems, it appears that the effects of thresholding may be of minor importance in many applications. However, it must be pointed out that the extent of these effects will depend on the characteristics of the preceding dynamic transformation and the stimulus conditions (i.e., power and mean level).

The issue that appears to be of greater practical importance, is the fact that the Wiener kernels of a neuronal system whose response is recorded extracellularly (at the axon, in the form of spike trains) may deviate considerably from the kernels of that system when the response is recorded intracellularly. This is due to the differentiating (band–pass) characteristic of the spike generation mechanism demonstrated in the previous sections. Some experimental evidence of this effect can be found in Sakai and Naka (1987) in first–order kernel measurements of catfish retinal ganglion cells (obtained both intracellularly and extracellularly), although the observed effect in that case is not dramatic. We believe that this point deserves further study, and caution should be exercised in assuming the spike generation mechanism as a static threshold in neurophysiological investigations.

CONCLUSIONS

The H–H equations were simulated for GWN inputs of various power and mean levels. The zero, first and second order Wiener kernels were estimated and shown to exhibit strong dependence on these input parameters in the suprathreshold region. This dependence reflects the presence of strong nonlinearities in this region (necessary for the generation of action potentials) and indicates the existence of higher order kernels. The primary effect of increasing GWN input power level P on the first–order kernel waveform is an increase of the undershoot and gain (kernel size) and a decrease of the resonance frequency and damping, as the GWN input power level P increases. Similar effects are observed when a positive mean level μ (depolarization) is introduced at the input, while a negative mean level (hyperpolarization) causes a significant decrease of the resonance frequency and increase of damping (for the same values of P). Similar results were obtained in experimental studies of the squid axon (Guttman et al., 1974). The effects on the second–order kernels are more difficult to describe (for lack of established terminology in describing their shape) but a definite increase of kernel size is observed in the transition from the subthreshold to the suprathreshold region. When a positive input mean level is introduced, the size of the negative lobes and the secondary positive lobes increases relative to the primary positive lobe. The introduction of a negative input mean level reduces considerably the size of the second–order kernel.

The first–order Wiener model was able to reproduce the system output with considerable accuracy in the subthreshold region. The second–order kernel contribution to the Wiener model prediction was found important in the suprathreshold region, although insufficient to reproduce the output spikes with great accuracy. Nevertheless, the second–order Wiener model was able to provide a significantly better prediction of the timing of the output spikes.

In spite of the inadequacy of the first two Wiener kernels (h_1 and h_2) to reproduce accurately the output waveform in the suprathreshold region, it was demonstrated that a considerable amount of information resides in these kernel measurements and in the changes they undergo for different conditions of stimulation (i.e., GWN input power and mean level). An approach was outlined for utilizing this information in order to test modeling hypotheses and possibly develop "block–structured" models that mimic the essential functional characteristics of the H–H equations. Since simple cascade models cannot explain the obtained kernel measurements, a broad class of nonlinear feedback models was explored and shown to reproduce certain important characteristics of the kernels of the H–H model. Although this particular modeling effort was not conclusive, the outlined approach (based on a methodology presented in Marmarelis, this volume) may hold the promise of developing a simplified model of the H–H equations in a "block–structured" form involving nonlinear feedback. Quantitative evidence of nonlinear feedback, utilizing the output signal and its derivative, was found in the obtained kernel measurements. A more extensive set of kernel measurements (i.e., for more values of P and μ) will be required before the aforementioned goal can be achieved.

The effect of replacing the continuous output signal with a point process (attaining binary values on the basis of a numerical threshold that assigns nonzero values to the output signal only when a spike occurs) was examined by computing the first and second order kernels for both output modalities. The obtained results show certain differences in the resulting kernel waveforms, albeit limited in extent. There are, of course, scaling effects dependent on the value of the numerical threshold. The effects on the kernel waveform are expected to be mitigated when a dynamic transformation of narrower bandwidth precedes the generation of action potentials at the axon hillock, as is often the case in actual neurophysiological studies. However, a more serious practical problem may arise in comparing experimental results obtained through intracellular and extracellular recordings, because of the changing dynamic (band–pass) characteristics of the spike generation mechanism for different conditions of GWN stimulation (i.e., power and mean level).

It is hoped that these results will assist neurophysiologists in the more accurate interpretation of experimental measurements involving neuronal spikes.

Acknowledgements

This work was supported by Grant No. RR–01861 from the Division of Research Resources of the National Institutes of Health.

REFERENCES

Cronin, J., 1987, "Mathematical Aspects of Hodgkin–Huxley Neural Theory," University Press, Cambridge.

Deutsch, S. and Micheli–Tzanakou, E., 1987, "Neuroelectric Systems," New York University, New York.

FitzHugh, R., 1969, Thresholds and plateaus in the Hodgkin–Huxley nerve equations, J. Gen. Physiol., 43:867–896.

Guttman, R., Feldman, L., and Lecar, H., 1974, Squid axon membrane response to white noise stimulation, Biophys. J., 14:941–955.

Hodgkin, A.L., and Huxley, A.F., 1952, A quantitative description of membrane current and its application to conduction and excitation in nerve, J. Physiol., 117:500–544.

Marmarelis, P.Z., and Marmarelis, V.Z., 1978, "Analysis of Physiological Systems: The White Noise Approach", Plenum Press, New York.

Marmarelis, V.Z., ed., 1987, "Advanced Methods of Physiological System Modeling: Volume I", Biomedical Simulations Resource, University of Southern California, Los Angeles.

Marmarelis, V.Z., 1988, Coherence and apparent transfer function measurements for nonlinear physiological systems, Ann. Biomed. Eng., 16:143–157.

Marmarelis, V.Z., 1989, Signal transformation and coding in neural systems, IEEE Trans. Biomed. Eng., 36:15–24.

Marmarelis, V.Z., Citron, M.C., and Vivo, C.P., 1986, Minimum–order Wiener modeling of spike–output systems, Biol. Cybern., 54:115–123.

Marmarelis, V.Z., and Herman, N., 1988, LYSIS: An interactive software system for nonlinear modeling and simulation, in "Modeling & Simulation on Microcomputers," Society for Computer Simulation, San Diego, pp. 6–10.

Sakai, H.M., and Naka, K–I, Signal transmission in the catfish retinal (IV), J. Neurophysiol., 58:1307–1328.

LIST OF CONTRIBUTORS

German Barrionuevo, Department of Neurological Surgery, University of Pittsburgh, Pittsburgh, PA 15260.

Theodore W. Berger, Departments of Behavioral Neuroscience & Neurological Surgery, Center for Neuroscience, University of Pittsburgh, Pittsburgh, Pennsylvania 15260.

David R. Brillinger, Department of Statistics, University of California, Berkeley, California 94720.

William D. Burton Jr., Epilepsy Research Center, Baylor College of Medicine, Houston, Texas 77030.

Mark C. Citron, Department of Neurology Research, Children's Hospital of Los Angeles, P.O. Box 54700, Terminal Annex, Los Angeles, California 90054.

Spiridon Courellis, Department of Biomedical Engineering, University of Southern California, Los Angeles, California, 90089–1451.

Robert C. Emerson, Department of Ophthalmology Box 314, and Center for Visual Science, University of Rochester, Rochester, New York 14642.

Andrew S. French, Department of Physiology, University of Alberta, Edmonton, Alberta, Canada T6G 2H7.

Patrick Harty, Department of Psychiatry, University of Pittsburgh, Pittsburgh, PA 15260.

Hugh King, Private Medical Practice, 1909 Ivanhoe Street, Denver, Colorado 80220.

Michael Korenberg, Department of Electrical Engineering, Queen's University, Kingston, Ontario, Canada K7L 3N6.

Donald N. Krieger, Department of Neurological Surgery, School of Medicine, University of Pittsburgh, Pittsburgh, Pennsylvania 15213.

Steven Levitan, Department of Electrical Engineering, University of Pittsburgh, Pittsburgh, Pennsylvania 15260.

Larry S. Liebovitch, Department of Ophthalmology, Columbia University, 630 West 168th Street, New York, New York 10032.

Vasilis Z. Marmarelis, Departments of Biomedical & Electrical Engineering, University of Southern California, Los Angeles, California 90089–1451.

Gottfried Mayer–Kress, Center for Nonlinear Studies, Los Alamos National Laboratory, Los Alamos, New Mexico 87545.

Aage R. Møller, Department of Neurological Surgery, School of Medicine, University of Pittsburgh, Pittsburgh, Pennsylvania 15213.

Paul L. Nunez, Department of Biomedical Engineering, Tulane University, New Orleans, Louisiana 70118.

Michael O'Toole, Departments of Cardiology and Physiology, Loyola University Medical Center, Maywood, Illinois 60153.

Arturo E.C. Pece, Department of Physiology, University of Alberta, Edmonton, Alberta, Canada, T6G 2H7.

Bernard Saltzberg, Department of Psychiatry and Behavioral Science, University of Texas Health Science Center at Houston, Medical School, Houston, Texas 77030.

Joseph Samosky, Department of Electrical Engineering, University of Pittsburgh, Pittsburgh, Pennsylvania 15260.

Robert J. Sclabassi, Departments of Neurological Surgery, Electrical Engineering & Behavioral Neuroscience, University of Pittsburgh, Pittsburgh, Pennsylvania 15260.

J.H. Shi, Department of Electrical and Computer Engineering, Drexel University, Philadelphia, Pennsylvania 19104.

Paul A. Sobotka, Cardiology and Physiology, Loyola University Medical Center, Maywood, Illinois 60153.

Jackie Solomon, Department of Electrical Engineering, University of Pittsburgh, Pittsbrugh, Pennsylvania 15260.

Hun H. Sun, Department of Electrical and Computer Engineering, Drexel University, Philadelphia, Pennsylvania 19104.

John X. Thomas, Jr., Department of Physiology, Loyola University Medical Center, Maywood, Illinois 60153.

Jonathan D. Victor, Department of Neurology, Cornell University Medical College, 1300 York Avenue, New York, New York 10021.

Joseph P. Zbilut, Surgical Nursing and Physiology, Rush University, Chicago, Illinois 60612–3864.